Biomedical Devices

Raymond H. W. Lam • Weiqiang Chen

Biomedical Devices

Materials, Design, and Manufacturing

Springer

Raymond H. W. Lam
Department of Biomedical Engineering
City University of Hong Kong
Hong Kong, Hong Kong

Weiqiang Chen
Department of Mechanical and Aerospace Engineering, Department of Biomedical Engineering
New York University
Brooklyn, NY, USA

ISBN 978-3-030-24236-7 ISBN 978-3-030-24237-4 (eBook)
https://doi.org/10.1007/978-3-030-24237-4

This Springer imprint is published by the registered company Springer Nature Switzerland AG
The registered company address is: Gewerbestrasse 11, 6330 Cham, Switzerland

Dedicated to
Guannan Luo

Preface

With the increasing public health awareness and a rising aging population, there is an increasing demand for high-quality, low-cost, and safe biomedical devices globally. Biomedical devices have been widely used in the healthcare industry and play a crucial role in diagnosing, assessing, and treating various diseases and injuries that affect various parts of our body. The huge and increasing demand in biomedical devices requires a significant investment in research, development, and manufacturing of new biomedical technologies and products.

Like any type of design engineering, biomedical device design must first begin with identifying a problem to solve. The ideation of a new device typically arises from medical problems in which people express frustration in current medical solutions or at the lack thereof. There exist numerous medical problems to be solved within medicine ranging from needs for improved small wound care bandages up to cancer diagnostic devices which can be shrunk down in size to make them more portable and accessible to a wider patient group. Thus, a biomedical device can be an apparatus, machine, implant, in vitro reagent, software, or similarly related article that are purposed for safe and effective prevention, diagnosis, treatment, and rehabilitation of illness and disease. They include complex devices, such as programmable pacemakers or microchip implants, and simple ones, such as simple surgical scalpels, medical syringes, and bedpans. Biomedical devices also include in vitro diagnostic products, such as general purpose lab equipment, reagents, and test kits. A few electronic products, such as diagnostic ultrasound products, X-ray machines, and medical lasers, are also classified as biomedical devices. In the biomedical device design and development process, medical professionals often voice their opinions on medical device designs as end user to avoid product difficulties or design flaws that may lead to serious life-threatening consequences. On the other hand, engineers must consider their ideal end users' need to propose technical solutions to develop appropriate biomedical devices. Medical device design is an incredibly expensive process which is closely monitored by regulatory agencies for years to ensure quality and safety concerns are met. Regulatory concerns begin even during this early conceptualization stage of design. The trend biomedical device manufacturing emerged and played an important role in biomedical device industry. Through the application of different manufacturing technologies, the safety, quality, quantity, cost, efficiency, and speed can be advanced for healthcare service and biomedical device production. After conceiving a device based on a specific need and regulatory specifications, the device will be designed and honed over time to maximize its potential usefulness and problem-solving capacity. This testing phase is crucial to successfully pass a device through regulations and often requires copious amounts of scientific data providing evidence of a device's safety and effectiveness at performing its intended task. Once a device clears this phase, it may finally enter production and be sold, though its producer is often required to conduct periodic follow-up evaluations to ensure the real-world safety of the product.

Biomedical device industry involves a variety of conventional as well as emerging fields including biomedical product design, manufacturing, process optimization, material characterization and processing, computer and data science, tissue engineering, medicine, etc. This book aims to provide the essential knowledge in the biomedical product design and development in order to provide ways to speed up the product development cycle. The content in this book is multidisciplinary and covers the principles in mechanical, chemical, biological, and physiological aspects. In general, we will introduce the biomedical device development through three main knowledge clusters: (1) material properties and selection (Chaps. 2, 3, 4, and 5), (2) fabrication processes (Chaps. 6, 7, and 8), and (3) design principles and techniques (Chaps. 9, 10, 11, and 12) for different biomedical applications. Biomedical engineers can learn the critical principles and techniques to apply the acquired knowledge of biomedical device design, prototyping, and manufacturing for particular applications.

While the book introduces many topics relevant to biomedical device development and industry, it is not meant to be an exclusive review of all the details of relevant areas. Instead, it intends to help biomedical engineers to develop basic principles involved in biomedical device design, manufacturing, development, regulations, and applications. Many references are provided as springboard for enthusiasts to launch their own in-depth research on topics of interests. We expect the introductory and interdisciplinary text will attract strong interests from biomedical engineers of a broad background in mechanical engineering, biomedical engineering, chemical engineering, electrical engineering, and materials science and prepare them with a familiarity and a clear understanding of the benefits of biomedical technologies and who will consider employment in the biomedical industry, universities, and the government.

Hong Kong, Hong Kong
Brooklyn, NY, USA

Raymond H. W. Lam
Weiqiang Chen

Acknowledgments

This book began as a project to fill an unmet need for the introductory course in biomedical device manufacturing and development that we started at the City University of Hong Kong and the New York University in the Fall 2012 and the Spring 2016, respectively. We embarked a journal to write a text to accompany our courses with the encouragement and support of our editors, Merry Stuber, Murugesan Tamilselvan, and Bibhuti Sharma at Springer Nature. There are many who have contributed their time in reviewing and editing the text, including our teaching assistants and students, Jacob Charles Harris, Apratim Bajpai, Jimmy Yau, and Kevin Guan, and to draft figures, including Yi Liu, Wei Huang, Lelin Liu, Lok Hin Law, and all the students of the course Manufacturing of Biomedical Devices at City University of Hong Kong in Fall 2017. We thank the support of Teaching Development Grant (project no. 6000613) from the City University of Hong Kong and the support from the National Science Foundation (CBET 1701322 to W. C.).

Contents

About the Authors

Raymond H. W. Lam is an Associate Professor in the Department of Biomedical Engineering at City University of Hong Kong and City University of Hong Kong Shenzhen Research Institute. He holds a first honor B.Eng. degree (2003) and an M. Phil. degree (2005) in Automation and Computer-Aided Engineering at Chinese University of Hong Kong and a Ph.D. degree (2010) in Mechanical Engineering at Massachusetts Institute of Technology. He was a Postdoctoral Fellow in the Department of Mechanical Engineering at the University of Michigan from 2010 to 2011, before he joined City University of Hong Kong as an Assistant Professor in 2011. He is also a Scholar of Croucher Foundation. His overall research objective is to bridge science and engineering knowledge, and currently, he aims at developing/applying microengineering techniques to advance the cell biology research.

Weiqiang Chen is an Assistant Professor in the Departments of Mechanical and Aerospace Engineering and Biomedical Engineering at New York University. He received his B.S. in Physics from Nanjing University in 2005 and M.S. degrees from Shanghai Jiao Tong University in 2008 and Purdue University in 2009, both in Electrical Engineering. He earned his Ph.D. degree in Mechanical Engineering from the University of Michigan in 2014. His research is focused on microfluidics, lab-on-a-chip and organ-on-a-chip systems, biomaterials, and biomanufacturing for new and improved solutions for emerging problems in cancer biology, mechanobiology, immune engineering, and stem cell-based regenerative medicine. He is the recipient of the Young Innovator Award of Cellular and Molecular Bioengineering from the Biomedical Engineering Society, the National Institute of Biomedical Imaging and Bioengineering Trailblazer Award, the American Heart Association Scientist Development Award, the New York University Whitehead Fellowship and Goddard Junior Faculty Award, the Baxter Young Investigator Award for distinguished research for critical care therapies, the American Heart Association Predoctoral Fellowship, the University of Michigan Richard F. and Eleanor A. Towner Prize for Outstanding Ph.D. Research, and the ProQuest Distinguished Dissertation Award.

Chapter 1
Introduction to Biomedical Devices

Abstract With increasing public health awareness and a rising aging population, there is an increasing market demand for high-quality, low-cost, and safe biomedical devices globally. Biomedical devices include a wide range of products with different complexities and purposes. Different regulatory agencies have different definitions of biomedical devices, which can be briefly classified as either diagnostic devices or treatment devices with different risk classes. This chapter discusses about the industry, e.g., the market growth and regulatory issues. It also includes an overview of biomedical devices such as surgical devices, skin devices, cardiovascular devices, skeleton devices, and tissue grafts.

1.1 Overview of Biomedical Devices

Biomedical/medical devices are instruments, machines, implants, in vitro reagents, software, materials, or other related articles that are purposed for the safe and effective prevention, diagnosis, treatment, and rehabilitation of illness and disease for human beings. These devices can achieve many aims such as diagnosing disease and injury, monitoring treatment, supporting life, investigating samples, replacing body parts, etc. Thus, biomedical devices actually include a wide range of products with different complexities and purposes, such as simple surgical scalpels, tongue depressors, medical syringes, contact lenses, medical thermometers, and blood sugar meters, to more advanced devices, such as X-ray machines, medical robots, programmable pacemakers, heart valves, microchip implants, and neuroprostheses.

Different regulatory agencies have different definitions of biomedical devices. The Food and Drug Administration (FDA) classifies a product as a medical device in section 201(h) of the Federal Food, Drug, and Cosmetic (FD&C) Act if it fits within the definition of "an instrument, apparatus, implement, machine, contrivance, implant, in vitro reagent, or other similar or related article, including a component part, or accessory," which is:

- Recognized in the official National Formulary, the United States Pharmacopeia, or any supplement to them

R. H. W. Lam, W. Chen, *Biomedical Devices*,
https://doi.org/10.1007/978-3-030-24237-4_1

- Intended for use in the diagnosis of disease or other conditions, or in the cure, mitigation, treatment, or prevention of disease, in man or other animals
- Intended to affect the structure or any function of the body of man or other animals, and which does not achieve its primary intended purposes through chemical action within or on the body of man or other animals and which is not dependent upon being metabolized for the achievement of any of its primary intended purposes

In addition, the most fundamental concept that differentiates medicines and biomedical devices is the idea of their mechanism of action, the means by which the principal intended purpose of the product is achieved. Medicines are designed to act largely by pharmacological, immunological, or metabolic means, with the assumption that they are interacting with the body at a cell receptor or molecular level. In contrast, according to the definition established in the general Medical Devices Directive 93/42/EEC, a “biomedical device” is any instrument, apparatus, appliance, software, material, or other article, whether used alone or in combination, including the software intended by its manufacturer, to be used specifically for diagnostic and/or therapeutic purposes and, necessary for its proper application, intended by the manufacturer to be used for human beings for the purpose of:

- Diagnosis, prevention, monitoring, treatment, or alleviation of disease
- Diagnosis, monitoring, treatment, alleviation of, or compensation for an injury or handicap
- Investigation, replacement, or modification of the anatomy or of a physiological process
- Control of conception

According to the World Health Organization definition, it must also not achieve its principal intended action in or on the human body by pharmacological, immunological, or metabolic means but may be assisted in its function by such means.

There are different ways of classifying biomedical devices. For instance, the rules by which biomedical devices are classified by the Medical Device Directive have the effect of stratifying them in terms of the amount of risk their use poses, although the terms “high risk” and “low risk” are not used in the classification rules. Devices are grouped into classes by aspects such as route of exposure (intact skin, breached or injured skin, natural body orifices, or introduction into the body via a surgical procedure) and the way in which they work (e.g., supply of ionizing radiation, delivery of medicinal products, intended to be absorbed by the body). Additional “special” rules apply, and these have the effect of placing the device into a specified class regardless of its other attributes. Furthermore, based on the purpose of the device, biomedical devices can also be classified as either diagnostic devices or treatment devices. Diagnostic biomedical devices help in figuring out the causality behind a malaise, whereas treatment devices can help in fixing the issues that crop up in patients.

Biomedical devices are expected to achieve their main purpose principally in physical, structural, or mechanical ways. Furthermore, it is possible that the manufacturer may wish to coat the device surface with a material that will enhance cell proliferation. Such material may include derived proteins such as growth factors or materials such as collagen or fibronectin which may be human- or animal-derived. The inclusion of such materials can have a huge impact on the way the product is classified for the purpose of regulations. Biomedical devices may be combined with pharmacologically active materials to assist their function provided that the pharmacological action is secondary to the main (physical) function of the device. All devices that contain a medicinal substance that has an integral action ancillary to that of the device itself, such that the device function is assisted by the ancillary action of the medicinal substance, are categorized in the highest-risk class. These devices require prior examination of their technical data by a notified body, and in addition, the safety and usefulness of the ancillary medicinal substance must be verified by a competent authority for medicines before the notified body can issue its approval.

On the other hand, "tissue engineering" is generally accepted to be the manipulation and culture of human cells or tissue ex vivo to produce therapeutic products. In December 2007, the advanced therapy medicinal products regulation (EC) 1394/2007 (ATMP) established a formal definition for "tissue engineering," bringing products that meet this definition firmly within the scope of medicinal product regulation. Article 2.1(b) "tissue-engineered product" means a product that:

- Contains or consists of engineered cell or tissues
- Is presented as having properties for, or is used in or administered to human beings with a view to regenerating, repairing, or replacing a human tissue

A tissue-engineered product may contain cells or tissues of human origin, animal origin, or both. The cells or tissues may be viable or nonviable. It may also contain additional substances, such as cellular products, biomolecules, biomaterials, chemical substances, scaffolds, or matrices.

Products containing or consisting exclusively of nonviable human or animal cells or tissue which do not contain any viable cells or tissues and which do not act principally by pharmacological, immunological, or metabolic action shall be excluded from this definition.

Cells or tissues shall be considered "engineered" if they fulfill at least one of the following conditions:

- The cells or tissues must have been subject to substantial manipulation so that biological characteristics, physiological functions, or structural properties relevant for the intended regeneration, repair, or replacement are achieved. The manipulations list in Annes 1 of Regulation EC No. 1394/2007, in particular, shall not be considered as substantial manipulations.
- The cells or tissues are not intended to be used for the same essential function or functions in the recipient as in the donor.

The reach of this definition is quite extensive, and the majority of products containing viable cells will be classified as an advanced therapy medicinal product (ATMP). Scaffolds do not in themselves meet the definition of an ATMP because they contain no engineered cells. However, if one is sold already seeded with engineered cells, the combined whole product will be regulated as a medicinal product under the ATMP regulation.

Like any type of design engineering, biomedical device design must first begin with identifying a problem to solve. The ideation of a new device typically arises from real-world cases in which people express frustration in current medical solutions or at the lack thereof. Medical professionals often voice their opinions on medical device designs as they cannot afford to encounter problems with their devices and tools when performing their work as product difficulties or design flaws may lead to serious life-threatening consequences. There exist numerous problems to be solved within medicine ranging from needs for improved small wound care bandages up to cancer diagnostic devices which can be shrunk down in size to make them more portable and accessible to a wider patient group. Engineers must also consider their ideal end user base which can vary between trained doctors and the general population and must take into account the traits of these users. For example, the accelerating rate of growth in the aging population around the world is greatly emphasizing the design of new devices which can be used for geriatric care. However, regulatory concerns begin even during this early conceptualization stage of design. Medical device design is an incredibly expensive process which is closely monitored by regulatory agencies for years to ensure quality and safety concerns are met. After conceiving a device based on a specific need, the device will be designed and honed over time to maximize its potential usefulness and problem-solving capacity. Then, according to regulatory specifications, testing of the proposed design will commence with detailed technical attention. This testing phase is crucial to successfully passing a device through regulations and often requires copious amounts of scientific data providing evidence of a device's safety and effectiveness at performing its intended task. Companies must be fully invested in their device's potential for marketability as this process may take years and millions of dollars to fully complete after multiple revisions. Once a device clears this phase, it may finally enter production and be sold, though its producer is often required to conduct periodic follow-up evaluations to ensure the real-world safety of the product. Each of these regulatory steps have their own nuances depending on a variety of factors, namely, the country in which they are being registered and the nature of the product being tested.

The trend of biomedical manufacturing emerged and played an important role in biomedical device industry. Through the application of different manufacturing technologies, the safety, quality, quantity, cost, efficiency, and speed can be advanced for healthcare service and biomedical device production. On the other hand, biomedical manufacturing can apply to many disciplines, but biomedical device is the core of it because it includes a lot of manufacturing operations. Compared with the conventional manufacturing, biomedical manufacturing has more potential, because as the technology develops, biomedical engineering is becoming more and more important in engineering field.

1.2 Biomedical Device Industry

With increasing public health awareness and a rising aging population, there is an increasing market demand for high-quality, low-cost, and safe biomedical devices globally. In the past decades, the demand for biomedical devices was booming, which led to explosive growth in the global biomedical device industry. The leading industrial countries such as the United States (USA), German, and Japan remain the standard setters in the biomedical device industry, producing the most competitive products and sharing the largest market shares worldwide. As statistics shows that the national health expenditures in United States in 2015 had reached 3205.6 billion USD, which account for about 10% of the US annual gross domestic product (GDP) and the German medical device industry also shares 11.9% of its GDP in 2012. In addition to these developed industrial countries, emerging markets like China have become new leading markets for biomedical device industry. Although the market is still relatively small in size and immature compared to developed countries, China's medical device industry was booming rapidly with an average annual growth rate of above 20% in the past decade and, no doubt, will continue to grow and even overwhelm some of these traditional markets in the near future.

Currently, imaging diagnostic instruments (magnetic resonance imaging, spiral computerized tomography, etc.), implantable medical device products (pacemakers, vascular stents, neural devices, urology department implements, etc.), orthopedic and dentistry devices, common surgical instruments, veterinary apparatus, and all kinds of consumables are the largest fields of biomedical device needs and also the important profit sources of medical device industry. In recent years, the biomedical device industry is rapidly expanding to advanced fields such as imaging diagnosis, telemedicine, molecular diagnosis, medical robotics, minimally invasive surgery, point-of-care devices, etc. The huge and increasing demand in biomedical devices requires a significant investment in research, development, and manufacturing of new biomedical technologies and products. Biomedical device industry involves a variety of conventional as well as emerging fields including biomedical product design, manufacturing, process optimization, material characterization and processing, computer and data science, tissue engineering, medicine, etc. With the rapid advancement of new technologies, product development demands ever-shortening life cycles and time to market. Besides, there is a clear trend toward greater functionality and miniaturization of biomedical products which is exemplified by more advanced manufacturing technologies. Overall, the market demand for biomedical manufacturing will be more and more needed and cover a broader range of healthcare issues, ultimately improving the quality and effectiveness of patient care and human life.

United States

Due to recent improvements in technology and health awareness, the US government has been putting more resources into different health categories in order to help its citizens lead healthier lives. In particular, with the Social Security Amendment of 1965, Medicare and Medicaid became the United States' first public insurance

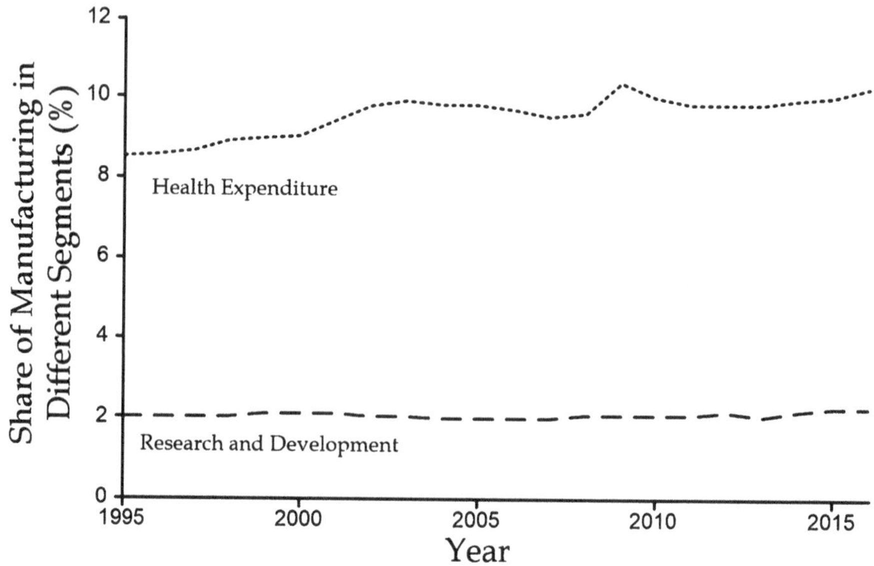

Fig. 1.1 Share of manufacturing GPD from different segments. (Public data from US Bureau of Economic Analysis)

programs. As shown in Fig. 1.1, the amount of national health expenditures in the United States in 2015 reached 3205.6 billion USD, which accounts for about 10% of the US GDP. It had been the highest share of the US economy in the past decade. More importantly, the amount of money spent on personal healthcare reached 2717.2 billion USD with an increase of 6%, the fastest increase of the three parts of health consumption expenditures, those being personal healthcare, administration, and net cost of private health insurance and public health. Besides, the proportion of personal healthcare over health consumption expenditures is almost 90%. All these figures show the fact that the health industry is taking a more and more important role in our lives and a bigger proportion of our GDP, and the public is beginning to be aware that their healthcare should come first when compared to other issues.

In the 30 years prior to 2008, the demand for medical devices in the United States was booming, which led to explosive growth in the American medical device industry. Medical device product developments in the United States have mainly been within cardiac devices, neural devices, diabetes, urological devices, and common surgical instruments. These areas are also the largest areas with medical device needs, especially implanted medical devices and diagnostic instruments as they are important profit sources. In recent years, the US medical device industry has rapidly expanded to advanced fields such as imaging diagnosis, telemedicine, molecular diagnosis, and minimally invasive surgery. American technology has been leading the field internationally, causing countries around the world to take notice. American companies have also been aggressively exporting medical devices to European and Asian markets in order to compete fiercely with their foreign counterparts. Europe has been America's largest export market for medical devices for years.

European Union

In 2012, the sales of medical devices from 11 Western European countries were valued at $ 82.4 billion, and this number has continued to rapidly grow in recent years. For example, Germany is second only to the US medical device industry's scale with more than 170 medical device manufacturers, most of them small- and medium-sized companies. Germany is Europe's largest producer and exporter of medical devices and is the world's leading medical device exporter. Over the past few years, German medical device exports have surpassed those of Japan to rank them second in the world. Current German medical device companies export about two-thirds of their total products, and the German medical device industry is equivalent to 11.9% of the country's GDP, significantly higher than the other major Western European economies. Due to the well-developed German medical device industry and the government's policy support, the product development costs for German companies are much lower than their foreign competitors. The average total cost to develop a new piece of medical equipment for a German company is between 8 and 10 million euros, which is much lower compared to American companies who must spend up to $80 million to develop a similar product. France is Europe's second largest producer of medical devices and also Europe's major medical device exporter. In 2012, the French medical device market totaled sales of up to 9 billion euros. France has a relatively well-developed medical device industry, imports medical equipment products, and exports medical device products. Imported products are mainly concentrated in magnetic resonance imaging (MRI), positron emission tomography (PET), spiral computerized tomography (CT), and other advanced electronic diagnostic imaging products along with implantable medical device products such as pacemakers and vascular stents.

China

In addition to the United States and Europe, China has emerged as one of the new leading markets in the medical device industry. For the past two decades, the total sales of medical devices have grown continuously in China, with an average annual growth rate of above 20%. In 2013, the market size of the medical device market in China (Chinese mainland) reached RMB 212 billion, an increase of 24.7% when compared to 2012 (Fig. 1.2). In 2016, medical device market sales reached about 370 billion RMB, an increase of about 75% in just 3 years of which medical device market is about 269 billion RMB. In 2015, the home medical device market exceeded the 100 billion RMB checkpoint at about 101 billion RMB, accounting for a growth of about 27.30%. In particular, the revenue growth in the manufacturing of medical equipment in the pharmaceutical industry above a designated size grew at a high rate of 13.25% (comparing to the year 2015).

Compared to the international market, although the Chinese total market size is still smaller than the United States, Europe, and Japan, China's medical device industry has maintained a relatively higher growth rate in the past decade and will properly continue to grow with a high speed. China has a large population, but most people still do not have enough access to medical devices. With an increasingly aging population, increasing health awareness, and strategic government support, there is a great potential for the medical device industry in China.

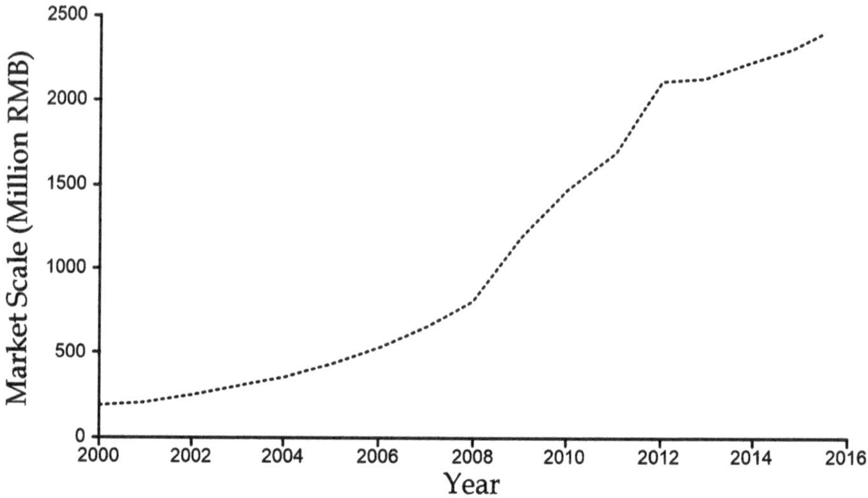

Fig. 1.2 Market scale of China medical device industry. (Public data from National Bureau of Statistics of China)

However, as the medical device industry in China started relatively late, there is still a certain gap between international medical device giants, especially for high-end medical equipment. Domestic medical institutions still largely rely on imported equipment. The high cost of imported medical equipment is one of the reasons for the high cost of medicine in China. Up to now, China's medical equipment manufacturing enterprises mainly concentrated on low-end conventional products, including small- and medium-sized equipment and consumable products, such as monitors, anesthesia machines, blood cell analyzers, biochemical analyzers, etc.

According to the National Bureau of Statistics of China, from the perspective of the product structure of China's medical device market in the past 3 years, imaging diagnostic equipment occupies the largest market share, which has been kept at about 40% in recent years and is still on the rise. The second largest share is all kinds of consumables, occupying about 20% market share. The market share of orthopedic and implantable medical devices is declining. The remaining market share is occupied by dentistry and other instruments. From the developing trend of the past few years, medical diagnosis, monitoring and treatment equipment, and medical (including surgical and veterinary) apparatus and instruments manufacturing industry have been occupying the major share (67% of total market share).

1.3 Regulatory Issues

The manufacturing and marketing of medical products are subject to extensive control via a range of methods. Legal instruments, such as European regulations and directives and the national laws of the member states, establish a framework that controls all aspects of the medical products business. These instruments are supplemented by several means, including guidelines and international standards. The classification of a product as a medical device or a medicinal product determines almost every aspect of development, approval for sale, and subsequent marketing, and therefore, an early and correct determination of how it will be regulated is of paramount importance.

The fundamental purpose or intended use of biomedical devices relates to prevention, diagnosis, or treatment of a disease or management of a physiological condition. Since most of these devices come in contact with the human body, their manufacturing has to be a zero-error process; and the guidelines regarding biomedical devices are very stringent. Failure of medical devices to adhere to these guidelines can lead to serious complications in patients such as in the case of metal-on-metal hip implants which during wear may release toxic metallic particles, causing serious health implications in the implant site and potentially throughout the body. A lot of advancements have happened in the field of biomedical devices, but it is still far from being an exhausted quarry and has a lot of scope for advancements and improvements upon existing designs. These products are subject to a complex framework of legislation that is primarily aimed to protect the public. The overall intentions of regulations are to ensure that products are not harmful under normal conditions of use, that their benefits exceed the potential risks, and that they have been shown to provide the benefits claimed by their manufacturer. New products are usually categorized by regulatory bodies based on their potential health risks, and the classification of a product will dictate the process it must undergo before being approved for marketing. Regardless of the complexity a new product, the fundamental health protection principles underlying their regulation and control can be summarized in terms of a few simple questions:

- Are the components and the final product acceptably safe for the intended use?
- Is the manufacturing process capable of delivering a consistent product?
- Does the product perform in the manner intended by the manufacturer?
- Are the risks associated with the use of the product outweighed by the clinical benefit of using it?

United States

Governments around the world have their own specialized agencies that help write and carry out regulations for new medical devices with each varying in its requirements for device submission and documentation requirements. Perhaps the most well-known federal medical device agency is the US Food and Drug Administration (FDA) which oversees all medical devices marketed within the United States. All of these items are subject to regulatory control under the FD&C Act and Title 21 of the Code of Federal Regulations (21 CFR). Twenty-one CFR includes a wide range of medical devices, ranging from tongue depressors all the way to X-ray scanners which include their own special set of regulations as a radiation-emitting electronic product. The FDA is notorious for its extensive regulatory prowess due to its extensive premarket application process.

A proposed medical device must first be classified based on its perceived degree of risk toward human health as illustrated in Fig. 1.3. According to the FDA, Class I is for the lowest-risk items which are subject to only general controls (e.g., manual toothbrush). Class II medical devices are those which have a moderate level of risk and are subject to both general and more extensive special controls (e.g., condom or noninvasive blood pressure monitor). These special controls are specified for Class II devices when general controls alone are insufficient to provide reasonable safety and efficacy assurance and when there is enough information to establish that these special controls would provide such assurance. Finally, Class III devices present the highest risk to users and are subject to general control and a premarket approval application (PMA) (e.g., artificial heart valves). These devices are meant to be used for supporting and sustaining human life or preventing impairment of health. They may also present a potential risk of illness or injury which general or special controls may not provide reasonable assurance of safety and efficacy. Most notably, a device may receive an automatic Class III designation if there is no such information to determine any of these factors until such information can be obtained.

The proposed device must undergo a certain premarket submission procedures. The most common of these is a 510(k) premarket notification, usually reserved for some Class I and most Class II devices which must be able to demonstrate that they are "substantially equivalent" to previous devices that also perform their intended tasks. PMA, the most stringent of approvals, is reserved for Class III devices and requires valid scientific evidence demonstrating reasonable assurance of safety and

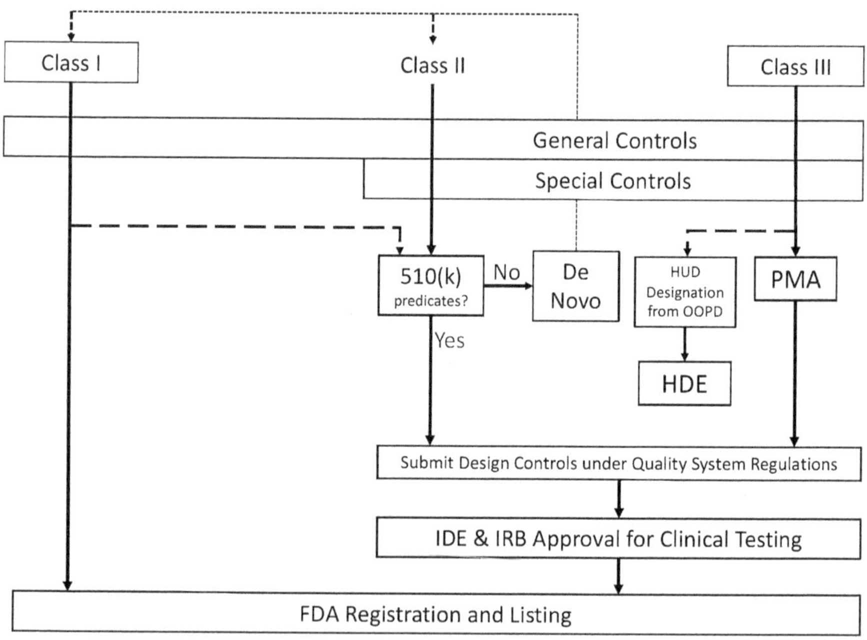

Fig. 1.3 Classification procedures of medical devices by FDA

effectiveness for the device's intended use. For novel devices without predecessors which automatically fall under Class III, a de novo submission is required to further classify a device into a Class I or II if it meets certain criteria. Finally, a Humanitarian Device Exemption (HDE) is a request to gain approval for a Class III device for use in patients with rare diseases or conditions. These devices require a designation as a Humanitarian Use Device (HUD) which must be granted through the FDA's Office of Orphan Product Development (OOPD). Once the correct premarket submission type has been chosen, the necessary information concerning the medical device must be collected, including design controls of Class II or III devices under Quality System Regulation (21 CFR 820.30). Any clinical testing will also require the granting of an investigational device exemption (IDE) from the FDA and the approval of an institutional review board (IRB). Finally, the premarket submission will require the payment of a fee for a 510(k) or PMA. If granted, the device can be listed in the FDA's device database, but this does not denote clearance or approval of establishment or the products by the FDA.

European Union
Biomedical device sold in the EU is subject to one of three Medical Device Directives: the general Medical Devices Directive (MDD), the Active Implantable Medical Device Directive (AIMDD), and the In Vitro Diagnostic Device Directive (IVDD), depending on the type of device under consideration. For the purposes of this discussion only, the general MDD will be considered. The scope of the other two directives is not likely to be relevant to tissue regeneration products. The authorization procedure for medical devices may take several different routes, ranging from "self-certification," in which the manufacturer makes no submission for external assessment but instead signs a declaration that their product is in compliance with the requirements of the relevant medical devices directive, to prior authorization of the product by one or more external assessment agencies before they can make the declaration that their product complies with the directives.

The central principle of the Medical Device Directives is the establishment of "Essential Requirements" (ERs) to which the device must conform, but the solutions the manufacturer may use to demonstrate conformity with the ERs are not specified in the legal documents. The use of ERs is a cornerstone of the "New Approach" adopted by the European Commission in order to establish requirements for the safety and suitability for purpose of a wide range of goods without being unduly restrictive. This approach is designed for fast-moving and innovative industries which develop and introduce new technologies rapidly and that would be disadvantaged by the imposition of fixed requirements upon a variable and rapidly evolving technical field. The manufacturer demonstrates their assertion that they have met all the ERs relevant for their product by affixing the "CE mark" ("CE" as an abbreviation of *Conformité Européenne*, meaning European Conformity) to the product or its labeling.

The EU Commission also uses a classification system to separate new medical devices based on their level of health risk as noted in Annex VII of the Medical Devices Regulations (MDR). The four major types of devices classified by this system fall into one of the following: noninvasive devices, invasive medical devices, active medical devices (i.e., electrically powered), and special devices (e.g., contraceptives, disinfectants, radiological diagnostic devices). Unlike the US FDA's system, there are two low-risk Class I subtypes: "self-certified" non-sterile devices with no measuring function and sterile devices which may or may not have a measuring function, including reusable surgical instruments. Class IIa is for medium-risk devices, while Class IIb is reserved for medium- to high-risk devices. Class III devices are considered the highest risk and receive the most strenuous testing.

In order to receive a "CE mark" of approval for marketing, all devices must undergo a series of regulatory steps. First, a "person responsible" for regulatory compliance within the development team must be appointed, and device classification should be determined using Annex VII of the MDR. Then, teams must implement a quality management system (QMS), usually based on ISO 13485, which can account for clinical evaluation, post-market surveillance (PMS), and post-market clinical follow-up (PMCF) plans along with making arrangements with suppliers about notified body audits. At this point, self-certified Class I devices can prepare a CE technical file based on Annex II, while every other device class must prepare a more extensive CE technical file (or dossier file for Class III devices) which includes details on the device's intended use, testing reports, Clinical Evaluation Reports (CER), risk management plan, and a unique device identifier (UDI). Next, all device teams must appoint an authorized representative located within the EU to handle regulatory issues. Nonself-certified Class I devices and up will then have their QMS and technical file (or design dossier for Class III) audited by a third-party notified body to review QMS and products. Once the audit is completed, the device is issued a CE marking certificate, and an ISO 13485 certificate is issued for the facility. A CE mark is valid for 3 years, but it will usually be reviewed annually during ISO 13485 audits. Then, all device types must prepare a Declaration of Conformity stating that the device complies with the MDR before placing the CE mark on the device and registering the device and its UDI in the European Databank on Medical Devices (EUDAMED). CE marks on self-certified Class I devices do not expire, but all device classes must ensure MDR compliance and perform clinical evaluation, PMS, and PMCF to keep their certifications. These standards apply across the whole of the European Union and help to smooth out the process of manufacturing and delivering medical devices throughout the continent.

A harmonized standard is a European standard, prepared by the European Committee for Standardization (CEN) (or the European Committee for Electrotechnical Standardization (CENELEC) for electrical products) under the mandate of the European Commission with the purpose supporting the Essential Requirements of a New Approach directive. Thus, for the medical device manufacturer, compliance with the harmonized standards provides a straightforward way of meeting the requirements of the directive: if the requirements of the standard are met, there is a legal presumption that the product will also meet the equivalent requirements of the directive. In the medical device field, standards cover specific aspects, for example, the quality of surgical gloves or safety and performance requirements for pulse oximeters. There are also standards covering general aspects applicable to all devices, such as requirements for biological evaluation of medical devices or the labeling of the devices.

Depending on the class of device and the route of conformity assessment chosen by the manufacturer, some combination of the design, manufacturing, and testing of medical devices should be performed in accordance with a quality system. The use of the harmonized standard ISO EN 13485, which adapts the general quality system requirements of ISO 9001 to specific requirements for medical devices, ensures that the manufacturer's processes and systems are compatible with the requirements relating to quality systems for CE marking of devices. The manufacturer may make use of also ISO 14971 for risk analysis procedures.

China

In comparison to the United States or the EU, China's regulatory process is in many ways alike to its Western counterparts while also possessing some unique quirks. There are three regulatory bodies within China tasked with monitoring compliance to medical device standards. The China Food and Drug Administration (CFDA) concerns itself with the quality of medical devices, drugs, and healthcare services. The Center for Medical Device Evaluation (CMDE) conducts dossier reviews on new medical devices during their registration processes, while the General Administration of Quality Supervision, Inspection, and Quarantine (AQSIQ) is responsible for the safety registration, certification, and inspection of certain medical devices. Needless to say, these bureaucratically overlapping duties can make the device regulatory process confusing, especially for Western companies. However, there is still a standard risk classification system to be found like in the United States or the EU. Class I devices present the lowest risk and the safety and effectiveness of which are tested through routine administration. Class II devices require further control during routine product testing to ensure their safety and may require clinical testing. Class III devices are the highest-risk items and include implantable devices or life-sustaining devices and therefore must face the most scrutiny during product safety and effectiveness evaluation.

The marketing registration process for medical devices in China first requires that any foreign-made device must have already received prior market approval in its country of origin. These can include an ISO 13485 certificate from the EU, an establishment registration from the US FDA, or a manufacturing license from Japanese or Korean companies. Along with this, the approval process for Class I devices is unusual compared to other countries. First off, a Class I device does not need to be registered with the central CFDA; rather, it can be registered at a local provincial branch using a notification application. Foreign-approved Class I devices also do not require any data acquired from testing performed in China. Data that was originally acquired during a company's foreign tests will suffice for Class I regulations. On the other hand, all Class II and III medical devices from abroad to be evaluated in China are tested at Chinese Medical Device Evaluation Centers. Class II and III will require extensive amounts of technical and clinical evaluation data before they begin their own testing. These devices are also always clinically tested at Evaluation Centers with a few exceptions. These include when the device has a clear working mechanism, finalized design, and mature production technology. Also, if a medical device of the same variety is already on the market and has been used in clinical practice for many years without any record of serious adverse events, the product may circumnavigate certain tests. Finally, if there will also not be any changes to the product's conventional usage or the safety and efficacy of product can be proven through nonclinical evaluations or through analysis of data from other clinical trials of other similar devices, then clinical tests will not be performed. Class I devices will require technical documentation and undergo administrative reviews, while Class II and III devices require much more extensive dossier preparations and technical and specification documentation, all of which must be in Simplified Chinese. After review, Class I devices will be issued a non-expiring voucher which allows for the sale of the device in China. Class II and III devices are given certificates which are valid for 5 years.

1.4 The Demands for Biomedical Engineers

The huge global demand in biomedical devices creates a variety of new job opportunities. The United States' medical device market value stood at approximately $156 billion, accounting for 40% of the worldwide medical device market, and exported more than $41 billion of products in 2017. More than two million jobs in the United States are based in the field, and many companies are up and coming start-ups which aim to invest revenues in the research and development of new products. Biomedical-related positions have been consistently ranked as top jobs in many developed countries and now are starting to boom in developing countries such as China as well. For instance, the job growth rate in biomedical engineer field (27%) is the highest among all occupations in United States, over petroleum engineering (25%), civil engineering (19%), and environmental engineering (14%).

Through these extensively arduous processes, medical device design is a highly collaborative endeavor, often bringing together the knowledge and skills of biologists, medical doctors, product engineers, manufacturing engineers, packaging specialists, quality control specialists, and many more professionals. The engineering process for these types of products requires breaking down the larger design problem into smaller, more manageable subproblems which can then be evaluated, solved, and combined into a complete device by multiple specialized teams. Medical device design is still a relatively new branch of engineering, and cooperation between more traditional product engineers focused on aspects such as device mechanics and medical specialists concerned with the biomedical implications of a device's functions must be combined effectively during the design process. Even software engineers have become intertwined within the medical device design field and are also subject to the same design regulatory standards. While these fields have previously had separate knowledge bases, the emergence of biomedical engineering has caused these fields to merge, requiring researchers to work together to solve unmet medical needs going forward.

When biomedical engineers start to learn biomedical manufacturing, they require studying a new set of knowledge, for example, the product design and optimization, material properties and selections, the manufacturing processes, control of the quality, the interaction between tissue and materials, committing for legal, the knowledge of anatomy and physiology, etc. Therefore, knowledge of biomedical manufacturing and regulations is a critical barrier to achieve successful development of new biomedical products. After getting this knowledge, biomedical engineers can design a product which has a well balance of different factors such as costs, material, machining process, etc.

Those who are interested in biomedical engineering normally need a bachelor's degree from a biomedical engineering program at least. An alternate option is to earn an undergraduate degree in a different type of engineering and either completing a graduate degree in biomedical engineering or receiving on-the-job training in biomedical engineering. Co-ops and internships can make a big difference for students in both understanding the work and finding jobs. Employers value work experience, so co-op programs where students earn academic credit for structured job experiences are valuable. Ph.D. programs and postdoctoral appointments also help applicants when applying for a job.

1.5 Human System Basics

Biomedical devices are intended to be developed specifically for diagnostic or therapeutic purposes for human beings. A human body is an intricately complex system. An average human body consists of around 37.2 trillion cells forming different organs in the body. Multiple organs that perform similar functions are grouped into the following different organ systems:

- Skin, hair, and nails make up what is called the integumentary system. Skin is the largest organ of the human body. It acts as the first line of defense against all manner of pathogens, like bacteria and viruses. In addition to this, the skin also helps in maintaining normothermia or euthermia, i.e., normal human body temperature by perspiration. Perspiration also assists in waste disposal.
- The skeletal system consists of bones, cartilages, connecting tissues like tendons and ligaments, and the teeth. The skeletal system supports our body and helps in movement. The bone marrow is involved in the production of blood cells.
- The lymphatic system includes lymph nodes, ducts, and vessels. Its main function is to manufacture lymph, which is a fluid that contains white blood cells and helps in fighting infections.
- The respiratory system consists of the lungs, diaphragm, and trachea. Cells need oxygen to perform their functions. This process releases carbon dioxide as waste. The respiratory system performs the function of breathing in oxygen and breathing out carbon dioxide.

- The muscular system is composed of three types of muscle tissue: the skeletal muscles or voluntary muscles, so named because of the fact that we can control them; the striated muscles or the involuntary muscles that are found inside organs; and cardiac muscles that make up the heart. They help in movement, the transport of substances through organs, and the pumping of blood through the body.
- The nervous system is made up of two parts: the central nervous system, i.e., the brain and spinal cord, and the peripheral nervous system, consisting of the nerves that connect the central nervous system to other parts of the body. It controls all our voluntary and involuntary actions.
- The digestive system consists of the mouth, esophagus, stomach, small intestine, large intestine, rectum, and anus, along with the liver and pancreas. It helps in transport, digestion, and absorption of food inside the body.
- The urinary system consists of the kidneys, ureters, bladder, sphincter muscles, and urethra and helps in eliminating urea, which is produced during the breakdown of food in the human body.
- The endocrine system consists of the pineal gland, pancreas, pituitary gland, testes, ovaries, thyroid gland, adrenal gland, and parathyroid gland. These organs secrete hormones that help cells in performing their functions. Furthermore, cells are organic engines, and like any engine, they need fuel and oxygen to perform their functions.
- The circulatory system functions as the carrier of nutrients and oxygen to the cells. Further, the waste material and carbon dioxide generated during cellular functions have to be removed from the cells. This task is performed by the cardiovascular system. Apart from this, the cardiovascular system takes care of the transport of other chemicals and enzymes to and from the cells. The cardiovascular system consists of the blood, blood vessels, heart, veins, and arteries.
- The reproductive system, as the name implies, is involved in the reproduction of human beings. Unlike other systems, the reproductive system is different in males and females. In males, it consists of the testes and penis. The female reproductive system consists of the vagina, uterus, and ovaries.

In reality, all these organ systems can be affected by disease and injury and thus require medical devices for diagnosis and treatment so as to facilitate their proper function and recovery. Dental implants, sutures, and casts for broken bones are some examples of such devices that have been in use for a very long time. There are also devices that help in determining the cause of malaises. Examples of such devices include sphygmomanometers, thermometers, stethoscopes, etc. In this chapter, we will specifically discuss a few representative examples of biomedical devices to help new biomedical engineers learn the basic concepts and principals that are involved in the design, development, and manufacturing of biomedical devices for different human organ systems.

1.6 Surgical Tools

1.6.1 Surgical Scalpel

The surgical scalpel is a small and extremely sharp bladed instrument used for surgery, the term for anatomical dissection. Scalpels may be single-use, disposable items or reusable. Reusable scalpels can have permanently attached blades that can be sharpened or have removable single-use blades. Disposable scalpels usually have a plastic handle with an extensible blade and are used once; then, the entire instrument is discarded. Scalpel blades are usually individually packed in sterile pouches but are also offered non-sterile. In surgery, doctors will choose blades with different shapes and sizes according to different needs. Scalpel blades are usually made of hardened and tempered steel, stainless steel, or high carbon steel; in addition, titanium, ceramic, diamond, and even obsidian knives are also chosen depending on the applications. For example, when performing surgery under the guidance of medical imaging such as magnetic resonance imaging (MRI), steel blades are unusable because the blades would be drawn to the magnets or may cause image artifacts. Historically, the preferred material for surgical scalpels was silver on account of its antimicrobial properties. Scalpel blades are also offered by select manufacturers with a zirconium nitride-coated edge to improve sharpness and edge retention.

The manufacturing of scalpels should consider the working environment of the human body. Thus, scalpels usually should have properties including corrosion resistance, high strength, no rust capability, high-load tolerance, and low weight. A scalpel must have very good comprehensive mechanical properties and process performance, so scalpels usually use stainless steel as their material. Stainless steel does not rust in corrosive media such as air, water, steam, etc. and in corrosive medium such as acid, alkali, and salt. Some of the steel that does not rust in weak corrosive media does not necessarily corrode in strong corrosive media, and steel which does corrode in strong corrosive media is generally good when not rusted. In order to make sure stainless steel has great corrosion resistance and good mechanical and other physical properties according to needed requirements, the producer will add a higher content of alloying elements in steel and chrome, including added nickel, molybdenum, manganese, nitrogen, and other alloying elements. This method can not only change the chemical composition of passive film by strengthening it in its harsh medium corrosion resistance but also give the steel sufficient strength, plasticity, and toughness, improving its process performance.

1.6.2 Surgical Sutures

The skin forms the first line of defense against the pathogens and injuries to the human body. However, wear and tear of skin due to cuts and scrapes is a common phenomenon. In the cases of deep cuts, surgeons have to close the wounds using devices called surgical sutures. Sutures for stitching the skin have been in use for a very long time. Early records of surgical sutures can be traced back to Egypt around 3000 BC.

There are two parts of a surgical suture: the needle that is used to penetrate the skin and the suture, i.e., thread that is used to hold it together. Needles have two basic end types: eyed with a hole at the end for the thread and eyeless where the thread and needle are joined together. Eyeless needles are better than normal-eyed needles in a few respects. Since eyeless needles are single use, the risk of infection is reduced. Also, since the thread is crimped within the needle, the hole size is equal to the thread size, and effective blocking can be achieved. Multiple uses can lead to a decrease in the sharpness of eyed needles, which is avoided in eyeless needles due to their single use.

There are different types of sutures used in closing wounds, usually divided into two categories: absorbable and nonabsorbable. Nonabsorbable sutures include sutures made of silk, cotton, stainless steel, polyester, nylon, and polypropylene. Absorbable sutures are made of materials that can be broken down by enzymes in the body. For example, plain gut can provide effective wound support for 8–9 days. Fast-absorbing gut can provide the support for 5–7 days; chromic gut can provide the support for 10–21 days. Vicryl can support for 21 days, and Vicryl rapide can support for 10 days. These types of sutures are preferred in different applications/wound sites based on the duration of effective wound support, e.g., fast-absorbing gut can be used in facial wounds, while plain gut and Vicryl rapide can be used in wounds of the chest and extremities.

1.7 Devices for Sensory Organs

1.7.1 Skin Devices

The skin is the largest organ as measured in terms of total body surface area and weight. The skin protects against pathogens, regulates and prevents water loss, and acts in mechanosensation. The skin accomplishes these tasks through the various proteins and receptors throughout its layers. It consists of the epidermis layer and the dermis layer. Beneath the dermis is the hypodermis or subcutaneous fatty tissues. As seen, the skin has many layers acting as a barrier to protect from mechanical touch (which is also sensed), chemical reactions, and radiation. The skin can also regulate body temperature by releasing heat through sweating and can retain heat by preventing water loss. Sweating can also affect peripheral circulation and fluid balance of other ions in the body. The skin can also synthesize vitamin D naturally. There also exists a large network of nerve cells and receptors that sense and relay changes in the environment to the brain to process. Since the skin is so important, it is not surprising that it is the most studied organ in tissue engineering. In fact, tissue-engineered skin has been used clinically for the last 25 years.

Cuts, scrapes, burns, and wounds are forms of skin injuries. Minor cuts, scrapes, and wounds can be naturally healed by the skin. Tissue engineers study burns as they are not naturally healed as effectively. Burns are classified into three categories in accordance to their severity in terms of depth and the layers that they affect: (1) epidermal injury (first degree), (2) superficial dermal injury (second degree), and (3) epidermal plus near-full to full dermal injury (third degree). Skin tissue engineers have invented autograft and artificial skins to provide epidermal cover and dermal replacement.

There are biomaterials in use for tissue engineering of skin. For epidermal wounds, covering layers are typically applied over the wound area in order to deliver cultured keratinocytes so that they take on the wound bed and form a new epidermal layer. Examples include cultured epidermal sheets (Epicell), cultured epidermal sheets from plucked hair follicles (Epidex), sub-confluent cells on a synthetic carrier (Myskin), and cells delivered in a fibrin spray. For dermal replacement, implants often provide a dermal alternative to promote wound healing or are used in a two-stage skin replacement protocol. Other than the donor skin, there are already a number of commercially available products such as Permacol, Dermagraft, Transyte, and Integra (the details are described in Sect. 11.6.1). Alternatively, split-skin grafts can be applied for epidermal/dermal replacements. The related commercial products include Apligraf, Permaderm, and Orcel. Tissue-engineered skin is also a possible option.

1.7.2 Contact Lenses

Contact lenses, also known as corneal contact lenses, are thin lenses worn directly on the surface of the eyeball in order to correct vision or protect the eye. The current global market for contact lenses is about \$8 billion, with annual growth of around 5%. Among consumers, the average age of contact lens wearers globally is 31 years old with two-thirds of wearers being female. It can be seen from these figures that the consumers are young and especially female for the reason that contact lenses are different from traditional glasses. Aesthetics and cosmetics are the main motivating factors for people who either want to avoid wearing glasses or change the appearance of their eyes. When compared with spectacles, contact lenses typically provide better peripheral vision and do not give vision which is influenced on rainy days or turns indistinct owning to liquefaction. Besides, their direct contact with the eye can make them preferable for sports and other outdoor activities. Contact lens wearers can also wear sunglasses, goggles, or other eyewear of their choice without having to fit them with prescription lenses or worry about compatibility with glasses.

Since contact lenses contact directly with the eyes, the material ought to have good oxygen conductivity. In the process of maintaining the normal physiological activity of the cornea, oxygen molecules and other ions and molecules are of vital importance. Under normal circumstances, our cornea needs to breathe as long-term hypoxia of the cornea can lead to corneal complications. So, lens oxygen transmission rate is an important index for assessing how good contact lens materials are.

Another important index on which we should focus is moisture content. High and low levels of moisture content affect the characteristics of lens. The higher the moisture content is, the softer the lens will be, but wet lens are also easier to deform and damage, causing them to more easily lose water. Conversely, the lower the moisture content, the more correct the shape of the lens will be as the deformation degree is small and the relatively high water-bearing trait of the lens will not cause them to be prone to dehydration. Moreover, moisture content can affect the refractive index. The higher the moisture content is, the lower the refractive rate will be.

1.8 Cardiovascular Devices

1.8.1 Artificial Heart Valve

Our cardiovascular system performs the function of transporting nutrients and waste products to and from the cells. A human heart works 24 hours a day and pumps around 5–7 L of blood per minute. An abnormality in the function of the heart can lead to serious medical ailments, even death. The human heart contains four heart valves, acting as one-way doors to guide blood flow into the proper direction in both the systemic and the pulmonary circulations. A heart valve is located between the atrium and the artery. It is a thin sheet structure formed by the endocardial process. The heart valve surface is covered with endothelium, which is dense connective tissue, and is attached to the fibrous ring. The function of the heart valve is to prevent reverse blood flow. The atrioventricular valves (tricuspid and mitral valves) prevent backflow of blood from the ventricles to the atria, while the semilunar valves (pulmonary and aortic valves) prevent backflow from the arteries into the ventricles during diastole. All valves are believed to function in a passive, pressure-driven manner. They open when pressure gradients force the blood forwards and close when backward pressure gradients push the blood backward. The atrioventricular valves are anatomically and functionally different from the semilunar valves, the former being supported by papillary muscles that prevent leaflet prolapse during ventricular contraction.

Heart valve disease causes either insufficient opening (stenosis) or closing (regurgitation) dynamics of the valve or a combination thereof, which ultimately will result in heart failure. Congenital heart disease affects 1% of all newborns and often has its origin in abnormalities of one of the valves or its function. A common cause for acquired heart valve disease is rheumatic fever, currently still persisting in developing countries, thereby affecting children and young adults. In industrialized nations, acquired heart valve disease is mainly considered a degenerative pathology, predominantly affecting the elderly. In general, heart valve diseases are considered to be a worldwide major public health problem causing significant morbidity and mortality. The left-sided valves (aortic valve and mitral valves) are most prone to degenerative dysfunction in adult patients as these are located in the systemic circulation and are thereby exposed to harsh hemodynamic conditions. When heart valve disease is diagnosed, the affected valve often has to be replaced. With the continuous growth and aging of the world's population, the social and economic impact of heart valve disease will increase. Therefore, the number of patients requiring a heart valve replacement is expected to be over eight million by 2050.

Degradation of the heart due to disease makes repair or replacement necessary. More than 45,000 heart valve replacements are implanted each year in the United States. In China, cardiovascular disease (coronary heart disease, stroke, heart failure, hypertension) has been the primary cause of death since 1990. The 2008 statistics bulletin of the Chinese health service development showed that cardiovascular disease accounted for 40.27% of the country's death toll; between ages 35 and 54, the death toll increases quickly in young adults. In China, the number of people with heart disease is at least 230 million, with 2 in 10 adults suffering from cardiovascular disease. Every year, three million people die from cardiovascular disease nationwide, and one in three deaths is from cardiovascular disease. Now, people with heart disease are getting younger, and the replacement heart valve has a greater market need. Nowadays, many valve replacement types are available, mainly classified as mechanical valves or bioprostheses. Mechanical valves as shown in Fig. 1.4 offer excellent structural durability but are prone to thromboembolic events, thereby committing patients to daily anticoagulants. Hence, mechanical heart valve prostheses are fabricated for durability and reduction in immunological reactivity.

The significance of the heart and its complexity as well as the risk of heart surgery determine which material will be used to manufacture the heart valve as it must have good durability and antithrombotic ability. Because of its special function and working environment, artificial heart valve materials have some special requirements. Initially, materials should be nontoxic. As artificial heart valves will be implanted into the human heart, the issue of inflammation needs to be avoided after implantation. The material also should have good blood and tissue compatibility. Second, the selected material should be able to avoid antiplatelet precipitation

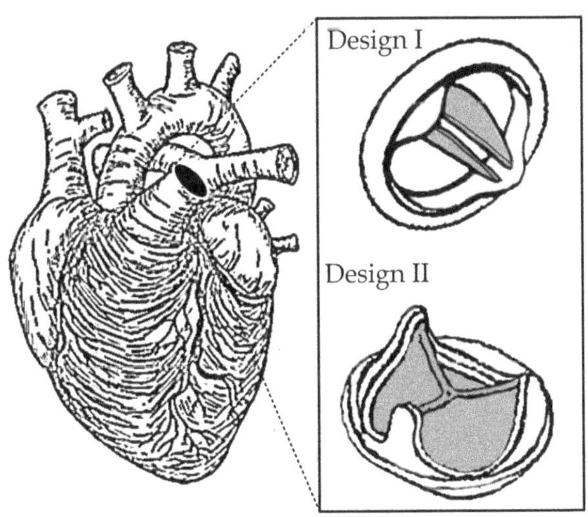

Fig. 1.4 Different designs of heart valves

and blood coagulation without causing degeneration or destruction of blood plasma. This can be evaluated by soaking the artificial heart valves in blood for a long time; if there are any adverse reactions, it will immediately affect the normal function of the valve and threaten the normal operation of blood circulation. Third, the selected materials should be chemically obtuse by not adsorbing blood components such as lipids and by not releasing foreign substances into blood circulation. In addition, the durability of materials is a factor which cannot be ignored. Both patients and doctors hope that artificial heart valves can be implanted into the body for a lifetime, requiring no secondary surgery. Other factors such as material source and price should seriously be considered.

1.8.2 Pacemaker

A pacemaker (Fig. 1.5) is a type of implant that monitors the patient's heart rhythm constantly. When an abnormal heart rhythm occurs, it sends an electric shock to defibrillate the heart muscle. Currently, the usefulness of the implantable pacemaker is defined not only by design constraints in achieving a shape appropriate to the physiological and biomedical requirements of the desired function but also by the properties of the materials of which the device is made. The use of alloplastic materials in replacement surgery has a long history and has mainly involved implants in the skeletomuscular system. The implantation of artificial parts in the cardiovascular system has only recently been developed. The major role played by implantable devices in modern medicine can be illustrated most simply by some statistics. Each year, more than 1,500,000 people worldwide are provided with vascular prostheses. Artificial heart valves are implanted in 100,000 patients, and about 220,000 receive an implantable cardiac pacemaker. In addition to the electronic requirements discussed previously, the biocompatibility of the implantable pacemaker is of great importance in the long-term success of arrhythmia treatment.

Manufacturers set their own quality standards for the pacemakers that they produce. Typically, the case of a pacemaker is made of titanium because it offers a

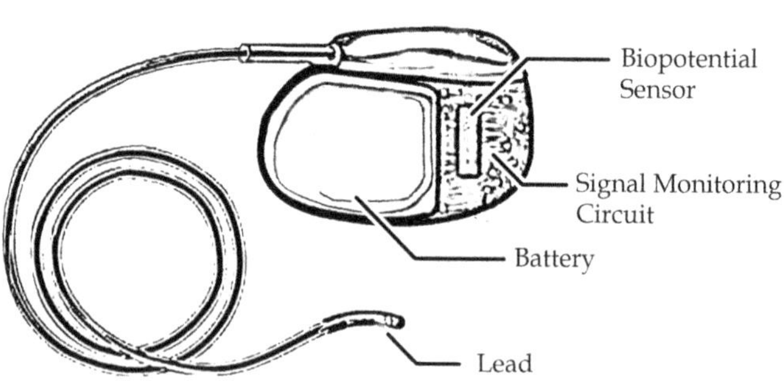

Fig. 1.5 A pacemaker

high modulus of elasticity, resistance to corrosion, durability, and strength. The electrodes are also made of titanium for these same reasons together with their electrical conductivity. The leads for transmitting electrical signals from the pacemaker to the heart are often insulated with surrounding silicone rubber or polyurethane, considering their low coefficient of friction and their capability of withstanding flexing induced by cardiac contraction. However, standards and performance recommendations are required by various medical organizations and governmental agencies. In the United States, pacemakers are classified as Class III biomedical devices, meaning that they require premarket approval from the United States Food and Drug Administration (FDA).

1.8.3 Vascular Stent

Blood vessels (usually arteries) of any size are affected by various diseases and may eventually become narrowed by being partly or completely blocked. This prevents normal blood flow with subsequent oxygen starvation of the tissues whose blood supply has been blocked. The patient develops symptoms such as angina or intermittent claudication. If the blockage is not relieved, it may progress to complete occlusion and cause a heart attack, stroke, or gangrene of a limb. Opening the artery before this critical stage is reached is clearly beneficial.

A stent (Fig. 1.6) is a tube inserted into a body passage to open it and/or keep it open, allowing normal flow of contents. A vascular stent is one placed into a blood vessel anywhere in the body. Vascular stents may be large or very small, e.g., 3 mm, and are usually expandable. Some stents are coated with slow-release drugs. Stents are most commonly used in the management of coronary artery disease, carotid artery narrowing, renal (kidney) artery narrowing, and peripheral vascular disease with narrowed arteries of the legs.

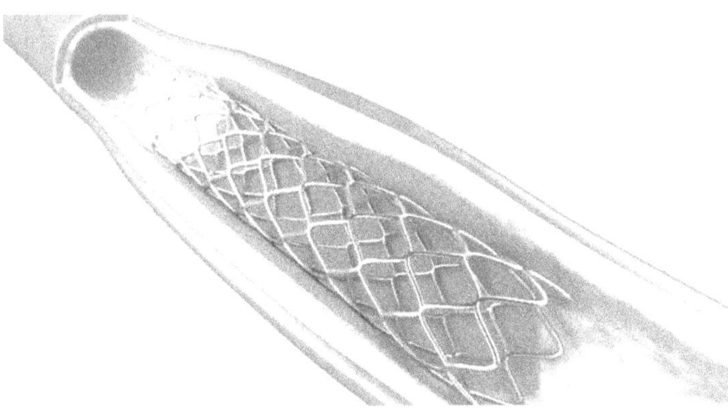

Fig. 1.6 Drawing of a vascular stent inserted in a vessel

1.9 Skeleton Devices

The skeletal system is the support structure for the entire body. As such, it is subjected to a lot of stress. This can lead to "cracks" in bones, called fractures. Most fractures occur due to stress and impact on the bones. However, some fractures can also occur because of the weakening of bones due to certain diseases like bone cancer. Athletes may suffer from fractures due to overuse of bones. Fractures can be classified according to the way in which the bones are broken, ranging from simple to multi-fragmentary fractures. In simple fractures, the break or crack is clean, i.e., in a straight line. In multi-fragmentary fractures, the bone is divided into three or more parts. In a case where the skin is pierced during the fracture, it is called an open or compound fracture; otherwise it is termed as a closed or simple fracture. Apart from fractures, joint degeneration due to old age is another malaise that plagues the skeletal system.

Internal fixation devices include plates, screws, rods, nails, pins, wires, and cables used in bone reduction, osteotomy, and arthrodesis procedures. The successful healing of long bone fractures (Fig. 1.7a) is often dependent on the mechanical environment created within the fracture, which in turn is dependent on the fixation strategy. These fixation devices are made with stainless steel and titanium, which are known for their durable, strong, and biocompatible properties. Internal fixation devices are reliable for stabilizing fractures and promoting healing without the risk of infection as compared to tradition methods using casts and splints. During the surgical procedure to set the fracture, the fragments are repositioned before being held together using specialized fixation devices. This reduces the risk of nonunion and malunion of the fracture site.

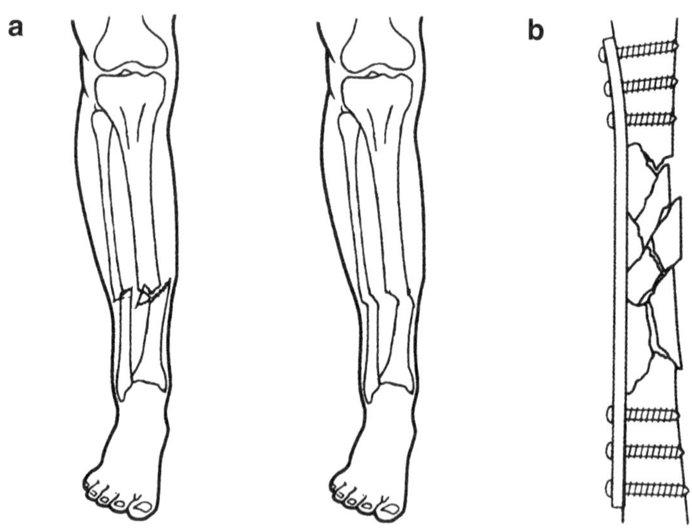

Fig. 1.7 (**a**) Nonunion (left) and angulated malunion (right) of fractured bones. (**b**) Plate and screw for stabilizing long bone fractures

Plates and Screws

Plates function similarly to splints, holding broken bones together. They are attached to the bones with screws and can be temporary or permanent (Fig. 1.7b). Plates are ideal for compression and can resist motion in three dimensions. However, the insertion of plates requires long and invasive operations, and such insertions often require a second operation for removal. This technique also introduces large amounts of foreign material into the body and might result in irritation.

This is the most common internal fixation device used. It often comes in various designs and sizes based on the specific needs. It can be used alone to set a fracture or used in conjunction with plates, rods, or nails. These devices can either be left or removed after recovery. They are good for interfragmental compression. However, inadequate surgical or fixation technique can increase risks of failure. Variations of an osteotomy (the cutting or removal of bone) might be necessary to accommodate the appropriate orientation of the screw. This insertion is a delicate process and might result in additional fractures if not done properly. The screw head might also cause irritation to the surrounding environment.

Nails or Rods

In the case of a long bone fracture, nails and rods are best used to hold the fracture pieces in place. A rod or nail (Fig. 1.8) is inserted through the hollow center of the bone that usually contains the marrow. At each end of the rod, screws hold the fracture in place, preventing the fracture from shortening or rotating. The rods and screws may then be removed or left in place after a full recovery. This method is ideal for fractures in the femur (thighbone) and tibia (shinbone).

Wires or Pins

Wires are commonly used to hold bones too small for screws back together. They are often supplemental to other internal fixation devices but can also be used alone for small bone fractures such as ones found on the hand or foot. They can also be temporary or permanent inserts. These are the easiest fixation devices that can be utilized and require minimal exposure. Removal of the device is also easy. However, with wires or pins (Fig. 1.9), they limit motion to only two planes, which might result in migration. They are also unable to provide compression for specific procedures. Furthermore, if percutaneous, pins could cause infection.

All of the devices in the previous sections are treatment devices used to help heal fractures. Some common problems faced when using these devices are:

Fig. 1.8 Intramedullary nail to treat a thighbone fracture

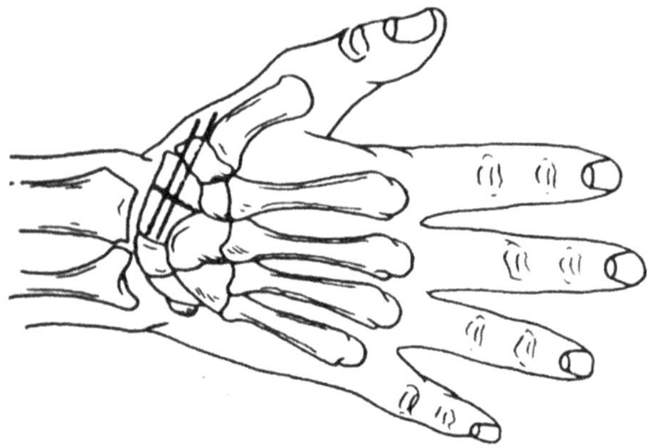

Fig. 1.9 Fixation wires (arrow) used to fix a fracture in the wrist

- Bio-incompatibility: Biocompatibility is a great concern when dealing with internal fixation devices since foreign material is introduced into the body. Since most available stainless steel implants are extremely inert, this problem mostly only arises with the use of rare metal devices.
- Allergic response: Reactions are usually secondary and based on sensitivity to one of the metal components of stainless steel such as nickel, chromium, molybdenum, and titanium. Eczema is the most frequent reaction.
- Infection: In spite of the success and increased use of internal fixation, postoperative infection is a significant problem. It could lead to delayed union, prolonged recovery, and increased morbidity. Hardware removal is commonly performed, but this might not eradicate the problem.
- Migration: This is motion occurring close to the fixation site. Smooth k-wires are the most susceptible to this problem, but the risk of migration happening can be reduced by bending the end of the wire or a prompt removal.

Joint Replacement

Synovial joints are mobile joints which include the human hip, knee, elbow, ankle, finger, and shoulder joints. Many elderly patients are subjected to degenerative bone and joint disease such as osteoarthritis and rheumatoid arthritis. Joint replacements are necessary to treat these diseases. Joint replacement medical devices are implantable medical devices intended to operate as a replacement for the shoulder, hip, or knee joints in whole or in part. They should substitute for the articulating surface of the joint, provide primary fixation to the bone, or act as a connector between bone and implantable medical devices.

Articular surfaces are surfaces of skeletal formation (bone, cartilage) that make normal direct contact with another skeletal structure as part of a synovial joint. Bony articular surfaces are usually covered with articular cartilage. These devices replace the ball and/or socket of joints such as the hip, knee, shoulder, and ankle joints (Fig. 1.10).

The risk of loosening of replacements is high for people with knee osteoarthritis and other non-traumatic diseases. Hence, fixations such as the stems of an artificial

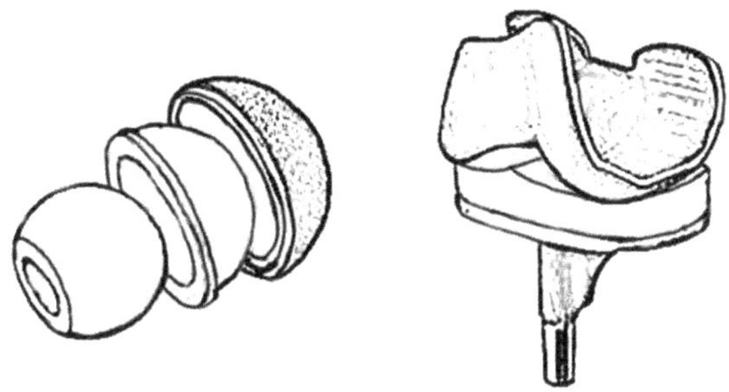

Fig. 1.10 Replacements for the articular surface of the joint

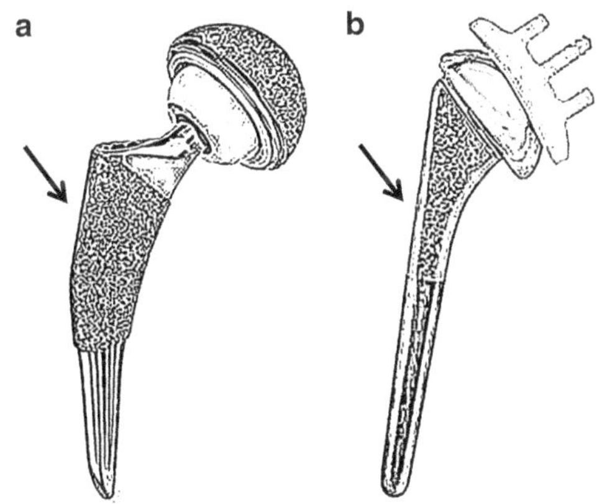

Fig. 1.11 (a) Fixation devices to keep replacements in place. (b) Intrinsic elements of joint replacement devices

hip (arrow in Fig. 1.11a) or of an artificial shoulder (arrow in Fig. 1.11b) come into play to keep replacements in place to serve its desired function. Joint replacement devices connect either directly or indirectly with the articular surface and/or the primary fixation. They function simply as an intrinsic element of the joint replacement, including the hip joint (Fig. 1.11a) and the shoulder joint (Fig. 1.11b).

The following example shows a hip joint replacement. The bipolar head replaces the articulating surface. The proximal femoral component acts as a direct connector to the bipolar head, and it also acts as an indirect connector to the cemented stem. The cemented stem provides primary fixation.

In the surgical procedures as shown in Fig. 1.12, the orthopedic surgeon makes an incision in the skin over the hip joint, cutting through skin, soft tissue, and muscle to provide visualization of the hip joint. The femoral head is then dislocated from the socket and removed from the end of the femur with a saw. The acetabulum (hip socket) is exposed, and a reamer (advanced grater) is used to create a perfect, hemispherical bone socket that matches the external shape of the acetabular cup. Next, the acetabular cup is carefully positioned into the socket. The cup is helped in place using a number of methods depending on surgeon preference and the patient's overall bone quality. These methods may include cement to hold the cup in place, but most cups are designed to allow the bone to grow into the metal for a permanent bond. The next step involves the acetabular insert, made of plastic, ceramic, or metal,

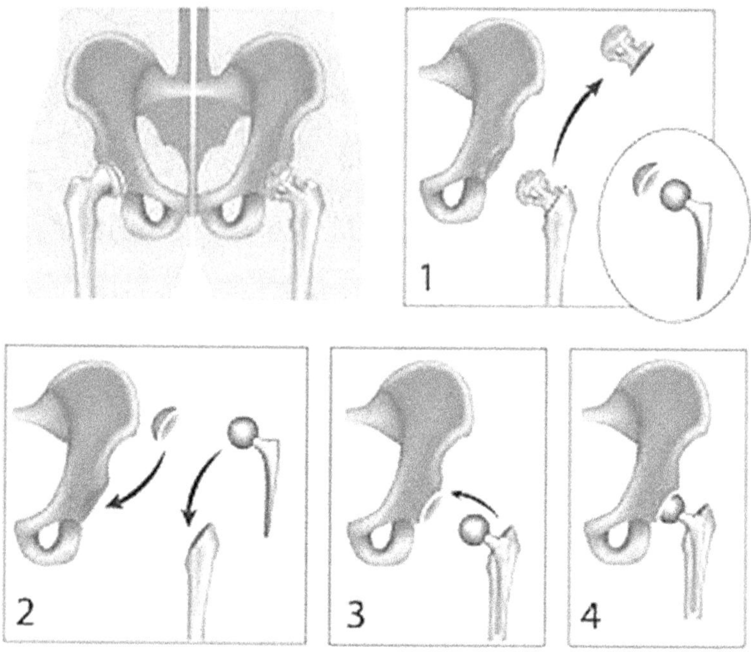

Fig. 1.12 Surgical procedures of an artificial hip joint

which is then positioned into the cup. The inside of the femur is cleared with various types of surgical instruments, and, at this point, a prosthetic stem is secured to the end of the thighbone. During surgery, a trial of mobility is evaluated with a test ball to determine the range of motion and stability of the joint. A final ball is placed onto the femoral stem, and the leg is again evaluated for mobility, stability, and leg length equality before the surgeon closes the incision site.

1.10 Tissue Grafts

Grafting is the procedure where tissue from one site is moved to another site without any blood supply from the donor site. Grafts can be classified into autografts, isografts, allografts, and xenografts. The former two are not rejected by the recipient, while the latter two are usually rejected and recognized as foreign by the receiving body.

- Autograft: These grafts are taken from an individual and transplanted onto another site in the same individual. This describes the process of simple skin grafts that can be used to treat severe burns.
- Isograft: This procedure involves grafts taken from an individual with the same genetic makeup as the recipient. This could be grafting between identical twins.

- Allograft: This is a graft taken from an individual and transplanted onto a body of the same species that is genetically nonidentical.
- Xenograft: This graft involves two individuals not from the same species such as an animal and a human.
- Synthetic materials: Graft material can be completely synthetic. Common materials used are ceramic or polymer-based materials.

Some common tissue grafting procedures involve skin, bone, nerves, tendons, neurons, blood vessels, fat, and even the cornea. Some examples are shown as the following:

- Skin grafting: Grafting is most commonly applied to the skin. It aims to treat wounds, burns, and infections. This technique can speed up recovery and improve function and appearance. The graft can be temporary or permanent. In the case of temporary grafts, the grafted material is gradually reabsorbed over time and replaced with new growth.
- Bone grafting: This is often used in dental implants where the bone can be harvested from other parts of the body, such as the iliac crest of the pelvis or the banked bone. For temporary grafts, the grafts are generally broken down over time and replaced with new bone.
- Vascular grafting: This surgical procedure is performed to redirect blood flow from one area to another by connecting blood vessels. This is often done to bypass a diseased artery.

There are also reasons for failure of grafting:

- Hematoma development: This occurs when the graft was placed at a site with active bleeding, resulting in a collection of blood beneath the graft. This in turn causes inadequate contact with the recipient's tissue bed and causes the site to not be well vascularized. This increases chances of failure since there is reduced blood and oxygen flow to the grafted tissue. Blood clots might also develop, increasing risk of pulmonary embolism.
- Infection: This is a risk faced by all surgical procedures. However, with grafts dealing with large areas, the risk increases when the air-skin barrier can be easily breached.
- Seroma development: Similar to hematoma development, collection of fluids under the graft could result in failure due to underdeveloped vascular systems under the graft.
- Inappropriate bed for new blood supply growth, such as cartilage, tendons, or bone: Since a new blood supply is required to grow tissue at the site to the new graft, when placed on an inappropriate bed, it could result in inadequate vascularization. This would then result in failure of the graft.
- Rejection from thc host body: Tissue or organ grafts between individuals of the same species are rejected with vigor proportional to the degree of the genetic disparity between them. Grafts between individuals of different species are rejected even more rapidly. Grafts between identical twins and from an individual himself survive indefinitely once vascular supply has been reestablished to the host.

1.11 Summary

In summary, biomedical devices play a crucial role in diagnosing, assessing, and treating various diseases and injuries that affect various parts of our body. Biomedical devices have been widely used in the healthcare industry. Biomedical manufacturing is currently a rapidly growing industry, brought by the globalization and the technology advancement. This book aims to provide the essential knowledge in the biomedical product design and development in order to provide ways to speed up the product development cycle. The content in this book is multidisciplinary and covers the principles in mechanical, chemical, biological, and physiological aspects. In general, we will introduce the biomedical device development through three main knowledge clusters: (1) biomaterials (Chaps. 2, 3, 4, and 5), (2) fabrication processes (Chaps. 6, 7, and 8), and (3) design techniques (Chaps. 9, 10, 11, and 12) for different biomedical applications. Biomedical engineers can learn the critical principles and techniques to apply the acquired knowledge of biomedical device design, prototyping, and manufacturing for particular applications.

References and Further Reading

1. World Health Organization: World Health Statistics, p. 2015. World Health Organization, Geneva (2015)
2. National Center for Health Statistics: *Health 2016, with Chartbook on Long-Term Trends in Health*. United States Government Printing Office (2017)
3. Hagedorn, T.J., Grosse, I.R., Krishnamurty, S.: A concept ideation framework for medical device design. J. Biomed. Inform. **55**, 218–230 (2015)
4. Schuh, J., Ann, C.L.: Medical device regulations and testing for toxicologic pathologists. Toxicol. Pathol. **36**, 63–69 (2008)
5. Medina, L.A., Okudan Kremer, G.E., Wysk, R.A.: Supporting medical device development: a standard product design process model. J. Eng. Des. **24**, 83–119 (2013)
6. Canaider, S.: An estimation of the number of cells in the human body. Ann. Hum. Biol. **40**, 463–471 (2013)
7. Yue, W.: Medical device registration system comparison between China and USA. Chin. J. Med. Instrum. **33**, 51–58 (2009)
8. Xu, X., Li, J., An, J.: Understanding the current regulatory landscape for companion diagnostic products in China. In: Companion and Complementary Diagnostics, pp. 335–364. Academic Press, Amsterdam (2019)
9. Khatik, I.: A study of various bone fracture detection techniques. Int. J. Eng. Comput. Sci. **6**, 21418–21423 (2017)

Part I
Biomaterials

Chapter 2
Basic Material Properties

Abstract Proper selection of material properties is significant for the processing of workpiece and for the functioning of biomedical device. Because the product design and manufacturing processes would largely depend on the chosen materials to utilize proper manufacturing processes and corresponding machines, a general understanding of materials and their properties is needed. The basic solid, thermal, and fluidic properties of materials are reviewed here to establish the foundation for a better understanding of the aforementioned issues mentioned later in this book. Considering that a product may also include multiple components assembled together, the surface and interfacial properties are also discussed.

2.1 Introduction

Product design engineers should have a deep understanding of materials and their properties for different biomedical device applications. Proper selection of material properties are significant for the processing of workpieces and for the functioning of biomedical device. In particular, for parts of biomedical devices with direct contact with human body, the materials used must be biocompatible that not only they support the device function but also they cause no severe harm problems to the body. Taking bone fixation implants as an example, the design engineers must make sure the implanted material can offer adequate physical support throughout the bone recovery period and cause no significant biological rejection or inflammation. Moreover, because the product design and manufacturing processes would largely depend on the chosen materials to utilize proper manufacturing processes and corresponding machines, a general understanding of materials and their properties is needed.

In most existing manufacturing processes, one key principle is to produce different components, each with a defined type of material with a desired shape, followed by assembling the different components into the final product. Hence, the manufacturing process that engineers develop needs to consider how the material geometry will change under a range of external physical factors. As the first conceptual thought, we may consider that when an external force is exerted on a solid object, the object will deform. If such force is sufficiently large, the material

R. H. W. Lam, W. Chen, *Biomedical Devices*,
https://doi.org/10.1007/978-3-030-24237-4_2

body will have permanent deformation, and therefore, the engineers can apply the appropriate range of force to change the material shape. Furthermore, if the applied external force is even higher, it can lead to separation of the material body into multiple parts. We can also think of another manufacturing strategy to remove small volumes form a raw material body bit by bit until the resultant material body converges to the shape of the target product component. Alternatively, a product component can be manufactured by first heating up a solid material until it melts, followed by transferring the molten material liquid into a prepared cavity. After the material cools down and solidifies, the material shape can then be defined. The basic solid, thermal, and fluidic properties of materials will be introduced here to establish the foundation for a better understanding of the aforementioned issues mentioned later in this book. Considering that a product may also include multiple components assembled together, the surface and interfacial properties are also discussed.

2.2 Solid Properties

2.2.1 Direct Stress and Strain

When a force is applied to a material body, the body deforms. For example, when a body is stretched by a tensile force, the body length increases, whereas a compressive force compresses a body through a decrease in the body length. Both the tensile and compressive forces are called direct forces. Here, we introduce a quantity called stress (or direct stress). We first consider that a cylindrical material body with a length L and a consistent cross-sectional area A is stretched from the circular ends on both sides along the axial direction with an external force F (Fig. 2.1). Stress is thus defined as the average intensity of the distributed forces per unit area and is often denoted by the Greek letter σ (sigma). The stress acting on a plane surface is first assumed to be uniformly distributed throughout the area. We consider a tensile force with a positive sign and a compressive force with a negative sign, considering that tension would increase the body length and compression would do the opposite. An equation for the magnitude of the stress is given by

$$\sigma = \frac{F}{A} \tag{2.1}$$

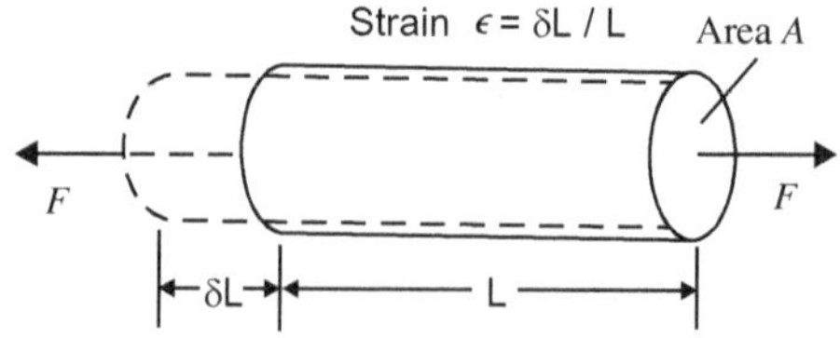

Fig. 2.1 Deformation of a cylindrical solid under stretch

The unit of σ is Pascal (Pa) or N/m^2, and most engineering fields often use kPa, Mpa, or GPa.

When a load is applied to a material body, it causes a change in the body length δL. We consider such change relative to the original body length L as the strain denoted by ε (epsilon), which is a dimensionless quantity with the expression

$$\varepsilon = \frac{\delta L}{L} \tag{2.2}$$

Again, the strain is called tensile strain with a positive value when the material body stretches, where a compressive strain has a negative value.

We often consider the original/undeformed body length and cross-sectional area for the stress and strain calculation in many engineering problems, and therefore, these are called engineering stress and engineering strain. In fact, the "true stress" is considered the instantaneous direct force divided by the instantaneous cross-sectional area, and the "true strain" is calculated by the instantaneous body length at any moment during elongation and the rate of increase in gauge length, i.e.,

$$\varepsilon = \int_{L_o}^{L} dL/L = \ln(L/L_o). \tag{2.3}$$

2.2.2 Stress-Strain Diagram

To determine the proper choice of material to be used in product manufacturing, the basic mechanical properties of materials need to be understood. These properties are tested by mechanical test experiments performed on specimens with defined dimensions. The experiments are conducted in laboratories equipped with testing machines capable of loading tension or compression (Fig. 2.2a). The main mechanism for testing is pulling the sample using the machine and analyzing the change of the sample by a data-processing system. The common relationship being tested is the change in sample length under a known external force. From the results of tensile testing, a curve of direct stress against direct strain is obtained, such as the sample plot for many metals as shown in (Fig. 3.1b). In the tensile test, for example, by continuously pulling until the sample breaks, a complete tensile profile about the sample is obtained.

From the origin to point A in the stress-strain diagram (Fig. 2.2b), the stress is proportional to the strain. The higher the stress, the higher the strain. The segment in this region is a straight line whose slope is called the modulus of elasticity. This line is no longer linear when the stress exceeds point A, and therefore, this point is defined as an elastic limit which is the upper limit to the linear line. This sweeping idealization and generalization applicable to all materials is known as Hooke's law.

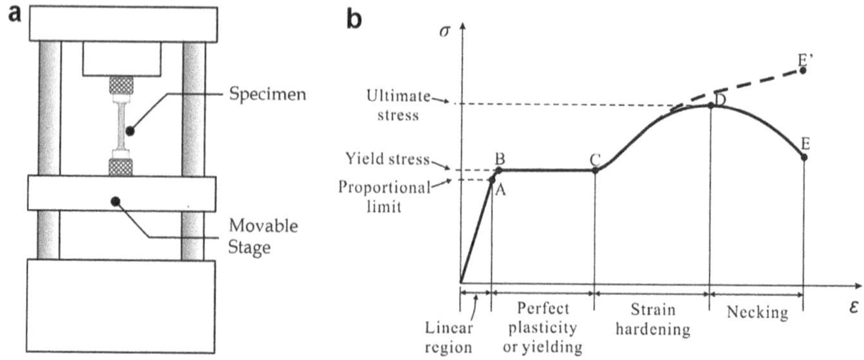

Fig. 2.2 (**a**) Configuration of mechanical tensile test. (**b**) Stress and strain diagram for mild steel

Hooke's law is the principle that elastic materials obey when they are released from stress and relax back into their original shapes. In the mechanical test, deformation of the material body is directly proportional to the force, $F = kx$ In which k is called the stiffness or the spring constant. It is always a positive real number and a common characteristic for springs but different values for different types and dimensions of material bodies. This stiffness equation can apply to both compression and extension cases, but the deformation (x) is positive at extension and negative at compression (this property is the same with force). Looking back to stress-strain perspective, the size can be eliminated by using stress and strain instead of force and deformation. Recall that in Eqs. 3.1 and 3.2, we obtain a relation $k = F/x = \sigma A/(\varepsilon L)$, and hence

$$\sigma = \frac{FL}{Ax}\varepsilon = E\varepsilon \tag{2.4}$$

where E is called the modulus of elasticity (or elastic modulus). The slope of the straight line in the elastic region of the stress-strain curve (Fig. 2.2b) is E. As the strain is dimensionless, as mentioned before, the unit of E is the same as the unit of stress.

A slight increase in stress above the elastic limit will result in breakdown of the material and cause it to deform permanently; this process is called yielding. The stress at yielding is called the yield stress. Once this yield point is reached, it delimits the elasticity from the plastic region. Beyond this point, plastic deformation begins, and the material in this state becomes perfectly plastic in which the specimen will elongate (strain) without any increase in load. In particular, yield strength is the stress that causing plastic deformation of the material. As shown in Fig. 2.3a (point B), it is located at the line offset where an arbitrary amount of 0.2% of strain is drawn parallel to the straight-line portion of the stress-strain diagram. Furthermore, the area enclosed by the loop in Fig. 2.3b corresponds to dissipated energy released through heat. Perfect elasticity of materials (between points B and C) occurs when there is no dissipation of any energy for deformations under a monotonic or cyclic loading.

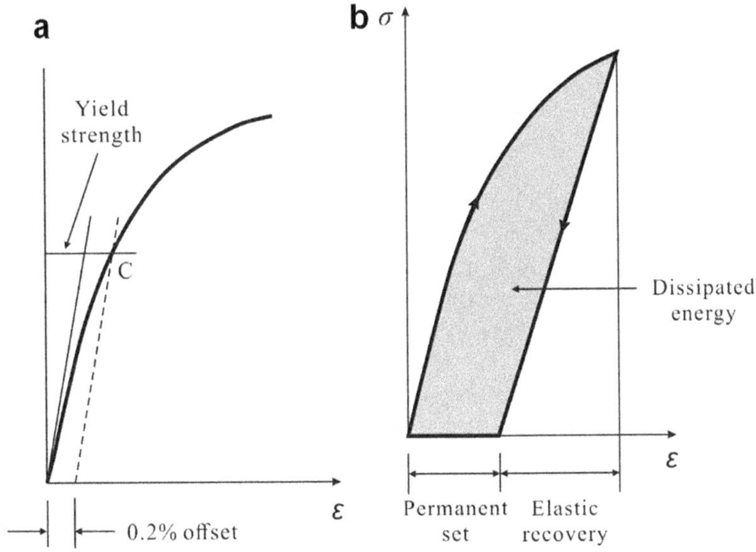

Fig. 2.3 (**a**) Offset method of determining yield strength. (**b**) Permanent deformation and energy dissipation after a cyclic loading

As the specimen is subjected to an increasing load beyond the point C in Fig. 2.2b, the curve rises continuously before becoming flat when reaching the ultimate stress, which is the maximum force that the specimen can endure before causing it to break or carry less load.

The rising of the curve between points C and D is called strain hardening, which occurs after the yielding ends when any further load is applied to the specimen. In strain hardening, the material crystalline structure changes as many crystalline dislocations propagate and interact with each other, resulting in the increased resistance of the material to further deformation. This phenomenon can be seen for tensile tests of some metallic and plastic materials. Also, while the specimen elongates, its cross-sectional area will decrease. Strain softening means that the stress required for further deformation of the specimen under loading beyond the yield point is smaller. Material effects, such as dynamic recrystallization, can cause strain softening. Strain softening can cause local deformation in strain and, hence, inhomogeneous strain profiles. For polymers, strain softening can be caused by the recombination of physical crosslinking points of macromolecular chains, resulting in the formation of supermolecular structures favorable to deformation development due to the larger strain. Elastic-plasticity means that the stress required for further deformation of the specimen is maintained.

In particular, in order to achieve the permanent material deformation, the loads and material stresses depend on the part geometry and flow stress of the material being formed. When the applied stress in uniaxial tension, without necking, reaches the yield stress, the material is considered to begin deforming plastically. Flow stress (σ) describes the stress-strain relation of a material undergoing uniaxial deformation in the strain hardening region, as a function of strain, strain rate, temperature, and the microstructure. The flow stress of most materials including metals at room temperature would increase with increasing strain. The phenomenon is known as strain hardening or work hardening. Strain hardening is the result of interaction of inclusions or dislocations in the crystalline structure of the materials.

The strain hardening region in the stress-strain curve is very important in the design of material deformation (e.g., forming and extrusion) processes, since they describe the material behavior during deformation. There are many different methods for mathematically representing the flow stress curves developed previously. The power law known as the Hollomon's law is a fairly good and simple approximation of the flow stress for materials during plastic deformation:

$$\underline{\sigma} = K_s \underline{\varepsilon}^n \tag{2.5}$$

where σ is the flow stress; K_s is the strength coefficient; n is the degree of strengthening; and ε is the strain during the material plastic deformation. For prestrained materials, the power law must be shifted as the known Swift's law, such that

$$\underline{\sigma} = K_s (\varepsilon_o + \underline{\varepsilon})^n \tag{2.6}$$

where ε_o represents prestrain of the material.

The ultimate stress or strength is the maximum stress level reached in the mechanical test (point D in the stress-strain curve). Beyond the strain at the ultimate stress, a particular cross-section of the specimen will have a rapid reduction in its area as shown in Fig. 2.4. As a result, a "neck" along the specimen tends to form; and such necking behavior can be observed for some metals. As the cross-sectional area of the neck gradually decreases, the smaller the area is, the less the load that this portion of specimen can carry. Therefore, the stress-strain curve goes downward (between points D and E in Fig. 2.2b) in the necking region, and, eventually, the specimen reaches its maximum possible strain and breaks at the fracture strain (point E).

Considering that the instantaneous load in tension is given by $F = \sigma A$ for a material in the strain hardening region during a tensile test. The criterion for the instability in tensile test (necking) can be formulated as the condition that F is just beyond the maximum (i.e., σ or σ is just beyond the strength), meaning that $dF/d\varepsilon \approx 0$, where ε is the flow strain. Near but slightly before reaching the maximum load, the uniform deformation conditions can be assumed. Recalling the true strain described in Eq. 2.3, and assuming the material volume during the strain hardening stage V_{sh} is roughly constant, we may have

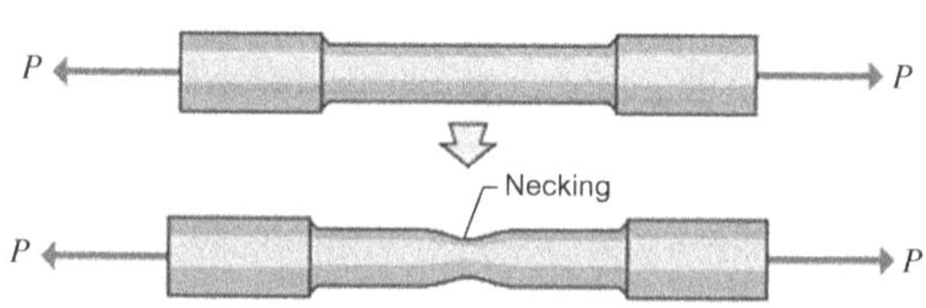

Fig. 2.4 Necking of a specimen in tensile test

$$F = \underline{\sigma A} = \frac{\underline{\sigma} V_{\mathrm{sh}}}{L} = \frac{\underline{\sigma} V_{\mathrm{sh}}}{L_{\mathrm{O}} e^{\underline{\varepsilon}}} = \underline{\sigma A_{\mathrm{O}}} e^{-\underline{\varepsilon}} \tag{2.7}$$

where $\underline{A}$ is the material area in the stain hardening region and $\underline{A}_{\mathrm{O}}$ is at the starting point of strain hardening. Hence

$$\frac{dF}{d\underline{\varepsilon}} \approx 0 = \underline{A}_{\mathrm{O}} \left(\frac{d\underline{\sigma}}{d\underline{\varepsilon}} e^{-\underline{\varepsilon}} - \underline{\sigma} e^{-\underline{\varepsilon}} \right), \tag{2.8}$$

and

$$\frac{d\underline{\sigma}}{d\underline{\varepsilon}} - \underline{\sigma} \approx 0. \tag{2.9}$$

Recalling Eq. 2.5, we have

$$\varepsilon_{\mathrm{neck}} \approx n \tag{2.10}$$

where $\varepsilon_{\mathrm{neck}}$ is the maximum allowable strain for stain hardening, or the strain required for necking.

In particular, in order to achieve the permanent material deformation, the loads and material stresses depend on the part geometry and flow stress of the material being formed. When the applied stress in uniaxial tension, without necking, reaches the yield stress, the material begins to plastically deform. Flow stress (σ) is simply the yield stress of a material undergoing uniaxial deformation as a function of strain, strain rate, temperature, and microstructure. The flow stress of most materials, including metals at room temperature, will increase with increasing strain. This phenomenon is known as strain hardening or work hardening.

The flow stress region in the stress-strain curve is very important in the design of material deformation (e.g., forming and extrusion) processes since they describe the material behavior during deformation. There are many different methods for mathematically representing the flow stress curves developed previously. The power law is a fairly good and simple approximation of the flow stress for materials during plastic deformation:

$$\underline{\sigma} = K_{\mathrm{s}} \underline{\varepsilon}^{n} \tag{2.11}$$

where σ is the flow stress; K_{s} and n are constants; and ε is the strain during the material plastic deformation.

2.2.3 *Shear Stress and Strain*

Shear force is a force applied sideways on materials (transversely loaded) as described in Fig. 2.5a, not only in the solid state but also in the fluid state or the gas state. Shear stress, denoted by τ, is the stress generated accordingly in the direction along the material surface and perpendicular to the tensile/compressive stress. Its unit is force per unit area carrying the load. The shear stress τ over a cross-sectional area A can be generated when the material is being cut or pinned with a shear force F:

$$\tau = \frac{F}{A} \tag{2.12}$$

Shear strain (γ) is the ratio of the distance deformed to the height of the material, which can be measured by the angular change. The side lengths of the specimen remain unchanged, but the shape changes, like from a rectangle to a rhombus as shown in Fig. 2.5a. For a cubic material block exerted under a shear force with a side length L, a dislocation of the sheared surface has a dislocation of x. When the angle of side surface is very small, we can say that

$$\gamma = \frac{x}{L} \tag{2.13}$$

For the range of small deformations, the shear stress and the shear strain can be related to a material property called the modulus of rigidity (G):

$$G = \frac{\tau}{\gamma} \tag{2.14}$$

We can conduct the mechanical test for determining the modulus of rigidity of a material with the known dimensions of a cross-sectional area A and the material thickness L by applying direct shear force (F) and simultaneously measuring the

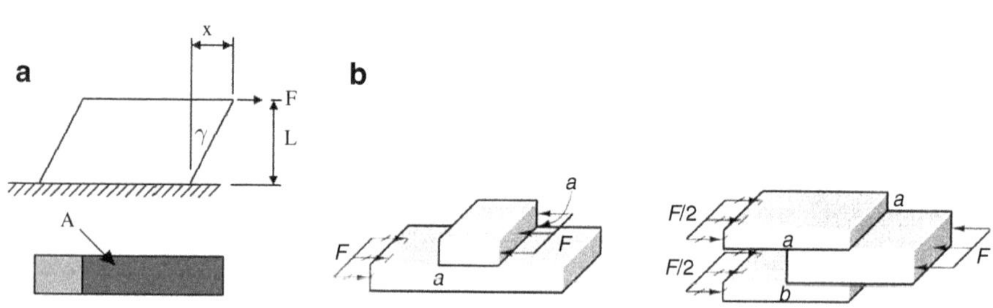

Fig. 2.5 (**a**) The deformation under shear stress. (**b**) Examples of loading conditions causing shear stresses between interfaces of glued blocks

surface displacement (x). For small deformations, the gradient in the graph of F against x is constant. This gradient can be converted to G by

$$\frac{F}{x} = \frac{L}{A} \times \frac{\tau}{\gamma} \tag{2.15}$$

Like the elastic limit in the tensile stress-strain curve, if a material is sheared beyond a certain limit, yielding occurs, and the material deformation becomes irreversibly twisted and cannot return back to its original exact shape. If the material is stressed to the ultimate shear strength, the material will break.

Shear stress can appear over interfaces between material bodies. To illustrate more details about interfacial shear stress, two cases of glued blocks under axial force are shown in Fig. 2.5b. In both cases, the blocks are in contact with other blocks by certain surface area a. In the left case, an external load F is acting on the side surfaces of the blocks, such that a shear force F is induced on the contact area. The right case contains three bodies with the side load of $F/2$ applied to the upper and lower blocks and the side load F in another direction. The load F on the middle block is distributed as two shearing forces over both the upper and lower interfaces, each with a shearing force of $F/2$.

2.2.4 *Poisson's Ratio*

If a solid body is subjected to axial tension, it contracts laterally at the same time. On the other hand, an object with an axial compression would have increases in its lateral side lengths as described in Fig. 2.6. This phenomenon is described by the material property called Poisson's ratio. It is the ratio of lateral contraction (extension) strain ($\varepsilon_{\text{lateral}}$) to axial extension (contraction) strain ($\varepsilon_{\text{axial}}$) in the direction of direct tensile (compressive) force. It is denoted by ν (nu) and is represented as

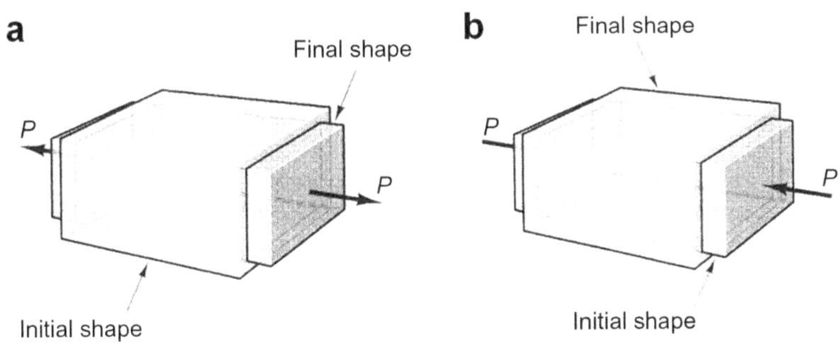

Fig. 2.6 (**a**) Lateral contraction and (**b**) lateral expansion of solid bodies subjected to an axial force

$$\nu = -\frac{\varepsilon_{\text{lateral}}}{\varepsilon_{\text{axial}}} \tag{2.16}$$

The equation of Poisson's ratio contains a negative sign to keep normal materials with a positive ratio because the signs of the lateral strain and axial strain are opposite to each other. The value of ν fluctuates for different materials over a relatively narrow range. Generally, it is on the order of 0.25–0.35. In extreme cases, values as low as 0.1 (some concretes) and as high as 0.5 (rubber) can occur. Some materials under the yielding with perfect plasticity may have $\nu \approx 0.5$, too.

Furthermore, considering when in the Cartesian coordinates, there is a direct stress σ_x along the x-direction applied on a cubic body with elastic and isotropic material properties, the strain along x, $\varepsilon_x = \sigma_x/E$ is induced based on the elastic modulus (or called Young's modulus) E, accordingly. Besides, considering the Poisson's ratio ν of the material, there are also strains along y- and z-directions as $\varepsilon_y = -\nu\sigma_x/E$ and $\varepsilon_z = -\nu\sigma_x/E$. For a more general consideration, a local cubic volume inside a material body may have stresses in all x-, y- and z- directions, so the Cartesian strains can be expressed as

$$\begin{aligned} \varepsilon_x &= \left(\sigma_x - \nu\sigma_y - \nu\sigma_z\right)/E \\ \varepsilon_y &= \left(\sigma_y - \nu\sigma_z - \nu\sigma_x\right)/E \\ \varepsilon_z &= \left(\sigma_z - \nu\sigma_x - \nu\sigma_y\right)/E \end{aligned} \tag{2.17}$$

Apparently, for the shear stresses and strains in the three Cartesian directions

$$\gamma_x = \tau_x/G, \quad \gamma_y = \tau_y/G, \quad \text{and} \quad \gamma_z = \tau_z/G \tag{2.18}$$

Both G and E are quantities for measuring the stiffness of materials under different modes of external forces. In fact, these moduli are dependent on each other and also the involvement of ν. For homogeneous isotropic linear elastic materials, we consider a cubic material element to be subjected only to shear as shown in Fig. 2.7a. The external applied shear stresses are τ_{xy} (i.e., the shear stress in the y-direction acting on the surface with an x-normal axis) and τ_{yx} (acting along the x-direction on the surface with a y-normal axis) where $|\tau_{xy}| = |\tau_{yx}|$. Recall Eq. 2.17:

$$\varepsilon_x = \frac{1}{E}\left[\sigma_x - \nu\left(\sigma_y + \sigma_z\right)\right] \tag{2.19}$$

In this case, the maximum stress ($=\tau_{xy}$) and strain ($\varepsilon_{\max}$) are along the direction of 45°, and the minimum stress ($= -\tau_{xy}$) and strain ($\varepsilon_{\min}$) are along the direction of −45° in the xy-plane. If we choose the new Cartesian coordinates by rotating the axes 45° counterclockwise about the z-axis such that the new x-axis (x') aligns with maximum stress and the new y-axis (y') aligns with the minimum stress:

$$\varepsilon' = \varepsilon_{\max} = \frac{1}{E}\left[\tau_{xy} - \nu\left(0 - \tau_{xy}\right)\right] = \frac{\tau_{xy}}{E}(1 + v) \tag{2.20}$$

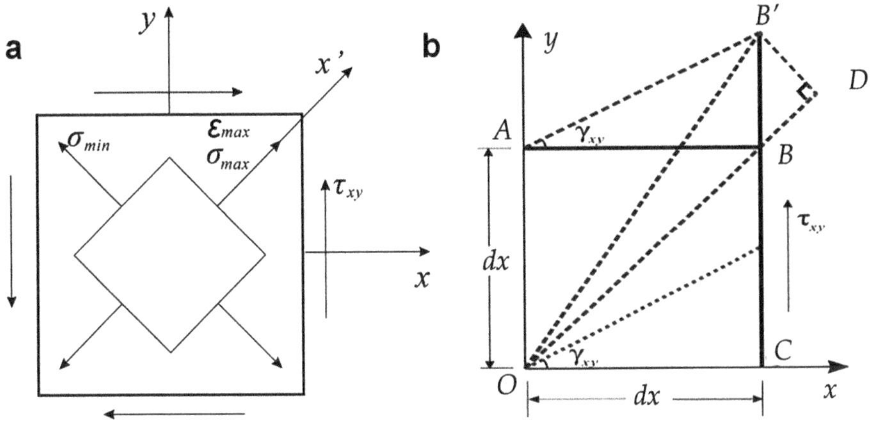

Fig. 2.7 (**a**) Pure shear state of plane stress. (**b**) Unit of stress under the shear deformation

This strain, which deforms the element along the new x' axis, can also be related to the shear strain γ_{xy}. Suppose the side length is dx in the plan element OABC, and the left side OA is fixed, as shown in Fig. 2.7b. Due to shear deformation, the right side BC with a length of dx dislocates with the distance $BB' = \gamma_{xy}dx$, where $\gamma_{xy} = \tau_{xy}/G$. If we now consider the new coordinate frame with OB as the principal axis with the strain $\varepsilon_{\max}$, the length of OB is $\sqrt{2}dx$, and its elongation is BD with a length as $1/\sqrt{2}$ times that of BB', i.e., $\gamma_{xy}dx/\sqrt{2}$. Therefore, the maximum strain is $\varepsilon_{\max} = \gamma_{xy}/2 = \tau_{xy}/(2G)$.

$$\varepsilon_{\mathrm{OB}} = \varepsilon_{\max} = \frac{\gamma_{xy}}{2} = \frac{\tau_{xy}}{2G} \tag{2.21}$$

Substituting Eq. 2.17 into Eq. 2.16 and rearranging the terms, we can obtain the relationship between Poisson's ratio (ν), modulus of rigidity (G), and modulus of elasticity (E) as

$$G = \frac{E}{2(1+\nu)} \tag{2.22}$$

2.3 Thermal Properties

2.3.1 Thermal Strain and Deformation

As the most obvious physical phenomenon, a body expands with an increase in temperature while contracting with a decrease in temperature. It has been found that thermal strain ε_{T} explains this change due to any temperature change for a body material that is homogeneous and isotropic. The thermal strain ε_{T} can be expressed as

$$\varepsilon_{\mathrm{T}} = \alpha \Delta T \tag{2.23}$$

where α is a property of the material referred to as the coefficient of linear thermal expansion. The units of α measure strain per degree of temperature. ΔT is the change in temperature of the material. This equation works well as long as the linear-expansion coefficient does not change much over the change in temperature ΔT and the induced thermal strain is small.

In general, substances expand or contract when their temperature changes with expansion or contraction occurring in all directions for isotropic materials. Recalling Eq. 2.17, thermal strain can be considered as an additional strain in the material deformation as

$$\begin{aligned} \varepsilon_x &= \left(\sigma_x - v\sigma_y - v\sigma_z\right)/E + \alpha\Delta T \\ \varepsilon_y &= \left(\sigma_y - v\sigma_z - v\sigma_x\right)/E + \alpha\Delta T \\ \varepsilon_z &= \left(\sigma_z - v\sigma_x - v\sigma_y\right)/E + \alpha\Delta T \end{aligned} \tag{2.24}$$

Sometimes, when a material body is constrained so that it cannot expand, then internal stress may be caused (or changed) by a change in temperature. This stress can be calculated by considering the strain that would occur if the body were free to expand and the stress required to compress the material back to its original size.

2.3.2 *Specific Heat Capacity*

Almost any substance can have a measureable specific heat capacity, such as chemical elements, compounds, alloys, solutions, and composites. The ratio of the absorbed heat (Q) to the mass (m) and the rising temperature (ΔT) of a substance of certain mass at elevated temperatures is called the specific heat capacity of this substance, denoted as c. The unit in the international unit system is the joule per kilogram of Kelvin [J/(kg·K)]. According to this theorem, we can get the following formula:

$$Q = mc\Delta T \tag{2.25}$$

where Q is the heat absorbed; m is the mass of an object; and ΔT is the change of temperature after the absorption of heat. There are also other definitions of the specific heat capacity, such as specific heat capacity at constant pressure C_p and specific heat at constant volume C_{v}. Furthermore, the quantity of $m \times c$ is called the heat capacity, denoted as C.

2.3.3 Changes of Moduli of Elasticity and Rigidity with Temperature

Material properties including the elastic modulus can be determined by bond strength, crystalline structure, chemical composition, molecular structure, and temperature. Often, the elastic modulus E decreases with temperature T, expressed as

$$E(T) = E_o(1 - \eta T) \tag{2.26}$$

where E_o is the elastic modulus at $T = 0$ K; η is the temperature coefficient proportional to the thermal expansion coefficient α as $\eta \approx \Psi\alpha$, where Ψ is a constant scaling factor.

On the other hand, the rigidity modulus G also reduces as the temperature rises. This trend can be described by an empirical model called the Mechanical Threshold Stress (MTS) shear modulus model, expressed as

$$G(T) = G_o - \Pi \frac{\exp(1) - 1}{\exp(T_o/T) - 1} \tag{2.27}$$

where G_o is the shear modulus at the temperature 0 K; T is the material temperature; T_o is the reference temperature; and Π is the reduction in the shear modulus at the temperature T_o. The values of T_o and Π are determined by experimental results.

2.4 Fluidic Properties

2.4.1 Viscosity

Viscosity is an important concept in fluids and is defined as the fluidic resistance to flow. It is also described as the momentum diffusion. Fluids with low viscosity are "air-like" – the flow of these types of fluids is very smooth and fast since there is weak internal friction of the moving fluid. Fluids with high viscosity are "honey-like" – their flow is very slow and resistive since the high internal friction between fluidic molecules resists motion. Viscosity is important for evaluating the quantity of fluids in transportation during a certain period of time and the energy losses due to transport of fluids in tubing, syringes, channels, and slits in manufacturing processes and even for vessels in the human body.

In order to examine the role of viscosity, we may first consider the Couette flow. It is a shear force induced laminar flow of fluid between two adjacent plates with the upper plate moving along one direction (Fig. 2.8a). If we assume a consistent viscosity μ over the fluid volume with a thickness H and impose a constant velocity of the moving plate U, we will have a consistent shear stress τ generated in the fluid volume between the plates, and the velocity profile inside the fluid $u(y)$ varying with

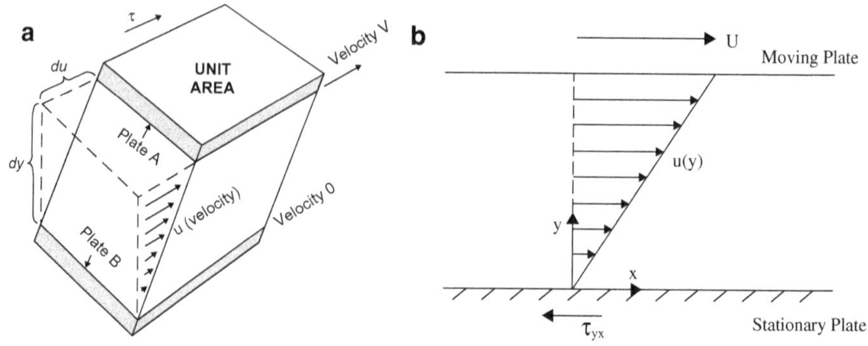

Fig. 2.8 (**a**) Configuration and (**b**) velocity profile of the Couette flow

the horizontal position y can be described as Fig. 2.8b. The role of viscosity in the associated shear stress can be explained via the flow between the two no-slip parallel plates as

$$\tau = \mu \frac{du}{dy} \tag{2.28}$$

We may further consider the no-slip condition where the boundaries of fluid adhering to the plates show zero velocity relative to the plate walls on which the fluidic adhesion force to the wall is stronger than the cohesion force by flow. Hence, we obtain $u(y = 0) = 0$ and $u(y = H) = U$. Therefore, $u(y) = Uy/H$ and $\tau = \mu U/H$.

Most common fluids, like water, air, and oil, exhibit a linear relation between shear stress (τ) and shear rate (du/dy) by the factor of viscosity μ. These fluids are called Newtonian fluids. Otherwise, fluids are considered non-Newtonian.

2.4.2 *Types of Flows*

The pattern of fluid flows are briefly classified into two major types. Laminar flow persists as unidirectional movement with moderate velocity. The parallel fluid layers move over each other up and down by diffusion, and no mixing occurs. Turbulent flow is irregular and random fluid motion, which is highly distorted with comparatively high velocity since the velocity of the fluid at different points shifts continuously in both magnitude and direction. There is large-scale flow perpendicular to direction of flow in turbulent flow. Also, the transfer of movement up and down by macroscale processes occurs. Eddies are one major trait of turbulent flow that can be commonly seen in daily life and are important to ecological balance in the ocean.

To predict whether a flow is laminar or turbulent, we calculate the Reynolds number (Re), which is a dimensionless number reflecting the ratio of viscous force to inertial force in a fluid. Re is the form of

$$\text{Re} = \frac{\rho U * L*}{\mu} \tag{2.29}$$

where ρ is the fluid density; $U*$ is the "characteristic velocity," which is sometimes considered as the average velocity; $L*$ is the "characteristic length" representing the dimension of the flow region; and μ is fluid viscosity. Roughly, Re < 1000 indicates dominance of the viscous force leading to laminar flow; and Re > 1000 reflects dominance of the inertial force leading to turbulent flow.

Flow in Slits/Parallel Plates

Considering that the characteristic length is smaller than a centimeter and the "characteristic velocity" is limited in a human body and in biomedical devices, corresponding flows are often laminar. In more extreme cases, inertial effects can be ignored as Stokes flow or creeping flow. For a liquid flow in a rectangular slit with length L, width W, and height H along the slit length direction driven by a fluidic pressure gradient along slit length as shown in Fig. 2.9a, further considering that $L \gg W \gg H$, the governing equation can be expressed as

$$\mu \frac{d^2 u}{dy^2} = \frac{dp}{dx} \tag{2.30}$$

where $u(y)$ is the flow velocity along the slit length, p is the fluidic pressure, x is the position along the slit length, and y is the position along the slit height. By letting the middle position in the slit be $y = 0$ such that $u(H/2) = 0$ and $u(-H/2) = 0$ (Fig. 2.9b), according to the no-slip boundary conditions, we obtain

$$u(y) = \left(-\frac{dp}{dx}\right) \times \frac{1}{\mu}\left(\frac{H^2}{4} - y^2\right) \tag{2.31}$$

The liquid should flow from a position with a higher pressure to a position with a lower pressure; therefore, the flow velocity u should have a direction along a negative pressure gradient $-dp/dx$. The fluidic pressure decreases along the flow direction caused by friction and kinetic energy change. Hence, the fluidic pressure at

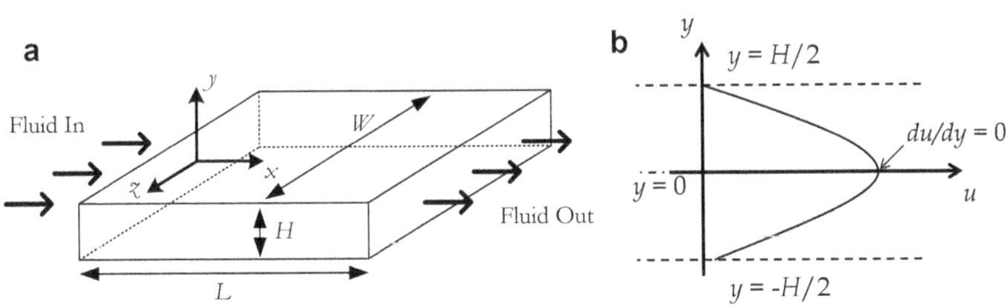

Fig. 2.9 (**a**) Key parameters and (**b**) flow profile of a slit flow

the slit inlet (P_{in}) is higher than the pressure at the slit outlet (P_{out}), and the negative pressure gradient can be estimated as $\Delta P/L$, where the gauge pressure is $\Delta P = P_{in} - P_{out}$.

Additionally, we assume that the fluid density is unchanged in the flow, i.e., incompressible, such that the fluid volume entering the slit is equal to that exiting the slit. Here, we suggest using the quantity of volumetric flow rate Q to describe the flow velocity along the slit. Q represents the volume of fluid flowing across a cross-section of the rectangular slit per time, which can be calculated by integrating $u(y)$ over the slit cross-section:

$$Q \approx W \int_{-H/2}^{H/2} u(y)dy = \frac{H^3 W}{12\mu L} \Delta P \tag{2.32}$$

Flow in Pipes

Compared to the slit flow, the only difference in flow in pipes is that it has a flow area with a circular cross-section as shown in Fig. 2.10a. Relations in this flow can be applied in components such as syringe needles, tubing, and even blood vessels. Considering the circular cross-section, we describe flow using cylindrical coordinates, with radial position r, angular position θ, and axial position x. If we consider a pipe with radius R and length L and $R \gg L$ driven by the pressure gradient dp/dx along the flow direction x, the governing equation of the velocity u as a function of r should be

$$\frac{\mu}{r}\frac{d}{dr}\left(r\frac{du}{dr}\right) = \frac{dp}{dx} \tag{2.33}$$

The boundary conditions are $du/dr = 0$ at $r = 0$ and $u = 0$ at $r = R$. u can then be expressed as a function of r (Fig. 2.10b):

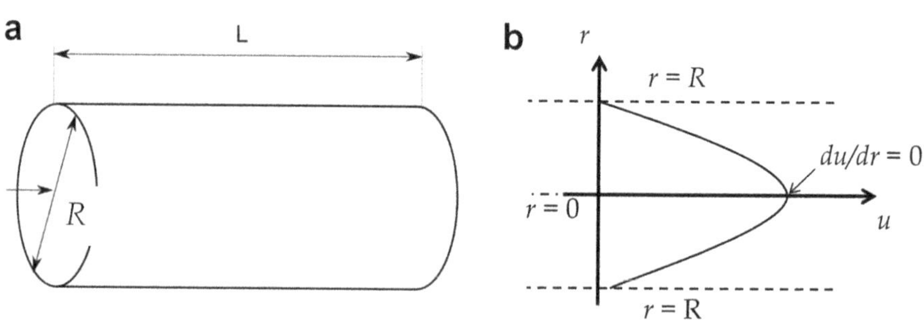

Fig. 2.10 (**a**) Key parameters and (**b**) flow profile of a pipe flow

$$u(r) = \left(-\frac{dp}{dx}\right)\frac{1}{\mu}\left(\frac{R^2 - r^2}{4}\right) \quad (2.34)$$

And the flow rate Q driven by gage pressure ΔP can be expressed as

$$Q = \int_0^R u(r) \times 2\pi r dr = \frac{\pi R^4}{8\mu L}(P_2 - P_1) \quad (2.35)$$

where P_1 is the inlet pressure and P_2 is the outlet pressure.

2.5 Surface and Interfacial Properties

2.5.1 Surface Roughness

Surface roughness (or roughness) is a property of the surface texture. It is quantified by the deviations in the direction of the normal vector of its surface. A rough surface is marked by finely spaced irregularities, protuberances, or ridges. The surface properties of a rough surface can be quantified by multiple measures as shown in Fig. 2.11a. Roughness width is the distance between the two successive peaks or ridges of the predominant surface pattern, which is measured in a direction parallel to the surface. Roughness width cutoff is the largest spacing between the irregularities. The measure of waviness, on the other hand, stands outside the value of roughness width cutoff.

On the other hand, waviness can be distinguished from roughness by the broader spacing between the surface irregularities. Heat treatment, residual stress, and vibrations are some the common causes for waviness of surface texture. Lay is defined as the direction of the predominant surface pattern. Roughness height is a measure of the height of the irregularities compared to a reference line. Waviness

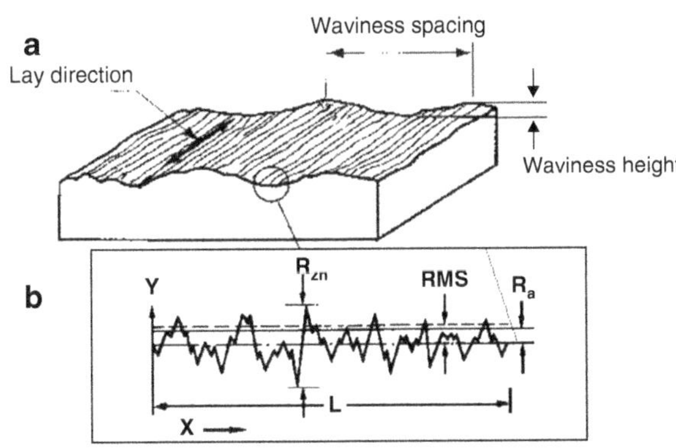

Fig. 2.11 Typical features of a rough surface. Inset: A scanned surface profile by a profilometer

width is the distance between two successive peaks or ridges of the predominant surface pattern, which is measured in a direction parallel to the surface.

Surface roughness or profile, R_a, is a quantitative calculation of the relative roughness measured across an area. R_a is determined via linear measurement of roughness through an area. Surface profile, R_a, is closely related to surface area roughness, S_a, in which several profiles are averaged and reported. In other words, S_a is an average of R_a though a surface. Several methods can be utilized in order to measure the surface roughness, which can be classified into contact and noncontact techniques. In a contact-based method, which is the basic function of a profilometer, surface roughness is measured via dragging a stylus across the surface; a sample profile is shown in Fig. 2.11b. Electron microscopy is a common example of a device using noncontact method for surface profile measurement.

Once a value of roughness is obtained, a relative value can be reported using several industry standard units. Surface profile, R_a, is generally reported as the arithmetic average of the peak heights and valleys from the mean line. The root-mean-square (RMS) roughness is a representation of R_a. This is often quantified by computing how a series (n) of height measurements at different spatial positions along the track $s(x)$ compares with the average height $\bar{s}(x)$. The following equation should be reminiscent of how the standard deviation on a sample surface is calculated:

$$R_a \approx \sqrt{\frac{1}{n}\sum_{x=1}^{n}[s(x) - \bar{s}(x)]^2} \tag{2.36}$$

Vessels and organs are composed of cells and molecules; hence, biological surfaces have a characteristic texture and a low, molecular-scale value of roughness. One obvious consequence of surface roughness occurs when an implant has a moving part. If surface roughness is high, then "high spots," called asperities, can make contact with the opposing surface. If the two surfaces are hard and rough, then the asperity contacts can result in significant friction. If the rough surface is harder than the opposing surface, then the soft surface will experience wear and become marred by scratches or plow tracks. Apart from the friction, surface wear, and reduced lifetime, another important consequence is that wear particles can be shed. These small particles are believed to activate immune responses in macrophages. The immune response results in erosion of the surrounding bone that secures the implant. Bone loss (osteolysis) can, in turn, result in implant loosening. Consequently, the combination of wear and loosening limits the lifetime of hip implants. Minimization of clotting can be achieved by processing biomaterials with smooth surfaces. The rougher a surface is, the more surface area exists for molecular and cellular adhesion. Such interactions with blood constituents can initiate coagulation. In particular, artificial heart designers use smooth plastics to construct the pumping chamber materials and surfaces. The smooth surface in combination with the fluid flow pattern and resultant shear forces is intended to limit the deposition of biological material on blood-exposed surfaces.

2.5.2 Friction and Lubrication at the Tool-Workpiece Interface

Knowledge of friction and heat transfer at the tool-material interface should be expressed quantitatively in order to develop an adequate design of manufacturing processes. The mechanics of interfacial friction is very complex. One way of expressing friction quantitatively is through a friction coefficient, μ. In the static conditions, the coefficient of friction (μ_f) between two materials is the ration of frictional shear stress τ_f to the normal compressive stress to the surface σ_f. For small values of the compressive stress/pressure, which is usually the case in manufacturing processes such as sheet metal forming, friction force increases with increasing pressure. In this case, friction conditions are best characterized by Coulomb's law:

$$\tau_f = \mu_f \sigma_f \tag{2.37}$$

Typically, $0 < \mu_f < 0.577$. With increasing pressure, friction force cannot increase indefinitely. It approaches a finite value for very high pressures. At that time, sticking friction conditions occur when $\tau_f \geq S_s$, where S_s is shear strength. S_s has a relation with the flow stress strength σ_{yield}:

$$S_s^2 = {\sigma_{yield}}^2/3, \tag{2.38}$$

according to the von Mises criterion of plastic flow. This condition occurs often in bulk material forming operations when $\tau_f \gg S_s$. At this high pressure, since there is no relative motion between tool and the workpiece at the interface, the coefficient of friction is no longer meaningful. In this case, the friction factor, m_f (between 0 and 1), is used to model such friction conditions.

Engine oils are specifically designed to modify a region of low shear rate using lubrication. It is believed that proteoglycans and glycoproteins have evolved naturally to reduce frictional effects in this way in a wide variety of biological processes, e.g., ocular mucins in the eye. Tool temperature may increase at high stroking rates due to friction and material deformation during manufacturing. Sometimes, this affects manufacturing processes by causing heat transfer between the tool and material and reduces the effectiveness of the lubricant used. Heating of the material also tends to decrease its flow stress and affects the manufacturing outcome.

2.5.3 Adhesion/Binding Strength

Adhesion or binding strength describes the effect of dissimilar particles or surfaces holding stably to each other by intermolecular forces. When an adhesive interface is formed between two materials, the adhesive forces can be caused by multiple means: chemical adhesion, dispersive adhesion, and diffusive adhesion. In particular, adhesion is an important phenomenon between biomaterials. Adhesion may be carefully controlled by biomolecular structures. On the other hand, materials can be bound together physically. A typical way of material binding is to melt materials at the material interface wherein the binding will form after the interfacial material has cooled down and solidified. Thermal bonding and welding are examples of such process. Determination of the failure point of interfacial adhesion and binding is important. Generally, the strength of the adhesion/binding between two materials is quantified as the tensile force required to break the adhesion divided by surface area over which the two materials contact.

In addition, an adhesive is a substance applied to bind two separate materials together. Adhesives can be classified as reactive and nonreactive adhesives, which refer to whether or not the adhesion is achieved via chemical reactions. Bone cement is an example of an adhesive used in biomedical devices. In some cases, the use of adhesives can replace physical binding techniques such as sewing and mechanical fastening. However, adhesives may have some disadvantages such as decreased stability at a high temperature, limited adhesive strength, and incapability of separating objects once adhered.

Problems

Problem 2.1

While running, the foot of a 68kg man is momentarily subjected to a force which is five times his weight. Determine the average normal stress developed in the tibia of his leg at the midsection of the bone. The cross section can be assumed as a circular hollow shape having an outer diameter of 4.5 cm and an inner diameter of 2.5 cm (please note that a bone structure should be in the hollow shape). Assume the fibula does not support a load.

Problem 2.2

The σ-ε diagram for a collagen fiber bundle from which a human tendon is composed is shown in Fig. 2.P1 below. If a segment of the Achilles tendon at *A* has a length of 16 cm and an approximate cross-sectional area of 1.43 cm^2, determine its elongation if the foot supports a load which creates tension in the tendon of 156 kg.

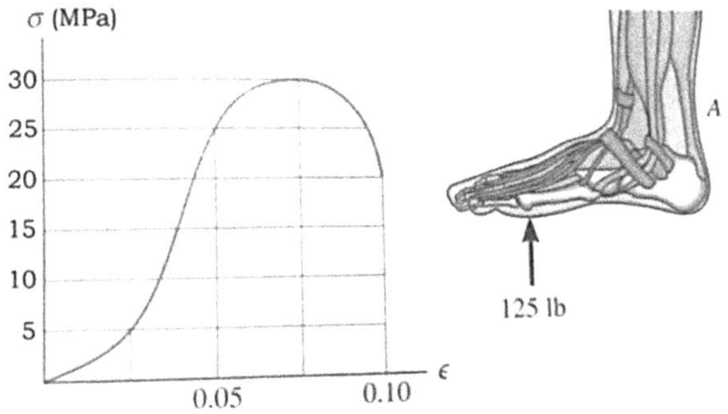

Fig. 2.P1 Stress strain curve for a collagen fiber bundle in a human tendon

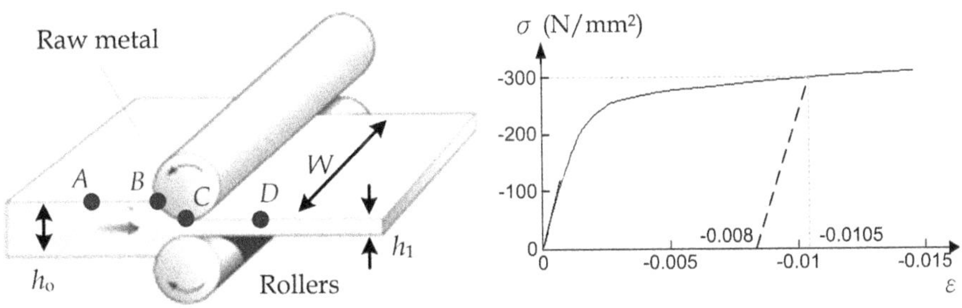

Fig. 2.P2 Rolling process on a plate made of stainless steel with the stress-strain curve shown on the *right*

Problem 2.3

Consider the rolling process on a stainless steel (304L/316L) plate with a width W of 50 cm as shown in the Fig. 2.P2 (*left*) above.

Two rollers clamp a metallic plate and keep rotating as indicated by the arrows on the roller cross-sections, such that the sandwiched metallic plate is forced to reduce its thickness and fit into the gap between the rollers. The material was compressed from an original thickness h_0 to a resultant thickness h_1. During the process, the material experiences variations in direct stress and direct strain. The *right* figure in Fig. 2.P2 shows the stress-strain curve during the process.

(a) Please indicate the points A, B, C, and D during the rolling process (*left* figure above) on the *right* stress-strain curve above.
(b) If the original metal plate before rolling is 5 mm thick, then what is the plate thickness after process?
(c) What are the vertical direct stress levels at Point B and Point C above?

The cross-section is shown in (Fig. 2.P3a) below, with the roller radius defined as $R = 10$ cm and the horizontal length of contact surface area L.

(d) What is Δh? Please estimate L by considering the contact surface is a plane with the cross-sectional view as the *green* line shown in Fig. 2.P3 below. Can you calculate for the value of L?

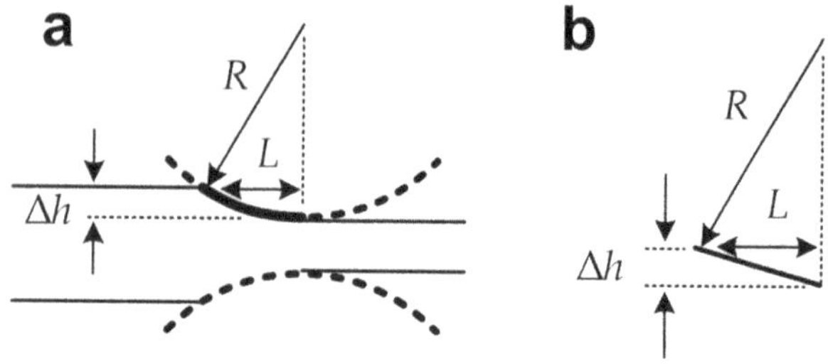

Fig. 2.P3 Cross section of the contact surface during rolling

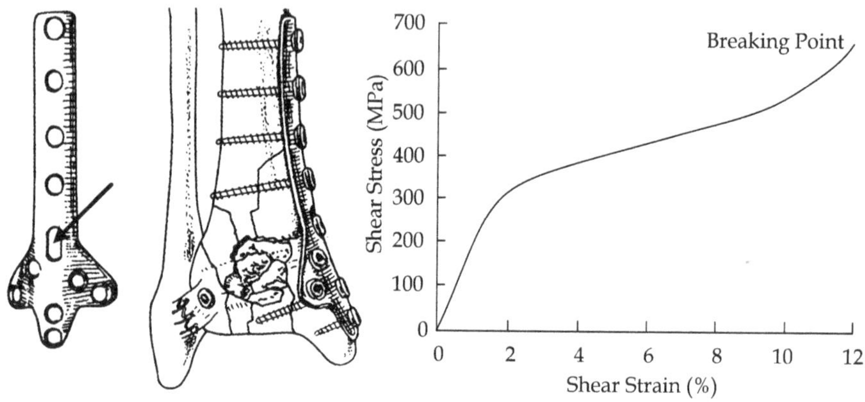

Fig. 2.P4 Configuration of a bone fixation plate made of stainless steel with the shear stress-strain curve shown on the *right*

(e) We may consider the average true stress over the contact surface as the average stress level of Point B and Point C as shown in Fig. 2.P3. Then, what is the compressive force on the plate?

Problem 2.4

The part illustrated in the Fig. 2.P4 is a stainless steel bone fixation plate which is mounted on a broken bone by screws for the fixation purpose. This medical tool is manufactured from a large stainless steel plate as the raw material.

(a) The thickness of the bone fixation plate is 3 mm, and a hole in the left figure above (arrow) has a diameter of 5 mm. Based on the shear stress-shear strain curve, estimate the force required to punch the hole on the fixation plate.
(b) The bone fixation plate should be bent in order to fit with the shape of a broken bone. Please specify a proper range of the shear stress during the bending process.

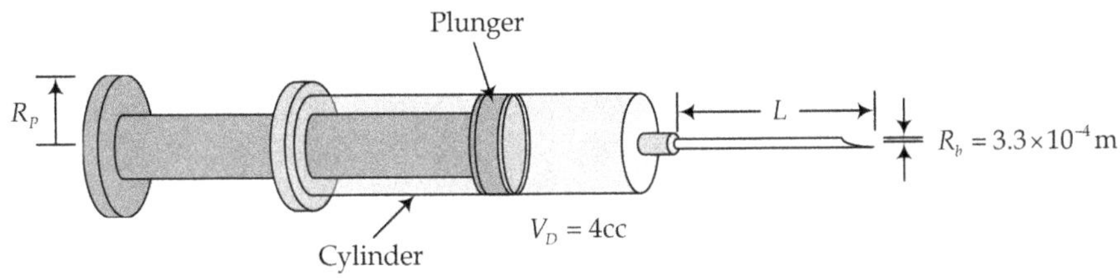

Fig. 2.P5 Dimensions of a clinical syringe

Problem 2.5

Consider the clinical requirement that 4 cc (or 4 ml) of a drug (V_D) needs to be injected in at least 2 s (t_D). As shown in Fig. 2.P5 above, the drug, whose viscosity is 3×10^{-3} kg/m s and density is 1 g/cm^3, will flow through a 5×10^{-2} m long (L) needle with a bore radius (R_b) of 3.3×10^{-4} m. The radius of the plunger (R_p), where the force for fluid motion is obtained through pressure exerted by the thumb, is 5×10^{-3} m.

Because viscous forces are high, let us presume that the flow is laminar. To simplify the calculation, we will also assume that the "back pressure" imposed at the needle's end within the body is similar to the background pressure that the atmosphere exerts on the plunger. Thus, the pressure drop associated with drug flow equals the pressure that must be generated by the person performing the injection. Then,

(a) What are the Re's of the flow along (i) the cylinder and (ii) the 20-gauge needle?
(b) What are the fluidic resistance ($\Delta P/Q$) along (i) the cylinder and (ii) the 20-gauge needle?
(c) Can you compare the pressure drop between the cylinder and that in the 20-gauge needle?
(d) What is the minimum required force applied by the thumb to support the required rate of drug injection?

Problem 2.6

3D printing technology can be applied in generating medical implants, such as the soft bone replacements as shown in Fig. 2.P6. Biomaterials such as polycaprolactone (PCL) can be printed line by line with a consistent radius of R_F in order to construct the target 3D structures.

The 3D printer can be configured as a horizontally (along z direction) movable stage holding the printed product, where the biomaterial is injected from a syringe mounted on a manipulator movable in both x- and y-directions (Fig. 2.P7).

The PCL prepolymer solution (60% volumetric ratio of PCL in acetone) of 10 ml (V_D) was prepared to be deposited on the substrate material. The solution had a

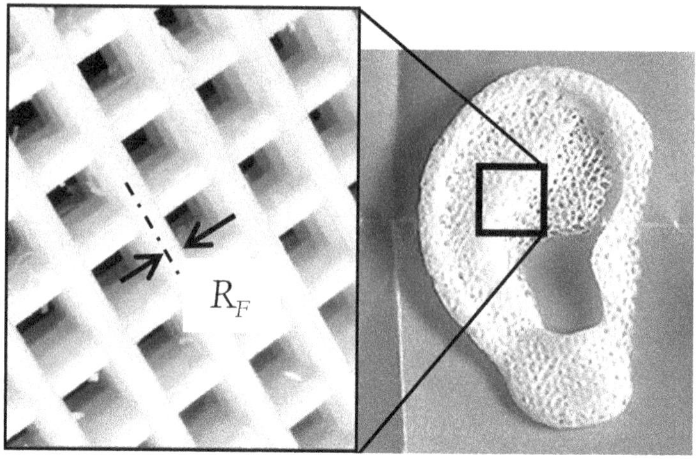

Fig. 2.P6 3D printed scaffold for soft bone replacement

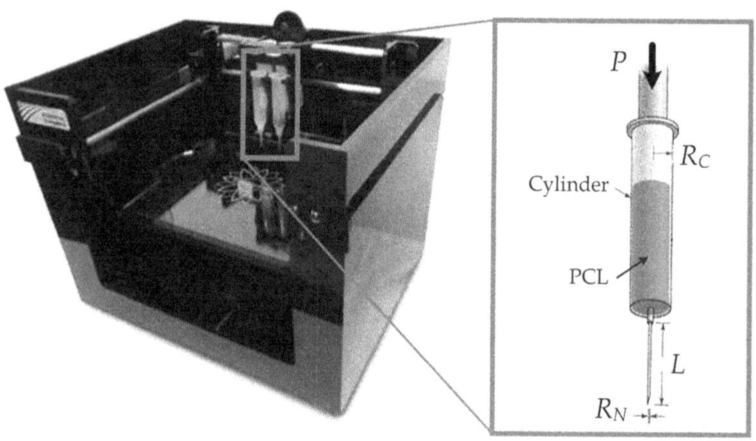

Fig. 2.P7 Configuration of a 3D printer

viscosity of 7.2×10^{-3} kg/m s and density of 1.0 g/cm^3. This solution should flow through a needle with a length of 1 cm (L) and an inner radius (R_N) of 0.3 mm. The inner radius of the cylinder (R_C) was 4 mm. The fluid motion was driven by a gas pressure (P) supplied via tubing connecting to the back of the cylinder. After injection of PCL solution and evaporation of acetone, the printed radius of material "line" (R_F) was expected to be ~0.23238 mm.

(a) Can you compare the pressure drop between the cylinder and that in the needle?
(b) Considering the target velocity of the syringe movement (either in x- or y-direction) is 50 mm/s, then what is the required gauge pressure (P) applied to support the required rate of the liquid injection?

(c) What are the *Reynolds* number of the flow along (i) the cylinder and (ii) the needle?

Problem 2.7

The injection molding process for plastics is described in Fig. 2.P8 below. A more comprehensive description is available in Chap. 8 of this book. Here, our focus is on the barrel, reciprocating screw, and nozzle.

(a) Please consider the nozzle as a cylindrical pipe with inner radius r and length l. Then, for the molten plastic with viscosity μ, what would be the upstream gauge pressure P_n of the nozzle inlet if a flow rate Q of the plastic is required for a sufficiently fast production speed? (Please assume the outlet pressure of the nozzle to be "0.")

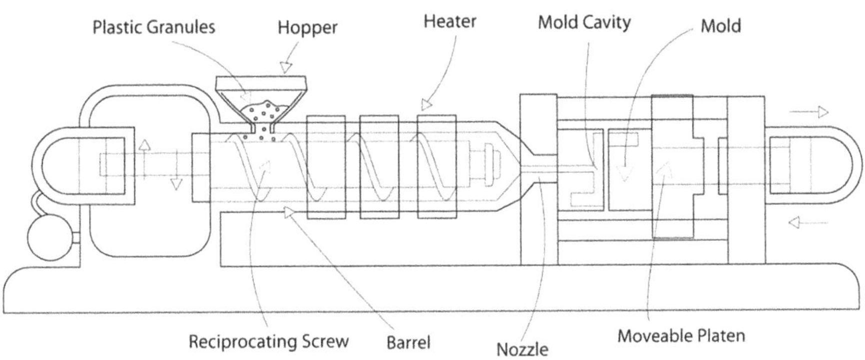

Fig. 2.P8 Structure of an injection molding machine

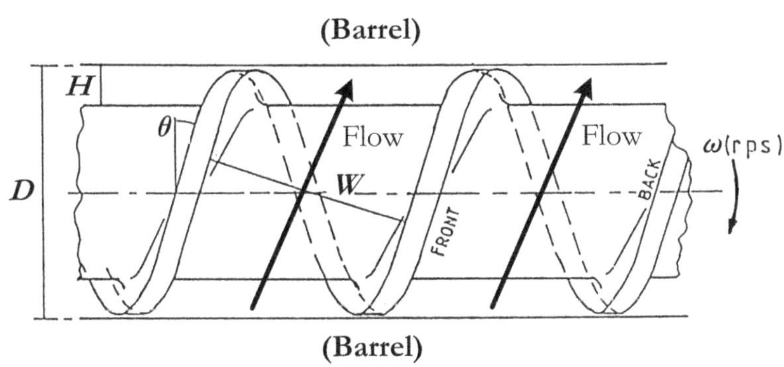

Fig. 2.P9 Key parameters of the barrel and the reciprocating flow

Please consider the key parameters of the barrel and the reciprocating flow as below (Fig. 2.P9). The inner diameter of the barrel can be considered to be the same as the outer diameter of the screw, denoted as D. There are gaps between the crest/tooth structures around the screw, inducing a flow region with a rectangular cross-section of flow with width W and height H ($H \ll W$). The crest/tooth structure has a tilt angle θ, and the overall flow direction follows the *straight* arrows. Therefore, in each turn (around the screw body) of the liquid flow, the flowing distance is about $\pi D/\cos\theta$ (given that $D \gg H$). And if there are N turns of the crest/tooth structure along the reciprocating screw, the total flow distance L passing the screw region would be about $N\pi D/\cos\theta$.

Now, please impose your imagination by considering the flow region to be "unfolded" as a straight rectangular channel with width W, height H, and length $L = N\pi D/\cos\theta$. For this "imaginary" rectangular channel, we may consider the outlet pressure as the nozzle pressure P_n as considered in Part (a) before. And for simplicity, we further consider the inlet pressure of this rectangular channel to be "0."

(b) Please first consider the side view of this rectangular straight channel along the flow direction as Fig. 2.P10 with the special case of $P_n = 0$. Here, we define the x-direction following the flow and $y = 0$ at the root surface on the screw (in other words the inner barrel surface is at $y = H$). Accordingly, under the reciprocating screw with a rotational speed ω, the inner surface of the barrel would have a moving speed U along the flow direction (the x-direction as indicated below), relative to the screw surface along $y = 0$. In this case, we may consider the flow as the Couette flow, with the side-view flow profile $u(y)$ as describe below.

Can you find an expression for U, the relative barrel surface movement along flow?

(c) Let's consider back a more general case that P_n is no longer zero but with the same expression as your previous answer in Part (a). Then, please sketch the side-view flow profile $u(y)$ which should not be the Couette flow anymore. Also, please identify the following three expressions in your flow profile sketch:

   (i) U at $y = H$
   (ii) $u(y)$ with $du(y)/dy = 0$
   (iii) The y-position of (ii)

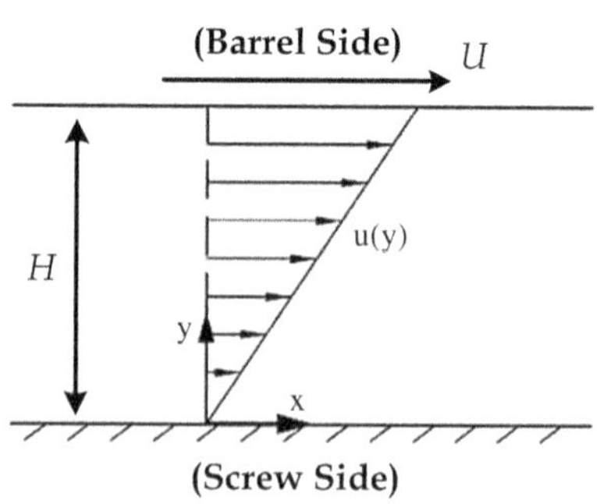

Fig. 2.P10 Couette flow between barrel and screw

Problem 2.8

Consider that the rough surface of an implanted bone fixation plate has steps that are 20 μm high and long; the steps repeat every 40 μm. The pattern is shown below (Fig. 2.P11). What is the ratio of surface area of the rough-to-perfectly flat surface? Assume that the material has a width that is way longer than the thickness.

Problem 2.9

In the following adhesive joints (Fig. 2.P12), assume that the adhesive tensile strength (E_s) and adhesive shear strength (G_s) are much smaller than the corresponding values of materials such that the maximum allowable stretching force for the following structures are determined by these adhesive parameters. Given that the depths (D) of the adhesion areas in all the configurations below are the same and $B \ll D$, L and W.

(a) Estimate the force F required breaking the adhesion joint in each of the cases above (i.e., design A, B, and C).
(b) Which one of the design A and C can provide larger adhesion strength? Please explain.
(c) Approximate for the expression of L in design C in order to induce an equivalent adhesion strength as design B with an angle θ under the same stretching force F.

Fig. 2.P11 Surface profile of a rough surface

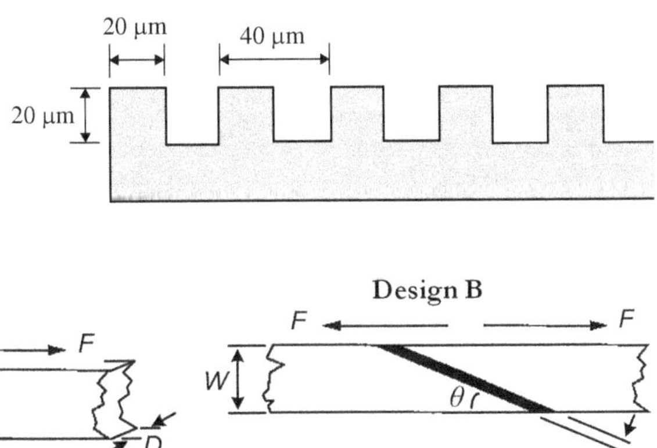

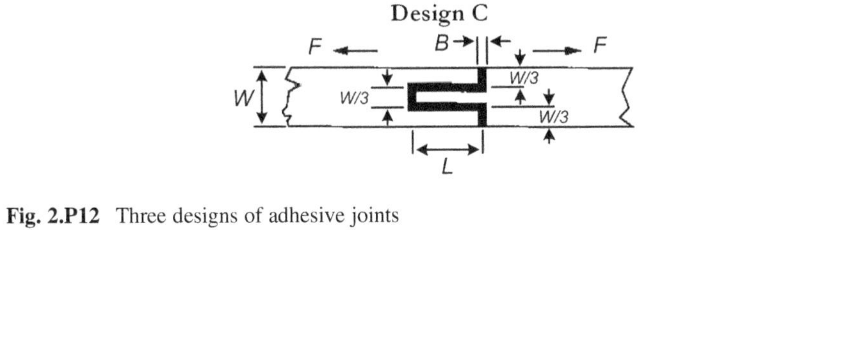

Fig. 2.P12 Three designs of adhesive joints

References and Further Reading

1. Boresi, A.P., Schmidt, R.J., Sidebottom, O.M.: Advanced mechanics of Materials. Wiley, New York (1985)
2. Hibbeler, R.C., Fan, S.C.: Statics and Mechanics of Materials. Prentice Hall, Upper Saddle River (2004)
3. Lemaitre, J., Chaboche, J.-L.: Mechanics of Solid Materials. Cambridge University Press, Cambridge (1994)
4. Kundu, P.K., Cohen, I.M., Dowling, D.W.: Fluid Mechanics. Academic Press, Boston (2008)
5. Desai, C.S.: Mechanics of Materials and Interfaces: The Disturbed State Concept. CRC Press, Boca Raton (2000)
6. Francis, L.F.: Materials Processing: A Unified Approach to Processing of Metals, Ceramics and Polymers. Academic Press, Amsterdam (2015)
7. Dowling, N.E.: Mechanical Behavior of Materials: Engineering Methods for Deformation, Fracture, and Fatigue. Pearson, Boston (2012)
8. Fung, Y.-C.: Foundations of Solid Mechanics. Prentice Hall, Upper Saddle River (2017)

Chapter 3
Metals and Alloys

Abstract The interatomic characteristics of metals determine the material properties. The metallic bonding gives metals most of their notable mechanical properties such as strength, malleability, and ductility. Metallic materials are good thermal and electrical conductors. Their electrical conductivity comes from the fact that metals are able to donate electrons. Metallic bonds are surrounded by positive ions because electrons form a "cloud" where electrons move freely. This chapter discusses on the crystalline characteristics, physical properties, and strengthening mechanisms of metals. It also introduces some commonly used metallic biomaterials (pure metals and alloys), such as stainless steels, cobalt-chromium alloys, and titanium-based alloys.

3.1 Overview

Metals are characterized by metallic interatomic bonding with valence shell electrons that form a "cloud" of electrons around their atoms/ions. In metals, the coordination number, i.e., the number of nearest neighboring atoms, is very high, which results in the valence shell electrons forming a cloud of electrons. As a result, the interatomic bonds are nondirectional, and electron movement within metal crystal lattices is easier than in ionic or covalently bonded materials. This fundamental distinguishing characteristic of metals results in the relative ease of plastic deformation as well as the high electrical and thermal conductivities of metals. In addition, metals can deform plastically at high loads to blunt sharp discontinuities, which results in relatively high fracture toughness, thereby reducing local stress concentrations. The high tensile and fatigue strength of metals, compared to ceramics and polymers in general, make them a widely used materials for many biomedical applications such as implants that carry mechanical loads.

The use of metals as biomaterials goes back thousands of years. Recently, archeologists in France have discovered a dental implant in a grave that may have been used to hold a decorative screw in place. The implant was an iron pin that was retrieved from the mouth of a skeleton in a Celtic burial site in La Chene, France, that has been dated back to the third century B.C. Metallic biomaterials continue to be

R. H. W. Lam, W. Chen, *Biomedical Devices*,
https://doi.org/10.1007/978-3-030-24237-4_3

used extensively for the fabrication of surgical implants, primarily for the same reason that led to their initial selection for these devices centuries ago. Metals belong to the category of synthetic biomaterials and are used as implant materials where implants are subjected to higher level of mechanical loading.

The metallic bonding gives metals most of their notable mechanical properties such as strength, malleability, and ductility. Metallic materials are good thermal and electrical conductors. Their electrical conductivity comes from the fact that metals are able to donate electrons. Metallic bonds are surrounded by positive ions because electrons form a "cloud" where electrons move freely. The electron cloud also makes metallic bonds nondirectional in nature. Because they are not fixed like ionic materials, metallic bonds can deform under pressure and force.

The high strength and resistance to fracture which this class of material can provide, assuming proper processing, gives reliable long-term implant performance in major load-bearing situations. The relative ease of fabrication of both simple and complex shapes using well-established metal fabrication techniques (e.g., casting, forging, machining) has promoted the use of metals in the fields of orthopedics and dentistry, the two primary areas in which highly loaded devices are most common. Metals have been widely used for forming cardiovascular devices (e.g., artificial heart valves, blood conduits, vascular stents, and other components of heart assist devices) and neurovascular implants (aneurysm clips) for similar reasons. Metals are also used in neuromuscular stimulation devices owing to their electrical conductivity. Alloys frequently provide improvements in material properties, such as strength and corrosion resistance, and are generally used in biomedical devices. At this time, three metals currently dominate the biomedical field: 316L stainless steel, cobalt-chromium-molybdenum alloy, and pure titanium and titanium alloys. Overall, the main considerations in selecting metals and alloys for biomedical applications are biocompatibility, appropriate mechanical properties, corrosion resistance, and reasonable cost.

In certain scenarios, metal prosthetics do present some problems. Corrosion of metals and how it should be stopped or at least minimized after implantation are always important concerns. Along with these, the higher moduli of elasticity and load-bearing capabilities of metals may sometimes cause problems. In the case of hip implants, the elastic moduli of common metals are at least seven times greater than that of natural bone. This mismatch of mechanical properties can cause "stress shielding," a condition characterized by bone resorption (loss of bone) in the vicinity of implants.

3.2 Crystalline Characteristic of Metals

3.2.1 Crystal Direction and Planes

Nearly all metals are crystalline solids. In crystalline solids, atoms bond with each other by forming a regular periodic pattern that constitutes a crystal structure as shown in Fig. 3.1a. The crystalline solid is essentially a periodic repetition of a small

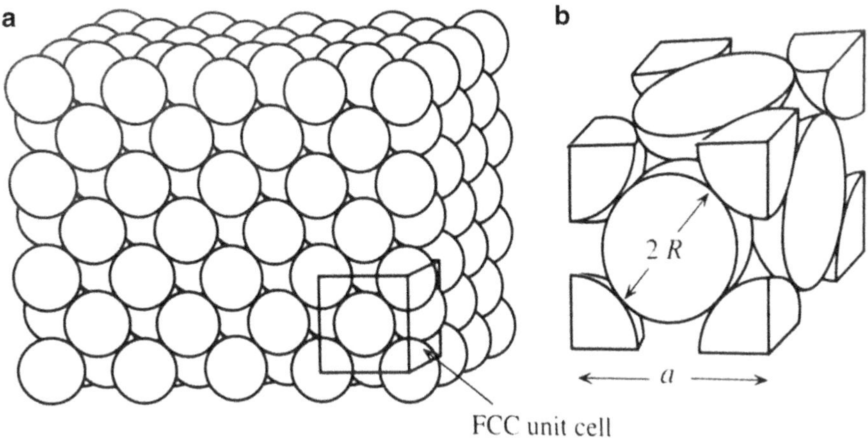

Fig. 3.1 (**a**) The crystal structure of nickel which is face-centered cubic (FCC). (**b**) An FCC unit cell with close-packed spheres

volume (or cell) of atoms in three dimensions. It is useful to identify the repeating unit so that the crystal properties can be described through this unit. For example, nickel has a face-centered cubic (FCC) structure (Fig. 3.1b); and annealed steel and iron have a body-centered cubic (BCC) structure. Crystal structure of metals can determine the material properties such as the flow stress. For instance, the degree of strengthening (n_s in Eq. 2.5) is relatively lower for FCC metals (0.1–0.15) than for BCC metals (0.5–0.55).

There are a number of possibilities for choosing a unit cell for a given crystal structure. As a universal convention, we generally represent the geometry of the unit cell as a parallelepiped with sides a, b, and c with angles α, β, and γ, referred to as the lattice parameters as depicted in Fig. 3.2. We insert an xyz coordinate system along a, b, and c directions as the reference frame. For nickel and iron as examples, their unit cell geometries have $a = b = c$, $\alpha = \beta = \gamma = 90°$, and cubic symmetry. The physical properties of crystalline solids should be specified along a direction or a particular plane of atoms. For example, the modulus of elasticity is directionally specific.

The crystal direction is specified by a vector from the (0, 0, 0) corner of a unit cell in the reference frame. Such vector is in the three dimensional space, and therefore, it has three numbers. It is preferable to have these numbers as integers. Some examples of the crystal directions for a cubic unit cell are shown in Fig. 3.3a. A direction is written in a square bracket without commas, e.g., [100] for a direction pointing to x. A bar would be placed above or below the number if the number is negative, e.g., [$\underline{1}$00] for the negative x direction. A group of the directions with the same absolute values (in other words, the underline notation removed) of the three numbers would call a family of directions, written inside a pair of triangular brackets. For instance, [100] and [$\underline{1}$00] are directions belonging to the family of directions <100>; <111> is demonstrated in Fig. 3.3b. Furthermore, to describe a particular plane in a crystal unit, we would consider the normal vector of the plane. Such plane vector also

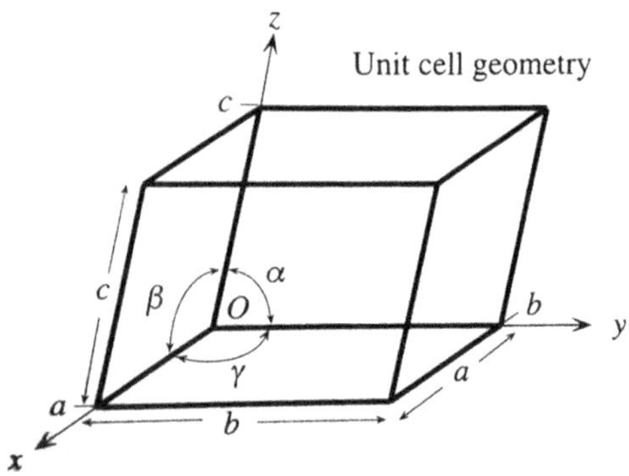

Fig. 3.2 A parallelepiped is chosen to describe the geometry of a unit cell

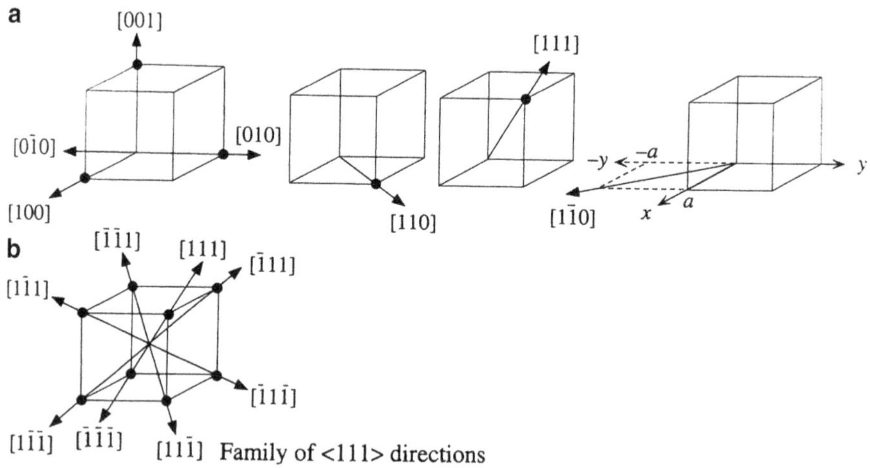

Fig. 3.3 Examples of (**a**) individual directions and (**b**) <111> in the cubic crystal system

consists of three numbers for the *x*, *y*, and *z* components of the vector, and these numbers are written in round brackets. Hence, the *xy* plane facing the negative *z* direction is represented as $(00\bar{1})$. A family of planes is denoted as three numbers inside a pair of curved brackets, e.g., {001}.

3.2.2 *Line Defects and Grains*

A line defect is formed in a crystalline solid when an atomic plane terminates within the crystal instead of passing all the way to the end of the crystal (Fig. 3.4a). The edge of this short plane of atoms is therefore like a line running inside the crystal. The planes neighboring this short plane are dislocated with respect to those below the line. We therefore call this type of defect an edge dislocation. The vertical line corresponds to the half plane of atoms in the crystal. It is clear that the atoms around

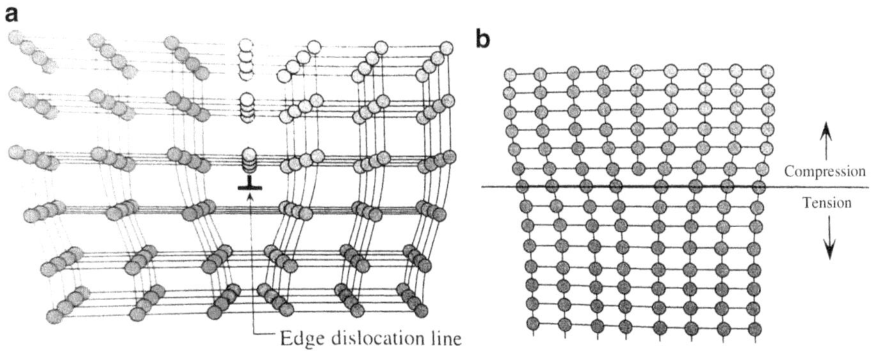

Fig. 3.4 (**a**) Dislocation is a line defect, which is at the mark pointing into the paper in this figure. (**b**) The line of strain field, with the atomic bonds compressed above and stretched below the dislocation line

the dislocation line have been effectively displaced from their perfect-crystal equilibrium positions, which results in atoms being out of registry above and below the dislocation. The atoms above the dislocation line (highlighted by the "reversed T" mark) are pushed together, whereas those below it are pulled apart, leading to regions of compression and tension above and below the dislocation line, respectively, as depicted by the shaded region around the dislocation line in Fig. 3.4b. Therefore, around a dislocation line, we have a strain field due to the stretching or compressing of the bonds.

Another type of dislocation is the screw dislocation, which is essentially a shearing of one portion of the crystal with respect to another by one atomic distance (Fig. 3.5a). The displacement occurs on either side of the screw dislocation line. The circular arrow around the line symbolizes the screw dislocation. As we move away from the dislocation line, atoms in the upper portion become more out of registry with those below; at the edge of the crystal, this displacement is one atomic distance.

The phenomenon of plastic or permanent deformation of a metal depends totally on the presence and motions of dislocations. Both edge and screw dislocations are generally created by stresses resulting from thermal and mechanical processing. A line defect is not necessarily either a pure edge or a pure screw dislocation; it can be a mixture (Fig. 3.5b). Screw dislocations frequently occur during crystal growth, which involves atomic stacking on the surface of a crystal. Such dislocations aid crystallization by providing an additional "edge" to which the incoming atoms can attach. To explain, if an atom arrives at the surface of a perfect crystal, it can only attach to one atom in the plane below. However, if there is a screw dislocation, the incoming atom can attach to an edge and thereby form more bonds, which is more favorable.

In fact, many naturally formed crystalline materials are polycrystalline, i.e., they are composed of many small crystals oriented in different directions. In fact, the growth of a flawless single-crystal form, called the melt, requires special, well-conditioned treatments. When a melt is cooled to below its freezing temperature, solidification does not occur at every point; rather, it occurs at certain sites at which small crystal-like structures form. The liquid atoms adjacent to the small crystal

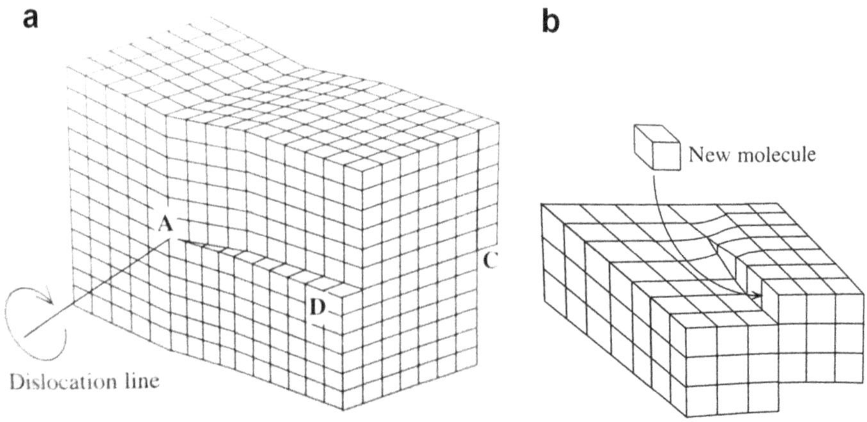

Fig. 3.5 (**a**) A screw dislocation in a crystal. (**b**) Screw dislocation facilitates crystal growth by providing more bonding sites

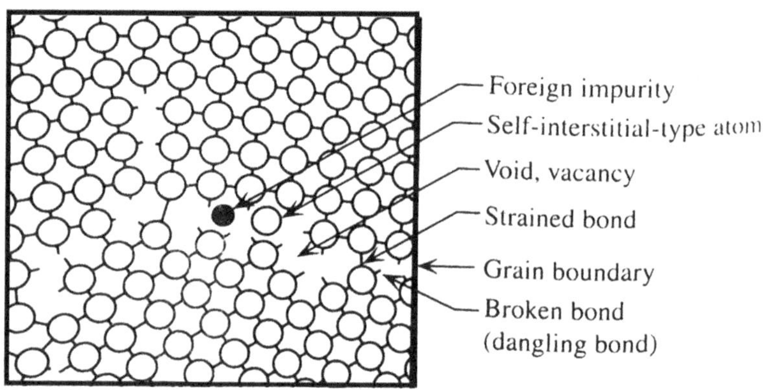

Fig. 3.6 The grain boundaries have defects such as broken bonds, voids, vacancies, strained bonds

structures contribute to the continuous growth of the crystal with a limited size, called a grain. Since these small crystal structures are randomly oriented when they are first formed, the grains have random crystallographic orientations during the crystalline growth. As the melt between the grains is consumed, some grains meet and obstruct each other. At the end of solidification, therefore, the whole structure has grains with irregular shapes and orientations.

Polycrystalline materials have grain boundaries where differently oriented crystals meet. The atoms at the grain boundaries obviously cannot follow their natural bonding habits because the crystal orientation suddenly changes across the boundary. Therefore, there are both voids and broken bonds at the grain boundary (Fig. 3.6). Consequently, the bonds at the grain boundaries are more easily broken, and so the entire bulk material will be weaker at these points. For instance, through various thermal treatment cycles, engineers can control the grain size and the mechanical properties of metals to suit different application needs.

3.3 Common Physical Properties

3.3.1 *Interatomic Attraction and Repulsion*

Metal atoms have only a few valence electrons. When many metal atoms are brought together to form a solid, these valence electrons are lost from individual atoms and become collectively shared by all the ions. If we consider only a simple case where two atoms are brought together, the valence electrons interact with each other and with the neighbor's positively charged nucleus. This interaction often results in the formation of a bond between the two atoms.

The formation of a bond requires that the energy of the system of two atoms together must be less than that of the two individual atoms. This general principle is demonstrated in Fig. 3.7a, showing two atoms brought together from infinity. As the two atoms approach each other, the atoms exert attractive and repulsive forces on each other as a result of mutual electrostatic interactions. The net force is the sum of the attractive force and repulsive forces. The variations of these two forces with distance differ in value. The attractive force decays relatively slower than the repulsive force with the increasing separation distance. When the atoms are placed so close that the individual atomic electron shells overlap, the repulsion force dominates. Accordingly, equilibrium will be reached at a defined position when the attractive force balances out the repulsive force and the net force is zero.

Let's first consider a simple case in which the atomic energy only consists of the potential energy of the two atoms which can be obtained by integrating the net force. Figure 3.7 shows the variation of the net force and the overall atomic energy with the interatomic separation r as the two atoms are brought together from infinity. The decrease in energy corresponds to an attractive interaction between the two atoms. In the state of force equilibrium, the atoms are separated by a certain distance (r_o as shown in Fig. 3.7a), which is called the bond length. At the bond length, the gradient

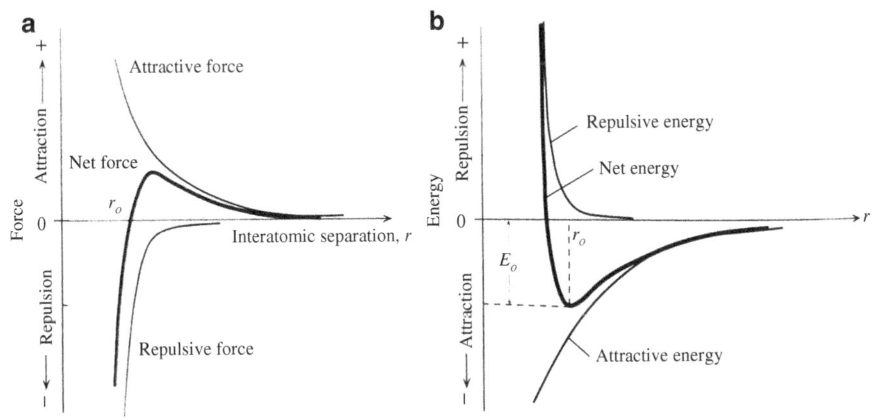

Fig. 3.7 (**a**) Force versus interatomic separation. (**b**) Atomic energy versus interatomic separation

of the net atomic energy is zero, reflecting a local energy minimum of E_o. Although we only consider two atoms in this example, similar arguments also apply to bonding between many atoms, even between millions of atoms as in a typical solid. In particular, metallic materials consist of mostly a limited number of repeating unit cells. More generally, even for a solid (not only for metals) in the presence of many interacting atoms, we can still identify a general potential energy and force curves per atom similar to the type shown in Fig. 3.7a. We can examine these curves for a better understanding of the properties of the solid, such as the thermal expansion coefficient and modulus of elasticity.

When a solid is under an external stretching force, the applied stress causes two neighboring atoms along the direction of force to be further separated. Their displacement δr ($=r - r_o$) results in a net attractive force δF_N between two neighboring atoms as indicated in Fig. 3.8, where F_N is the net interatomic force. δF_N attempts to restore the separation to equilibrium and is balanced by a portion of the applied force acting on these atoms with an effective area. Such area for each atom would be roughly r_o^2. For an ideal pure metallic body without defects and impurities, if we consider $\delta F_N/r_o^2$ as the average "atomic" stress and $\delta r/r_o$ as the average "atomic" strain, we may relate these quantities with the bulk elastic modulus E of a material body (defined in Eq. 2.4) as $\delta F_N/r_o^2 = E\delta r/r_o$ or

$$E \approx \frac{1}{r_o}\frac{dF_N}{dr}\bigg|_{r=r_o}. \tag{3.1}$$

Clearly, E depends on the gradient of F_N versus r curve at r_o in the plot indicated in Fig. 3.7a.

As mentioned in Sect. 2.3.1, nearly all materials expand as their temperature increases. This phenomenon is due to the asymmetric nature of interatomic forces and the increase in the amplitude of atomic vibrations with temperature as expected from the kinetic molecular theory. Recalling the potential energy curve as shown in Fig. 3.7b, in equilibrium, the atomic energy is at a minimum level of E_a, which has an absolute value identical to the potential energy $U(r) = U_{min}$ at $r = r_o$. Obviously, the atoms are separated by the equilibrium separation r_o. However, according to the kinetic molecular theory, atoms are vibrating at their equilibrium positions with a

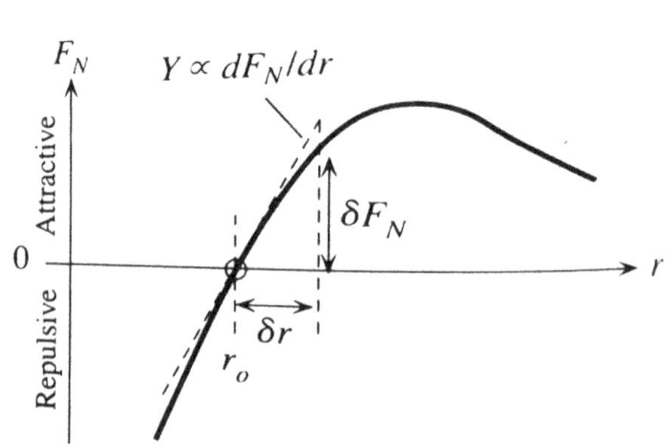

Fig. 3.8 In equilibrium, the applied force is balanced by the net force δF_N between the atoms as a result of their increased separation

mean vibrational kinetic energy that increases with temperature T as $(3/2)kT$, where k (=1.38×10^{-23} m^2 kg s^{-2} K^{-1}) is the Boltzmann constant. At any instant, the total energy of the pair of atoms is the sum of potential energy and kinetic energy without any external forces applied. Under a temperature T_1 above absolute zero, the atoms will be vibrating about their equilibrium positions, stretching and compressing their bond as depicted in Fig. 3.9. At positions B and C (and their corresponding atomic radial positions r_B and r_C, respectively), E is at a level above its minimum value due to the kinetic energy. The atoms are confined to vibrate with a radius position between r_B and r_C, executing simple harmonic motion and, hence, maintain their total energy level. However, the potential energy curve is "asymmetric," meaning that the separation position of point A (Fig. 3.9) with respect to the kinetic energy level $r_A = (r_B + r_C)/2 \neq r_o$. In fact, for most materials, $(r_B + r_C)/2 > r_o$. As the temperature increases to T_2 ($>T_1$), the kinetic energy and total energy increase. Subsequently, the atoms vibrate more rigorously, and the interatomic separation oscillates within a wider position range (i.e., the distance between B′ and C′, which is longer than that between points B and C as illustrated in Fig. 3.9). The new average separation at point A′ ($r_{A'}$) is even further, i.e., $r_{A'} > r_A$. The resultant effect is that the interatomic separation distance, and hence the bulk material strain, increases with temperature, referred to as the phenomenon of thermal expansion.

Furthermore, as higher temperatures in a material induce a longer equilibrium separation radius, the interatomic forces follow the curve as indicated in Fig. 3.7a. Typically, the gradient of the net atomic force (dF_N/dr) decreases with r. From Eq. 3.1, we know that the modulus of elasticity decreases with temperature, agreeing with our discussion in Sect. 2.3.3.

Importantly, as previously discussed in Sect. 3.1, the repeating crystalline unit cells along lines and planes enable the material deformation called "slip," especially for pure metals. A slip system describes the set of symmetrically identical slip planes and associated family of slip directions along which offsets of atoms can easily occur and lead to plastic deformation. The slip phenomenon offers high deformability and manufacturability of metals and alloys.

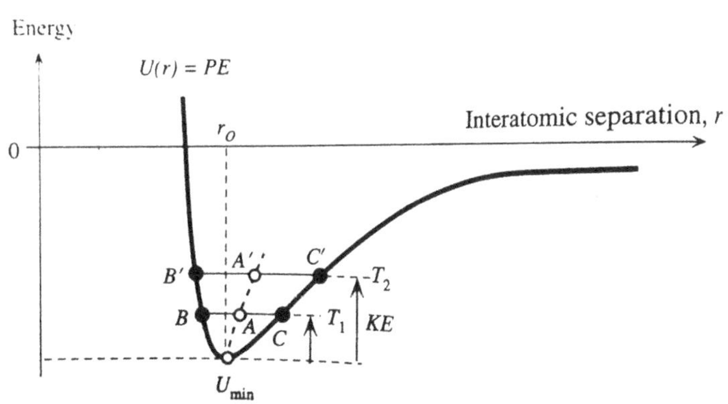

Fig. 3.9 In equilibrium, the applied force is balanced by the net force δF_N between the atoms as a result of their increased separation

3.3.2 Corrosion

As a biomedical device manufacturer, one has to provide solutions that will not only solve the problem at hand but also avoid any complications that may occur in future. The physiological environment in a human body can be modeled as a 37 °C aqueous solution at pH 7.4 with dissolved gasses (such as oxygen), electrolytes, cells, and proteins. This is a rather inhospitable environment for most metals in their natural state, and immersion of metals in this environment can lead to corrosion. Implant materials and alloys are influenced by the presence of various biological molecules like proteins and organisms like bacteria. Different fluids like saliva, blood, urine, etc. have different chemical concentrations and can have different effects on implant materials. Not only this, injury or illness can significantly change the environment surrounding the implant and can result in initiation of corrosion. Temperature fluctuation, dissolved gasses, mechanical loading cycles, pH of electrolytes, and bioelectricity are some of the factors that are important to consider while talking about in vivo corrosion in implant biomedical devices.

Metals in their pure form are generally unstable in the human physiological environment and tend to revert to their more stable mineral form. This is the primary reason for corrosion. Corrosion is derived from the Latin word *corrodere*, which means "gnawing into piece." During corrosion, metals may lose some important properties that are valuable for their medical applications. In general, corrosion can be classified on the basis of three factors: nature of the corrodent, corrosion mechanism, and appearance of corroded metal. In the case of biomedical implants, corrosion occurs in a "wet" medium. The most common types of corrosions are shown in Fig. 3.10.

Corrosion can also be grouped into two main categories based on whether the entire surface of the metal device is affected or if the corrosion is localized. Most

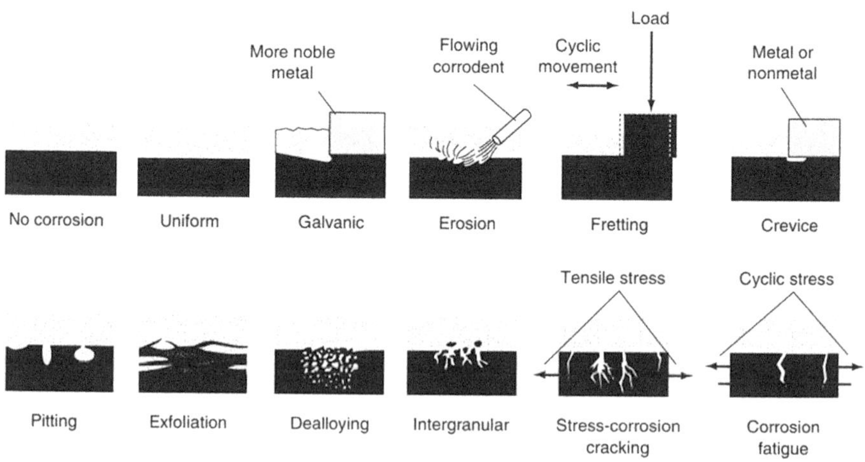

Fig. 3.10 Types of corrosion. (Davis, J.R.: Corrosion: understanding the basics, 2000, Chapter 1, Fig. 2.)

implants in a human body are protected from corrosion by passivation. Passivation is the process by which metals become less affected by corrosion in the environment. Passivation can be achieved by making the surface of the material that is exposed to the environment less reactive by applying a sacrificial layer of another metal or by either forced or natural reactions of the base metal. For example, the failure mechanisms of the Kuntscher cloverleaf intramedullary rod, which is a biomedical device that is used to treat fractures in long bones, found that the failure of the intramedullary nail was a result of localized stress corrosion caused by pitting or crevice corrosion. Hence, most of the corrosion that plagues the metallic implants inside the human body is of the form of localized pitting and crevice corrosion. In fact, there are many cases of implant failures related to corrosion and its adverse effects on biomedical implants.

During the electrochemical process of corrosion, metallic biomaterials can release ions, which may reduce the biocompatibility of materials and jeopardize the fate of implants. The released corrosion products may alter the functions of cells in the vicinity of implants as well as of cells at remote locations after transport via the circulation system. For example, nickel, an alloying element in steel, is released during dissolution of steel in blood. Nickel is an allergen to around 15–20% of the world's population and can result in the failure of steel implant materials.

As stated earlier, the majority of corrosion in the human body occurs from localized corrosion reactions. Pitting corrosion is a localized corrosion that results in the formation of pits and crevices inside materials. Pitting corrosion is more dangerous compared to uniform corrosions as it is harder to detect and prevent. The main cause of pitting corrosions is due to localized damage to the protective layer on a metallic surface. In a human body, this can be enhanced by proteins that absorb the protective layer and assist in its breakdown. Crevice corrosion is another type of localized corrosion of a metal surface which occurs in the vicinity of the gap between two joining surfaces. This gap can be formed between two metals or a metal and nonmetallic material. The presence of such gap may cause localized acumination of corrodents and induce accelerated corrosion. For example, such corrosion often occurs at the interface between the bone plate and the screw joining the plate.

In addition, corrosion testing of implant materials and devices is an important step prior to device manufacturing. There are two primary ways of testing for corrosion resistance. The in vivo methods of corrosion testing are qualitative in nature, while in vitro methods are more quantitative in nature. One major issue with in vitro simulation of corrosion in the human body lies in the complexity of the physiological environment. There are four ASTM standards used to determine the feasibility of use of biomaterial devices and materials as implant materials in human body: ASTM F746: Test Method for Pitting and Crevice Corrosion of Metallic Surgical Implant Materials; ASTM F897: Test Method for Measuring Fretting Corrosion of Osteosynthesis Plates and Screws; F1875: Practice for Fretting Corrosion Testing of Modular Implant Interfaces; and F1801: Practice for Corrosion Fatigue Testing of Metallic Implant Materials.

From the practical application point of view, even before implantation, an oxide surface film will spontaneously form on the surface of a metal through chemical reactions. Because ceramics are oxides, which are electrical and thermal insulators, the electrochemical reactions that lead to corrosion are reduced or prevented.

An alloy surface may be modified in some ways to improve tribological properties. There are a few general types of treatments available. For instance, the oxide layer may be enhanced by a suitable oxidizing treatment such as anodizing. The surface can also be hardened by the diffusion of interstitial atoms into surface layers. The flame spraying of a metal (such as molybdenum) or metal oxides onto the surface may be employed as an insulting layer. Other metals may be electroplated onto the surface. In the case of implant applications with biocompatibility requirements, the enhancement of the oxide layer is usually chosen as the most appropriate technique.

3.3.3 Biocompatibility

Metals are typically hard, opaque, shiny, and highly electrically and thermally conductive. Their high moduli, yield points, and ductility make them good materials for implants that need to bear large loads without deforming or changing in shape and size. Metallic implants are usually used to replace portions of the body such as joints, long bones, and skull plates. Metals used to manufacture these implants can usually be tolerated by the body in minute amounts but not in large amounts. The hostile environment inside the human body can lead to corrosion. Corrosion of the implant and release of corrosion products into the surrounding tissues is a major concern in developing metal implants as noted previously.

Another important consideration to take into account while working with metals is the biological tolerance of different metals in the human body. For examples, the human body contains approximately 4–5 g of iron. Single doses (as low as 10–20 mg/kg) of iron injected into the body may cause adverse symptoms. Doses higher than 40 mg/kg require medical attention. Cobalt exits in the human body in the form of vitamin B12. A daily intake of 3 μm of vitamin B12 is adequate. No obvious side effect of free cobalt has been clinically shown. Chromium(III) is an essential nutrient required for normal energy metabolism. The Institute of Medicine (IOM) of the National Research Council (NRC) has determined that an adequate intake of 20–45 μg chromium(III)/day is sufficient for adolescents and adults. Molybdenum with a concentration of 1–3 ppm exists in the liver. Molybdenum is essential for the functioning of the enzymes xanthine oxidase, aldehyde oxidase, and sulfite oxidase, but toxicity may be induced with large dosages. The presence of nickel is approximately 10 mg in adult human tissues. A normal blood level of nickel is around 5 mg/l. In humans, inhalation of nickel may lead to renal effects, but the most notable observations of toxicity are largely confined to carcinogenesis and hypersensitivity. Normal manganese levels are around 12–20 mg in a 70 kg man, and a normal blood level of the metal is 7.0–28.0 μg/ml. Manganese is a cofactor for a number of enzymes; among them are carboxylases and phosphatases. Manganese is one of the least toxic trace elements. Titanium is the preferred element in terms of biocompatibility as it is nontoxic to human beings.

Generally speaking, the properties of metals make them ideal candidates for use in orthopedic and orthodontic applications. However, there are a few issues that arise when we use metallic biomaterials. Stress shielding refers to the reduction in bone density (osteopenia) as a result of the removal of typical stress on bone by an implant. According to Wolff's law, bones in a healthy person remodel in response to the loads that they experience. The presence of metallic replacements leads to a decrease in the bone loading which leads to the bones becoming weaker and less dense.

3.4 Metal Strengthening

3.4.1 *Work Hardening*

Metal deformation generally takes place by means of a mechanism called slip whereby a plane of atoms slides across another plane of atoms. Slip occurs within a grain. The techniques that prevent or hinder slip can increase the strength of the material.

Work hardening occurs when the material is under a large enough compressive/tensile stress such that it is plastically deformed. In the work hardening processes (or cold work), lines of atomic defects (dislocations) are created to achieve material strengthening. We may first consider the case of a dislocation guide through part of a single crystal with length, width, and thickness of L, l, and h, respectively (Fig. 3.11). This dislocation would generate an additional shear strain γ to the atoms in the alloy with a top surface displacement of b and a partial ratio $\Delta L/L$ of effectiveness along the length direction. For N_d dislocations guide through part of the crystal, the developed shear strain would be

$$\gamma = N_\mathrm{d}\frac{\Delta L}{L}\frac{b}{h} = \frac{N_\mathrm{d} lb\Delta L}{V} = \rho bl\Delta L \tag{3.2}$$

where $\rho_\mathrm{d} = N_\mathrm{d}/V$ is defined as the density of dislocation. The relation described in Eq. 3.2 is also called the Orowan equation.

We may further consider that the dislocations are distributed randomly over the length-width plane. The average distance between dislocations (l_d) can then be roughly approximated from the spreading area point of view $N_\mathrm{d}{l_\mathrm{d}}^2 \approx hL$, implying that

$$l_\mathrm{d} \approx \sqrt{\frac{hl}{{N_d}^d}} = \sqrt{\frac{1}{L}\rho_\mathrm{d}}. \tag{3.3}$$

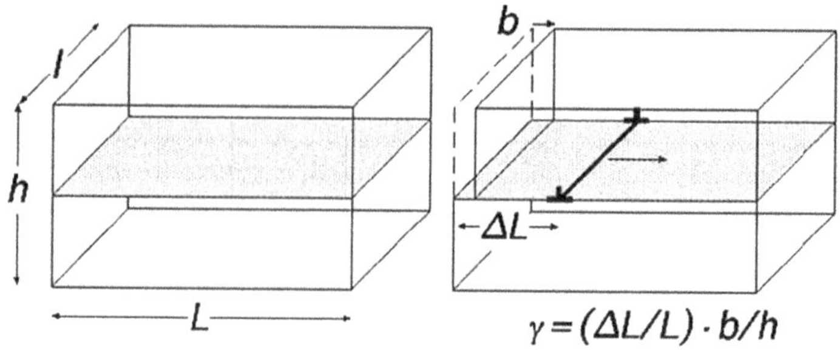

Fig. 3.11 Simplified condition of one dislocation lying in a single crystal

A better approximation of the additional shear stress τ lies under the assumption that material portions between dislocations would have purer shear effects. We may then express τ as

$$\Delta\tau = G\gamma = G(k''b/l_\mathrm{d}) = k''Gb\sqrt{L\rho_\mathrm{d}}, \tag{3.4}$$

where k'' is a constant. Substituting Eq. 4.2, we would have

$$\Delta\tau = k''Gb\sqrt{L\rho_\mathrm{d}} = k''Gb\sqrt{\frac{\gamma}{b\Delta L}} = k'\sqrt{\gamma}, \tag{3.5}$$

where k' is a constant. Hence, we would have effective internal stress caused by the interactions of dislocations as $\tau_\mathrm{d} = \tau_\mathrm{f} + \Delta\tau$, where τ_f is the baseline of stress required to move a dislocation in the absence of all other dislocations. τ_d also represents the threshold of stress level for large deformation, such that the dislocations within the material body can pass across each other.

$$\tau_\mathrm{d} = \tau_\mathrm{f} + k'\sqrt{\gamma}. \tag{3.6}$$

However, this relation can only describe the behaviors of many materials at large shear strains ($\gamma \gg 0$). It fails to depict the work hardening phenomenon quantitatively, largely due to factors including (1) the rare occurrence of regular dislocation distributions and (2) the ignorance of screw dislocations (and hence the cross slip).

Qualitatively, the tensile stress-plastic strain response of polycrystalline materials resembles the single-crystal scale of the same material. Considering an external force is applied on a single crystal, it induces tensile stress of σ over a crystal area A_c (Fig. 3.12). Yet, directions of the crystalline structure and the slip plane may not align perfectly with the external force. We denote the angle between the external force and the slip direction as λ and the angle between the force and the normal of slip plane as φ. Then, $F\cos\lambda$ is the effective shear force in the slip direction, and $A_\mathrm{c}/\cos\varphi$ is the effective shear area. Hence, the effective shear stress on the slip plane τ_s is

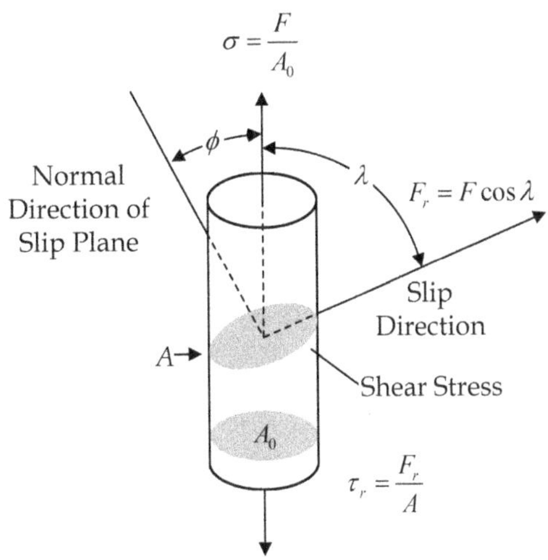

Fig. 3.12 Resolved shear stress on a slip plane under an external force

$$\tau_s = \sigma_c \cos \lambda \cos \varphi. \tag{3.7}$$

where $\sigma_c = F/A_o$. The minimum shear stress required for slip (yield) in a single crystal occurs when $\lambda = \varphi = 45°$. Clearly, slip can be easier or harder depending on the crystal orientation.

If we further consider the plastic strain rate $\dot{\varepsilon}_s$ for a slip to be similar to the stress by the slip strain $\dot{\gamma}_s$ multiplied by the orientation factors, then

$$\dot{\varepsilon}_s = \dot{\gamma}_s \cos \lambda \cos \varphi, \tag{3.8}$$

and

$$\frac{d\sigma_s}{d\varepsilon_s} = \frac{1}{\cos^2\lambda \cos^2\varphi} \frac{d\tau_s}{d\gamma_s}. \tag{3.9}$$

The above relations reveal the relationship between the modulus of rigidity and the modulus of elasticity in a crystalline structure. According to Eq. 3.4, the shear stress threshold and the modulus of rigidity can be increased by the higher density of dislocations. This means that we can strengthen crystalline metallic materials by inducing more dislocations via cold works. However, if there are too many dislocations generated and over an acceptable level, the material may crack; and therefore there is a limit to the amount of cold work which can be used. Often, dislocations and strengthening effects appear more frequently at the surface than at the center of the material upon an external force. It should not be a problem in most applications since it is desirable to have the enhanced properties at the surface to protect against wear. The percentage of work hardening is often expressed as a percentage of reduction in thickness of the material during processes such as rolling or as a percentage of reduction in cross-sectional area.

After the material is cold-worked, its yield strength, tensile strength, and hardness increase. Amounts of these increments can reach 50–100%. To achieve work hardening, the induced dislocations actually reflect irregularity of the material deformation/strains within the material body. Therefore, the mismatch of strains at different local sites will induce residual stresses inside the material body. This implies that a relative lower level of the additional external stress can cause the material to crack, i.e., the material will become more brittle.

It is important to mention that dislocations and brittleness can be removed by heating the material as annealing to relieve the residual stress/strain caused by cold work and to form the strain-free grains. The required temperature is called recrystallization temperature, which is approximately one-third to one-half of the melting point. Hot work is the cold work process at a temperature above the recrystallization temperature. When a material is hot-worked, the material recrystallizes immediately without the work hardening effect. The advantage of hot working is that less force is needed to permanently deform a material, yet heating may oxidize the material surface.

3.4.2 *Grain Size Control*

Grain boundaries can prevent slip because two adjacent grains have their planes at different orientations. If slip occurs along a plan in one atom at the grain interface, the change in orientation of the atomic arrangement in the adjacent grain makes slip more difficult. In principle, materials with more grains have more frequent orientation changes in their atomic arrangements, and therefore, material strength increases as the grain size decreases.

To investigate this mechanism, we first look into the resolved shear stress τ_s at a dislocation for each grain with N_d dislocations, which can be approximated as

$$\alpha_g \tau_s \approx N_d G \frac{L}{\pi(D_g/2)}, \tag{3.10}$$

where L is the length scale of the material; D_g is the average diameter of grains; α_g is a scaling factor with the value 1 for screw dislocations or $1 - \nu$ (Poisson's ratio) for edge dislocations. As the worst case scenario, we hypothesize that the leading dislocation in the pileup bursts through a grain boundary due to stress concentration. Then, the total resolved shear stress over the grain ($N_d\tau_s$) concentrates at a point. If such total shear stress exceeds the threshold stress level τ_c and induces cracking, based on Eq. 3.10,

$$N_d \tau_s = \alpha_g \frac{\pi(D_g/2)}{GL} {\tau_s}^2 > \tau_c. \tag{3.11}$$

The effective internal stress τ_d that a grain can withstand would be

$$\tau_{\mathrm{d}} \leq \tau_{\mathrm{f}} + \sqrt{\frac{2\tau_{\mathrm{c}}GL}{\alpha_{\mathrm{g}}\pi D_{\mathrm{g}}}} = \tau_{\mathrm{f}} + k'_{\mathrm{g}}\sqrt{\frac{1}{D_{\mathrm{g}}}}. \tag{3.12}$$

where τ_{f} is the baseline of stress required to move a dislocation in the absence of all other dislocations. This indicates that the strength enhancement is limited. Practical experiments reveal that the change in material strength can be inversely related to the square root of the grain diameter, and such enhancements have a typical level in change of 5–15% with a maximum of approximately 30%.

However, it is virtually impossible for all the stress to be concentrated at the leading dislocation. Equation 3.12 is valid only for larger D_{g}. The expression of k'_{g} would need to be further refined as proposed by Cottrell in 1958. Instead, we may consider the stress concentration to be caused by a pileup in one grain (grain "1"), activating a dislocation source in a neighboring grain (grain "2′") as shown in Fig. 3.13. For a grain with an average diameter of D_{g}, the possible pileup can be assumed to be a shear crack. If we consider the resolved shear stress along the interface between grain "1" and grain "2," recalling Eq. 3.11 by considering the length scale L as the displacement along the interface ΔL

$$\tau_{\mathrm{c}} \propto \frac{\alpha_{\mathrm{g}}\pi(D_{\mathrm{g}}/2)}{G\Delta L}{\tau_{\mathrm{s}}}^2 = \frac{\alpha_{\mathrm{g}}\pi(D_{\mathrm{g}}/2)}{G\Delta L}\left(\frac{r_{\mathrm{g}}}{r_{\mathrm{g}}}\right){\tau_{\mathrm{s}}}^2 \tag{3.13}$$

When τ_{s} is at the level where it is about to induce a crack along the grain interface,

$$\frac{\alpha_{\mathrm{g}}\pi(D_{\mathrm{g}}/2)}{r_{\mathrm{g}}}{\tau_{\mathrm{s}}}^2 \propto \tau_{\mathrm{c}}\frac{G\Delta L}{r_{\mathrm{g}}} \propto {\tau_{\mathrm{c}}}^2; \tag{3.14}$$

thus

$$\tau_{\mathrm{d}} \leq \tau_{\mathrm{f}} + k_{\mathrm{g}}\tau_{\mathrm{c}}\sqrt{\frac{r_{\mathrm{g}}}{D_{\mathrm{g}}}}. \tag{3.15}$$

where k_{g} is a constant.

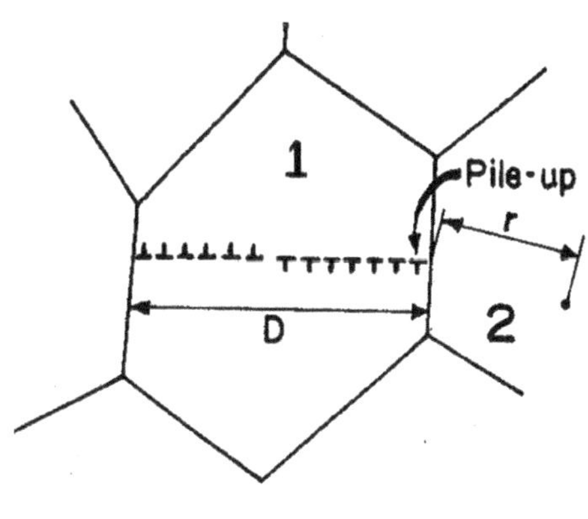

Fig. 3.13 Resolved shear stress along an interface between two grains

Often, the relation of Eq. 3.15 is expressed in the form as the Hall-Petch equation:

$$\sigma_y = \sigma_o + K_{HP} D_g^{-1/2} \tag{3.16}$$

where σ_y is the yield strength (the level of stress necessary to cause a certain amount of permanent deformation), σ_o is a constant for the metal, and K_{HP} is the grain boundary strengthening constant. Recall that the yield strength of a metallic material is about the minimum level of stress that is needed to initiate plastic (permanent) deformation. The Hall-Petch equation is not valid for materials with unusually large or ultrafine grains. An example of the Hall-Petch relationship is given in Fig. 3.14.

3.4.3 *Alloying*

In alloying, a different metal atom is added to a base metal. A pure metal or compound has a specific unit cell and crystal structure. Since all of the elements have different atom sizes, when a second atom is added, the unit cell will become distorted, and slip will be made more difficult. The second atom may substitute for the first atom at a lattice site. If the second atom is larger in size, it will distort the surrounding atom unit cells and make atomic slip of the alloy more difficult. If the atom is smaller in size, the adjacent unit cells will tend to be slightly distorted. On the other hand, if the second atom is much smaller, it may be an interstitial atom and locate itself in one of the void spaces in the unit lattice cell. Such difference will cause a lattice strain and also prevent the atomic slip.

Alloying can be achieved by increasing the strength of a metallic material via the formation of a solid solution, which is called solid solution strengthening. Substitutional or interstitial impurities in a solution produce point distortions in the crystal lattice. These solute atoms segregate around the dislocation core to occupy atomic sites suitable for their sizes (either substitutional or interstitial) in order to interact with dislocations and disrupt their movements. In fact, solid solution strengthening is a result of the elastic interaction between the stress fields of dislocations and solute

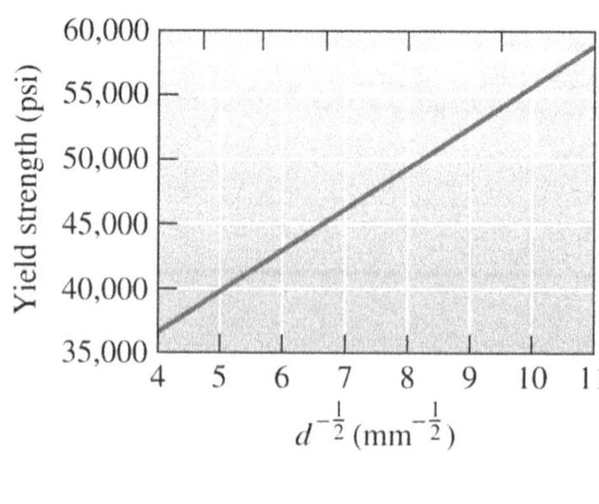

Fig. 3.14 The effect of grain size on the yield strength of steel at room temperature

atoms. A distortion core can be considered as an elastic sphere of radius $r_a(1 + \delta_a)$ and volume V_s inserted into a spherical hole of radius r_a and volume V_h in an elastic matrix. For our simple analysis, we assume both the sphere and the matrix are isotropic with the same/very similar modulus of rigidity G and Poisson's ratio ν. On inserting the sphere in the hole, V_h changes by ΔV_h to leave a final defect of radius $r_a(1 + \varepsilon_a)$ as shown in Fig. 3.15. Since radii of the sphere and the hole can be different, there is a mismatch in volume ($=V_s - V_h$). Thus for a very small ε_a

$$\Delta V_h = \frac{4}{3}\pi r_a^3 (1 + \varepsilon_a)^3 - \frac{4}{3}\pi r_a^3 \approx 4\pi r_a^3 \varepsilon_a \tag{3.17}$$

Likewise, the change in volume of the sphere is

$$\Delta V_s = \frac{4}{3}\pi r_a^3 (1 + \varepsilon_a) - \frac{4}{3}\pi r_a^3 (1 + \delta_a)^3 \approx -4\pi r_a^3 (\delta_a - \varepsilon_a) \tag{3.18}$$

The degree of solid solution strengthening depends on two factors. First, a large difference in atomic size between the original (base) atom and the added (substitutional or interstitial) atom increases the strengthening effect. A larger size difference produces a greater disruption of the initial crystal structure, roughening slip planes and making slip more difficult. Additionally, the greater the amount of alloying element added, the greater the strengthening effect. The effects of alloying vary approximately with the square root of the concentration of the second element.

We consider that the threshold stress level τ_c for solid solution strengthening would have a relation

$$\tau_c \propto \frac{F_{max}}{d_f l_f} \quad \text{or} \quad l_f \propto \frac{F_{max}}{\tau_c d_f} \tag{3.19}$$

where F_{max} is the maximum possible interaction force; l_f is the average obstacle (dislocation) spacing referred as the Friedel length; and d_f is the distance between F_{max} and the dislocation site. Furthermore, the area of the glide (slip) plane per

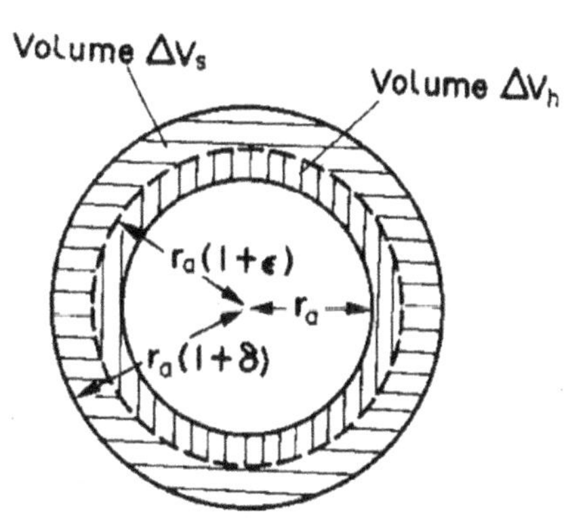

Fig. 3.15 Mismatch volume between the dislocation core and the corresponding void of the base material

dislocation should be inversely proportional to the concentration C_A of the substitutional (or interstitial) atoms added to the base metal. C_A can be approximated as

$$\frac{1}{C_A} \propto \frac{\tau_c\, l_f^3}{G\, d_f}, \quad \text{or} \quad l_f^3 \propto \frac{G d_f}{\tau_c C_A} \tag{3.20}$$

where G is the modulus of rigidity close to both the base and substitutional atoms. Substituting Eq. 3.19 into Eq. 3.18,

$$\tau_c \propto \frac{{F_{max}}^{3/2}}{G^{1/2}} \frac{1}{{d_f}^2} \sqrt{C_A} \tag{3.21}$$

According to Eq. 3.20, we may roughly convert the shear stress to the tensile stress. The alloy strength σ_S can then be approximated expressed as

$$\sigma_S = {}^{S}\sigma_{SB} + K_A \sqrt{C_A} \tag{3.22}$$

where σ_{SB} is strength of the base metal; K_A is the solid solution concentration constant depending of the alloying process (e.g., the interaction force, which depends on the size difference between the base atom and the substitutional atom); and C_A is the volumetric ratio between the second metal and the base metal. Typically, the increase in strength by alloying tends to be limited to be ~5–20%.

3.5 Common Metallic Biomaterials

Stainless Steels

In biomaterials, stainless steel is the favorite choice for implant alloys because of its ease of manufacturing, high wear resistance, and ability to handle mechanical load. The Young's modulus of stainless steel is ~190 GN/m^2, a high ratio for a stiff and rigid material. Compared cobalt-chromium alloy and titanium alloy, stainless steel is more likely to corrode as it has a thin passive film that can be permanently damaged frequently from interface corrosion due to various environmental factors such as pH and temperature. The elements that combine to form stainless steels are manganese, iron, nickel chromium, nickel, and molybdenum.

Stainless steel should have at least 11% of chromium out of its total alloy concentration to resist corrosion because chromium reacts with oxygen in order to form chromium oxide which is able to resist corrosion. Stainless steel with 3% molybdenum, 1% silicon, 10–14% nickel, 16–18% chromium, and a low carbon percentage can form pitting corrosion resistance. Nitrogen can also increase resistance to pitting corrosion and increase mechanical strength.

Stainless steel can be split into four categories: austenitic, duplex, ferritic, and martensitic ones. Austenitic stainless steel has a face-centered cubic structure with a chemical composition of 15–20% chromium, 1–7.5% manganese, and 3–14% nickel. Mixing aluminum, molybdenum, niobium, and silicon can increase pitting corrosion and oxidation. Heat treatment and cold working can both harden this stainless steel. It has cryogenic properties, oxidation resistance, strength at high temperatures, and formability. Nickel increases this formability significantly. *316L*, the most common austenitic stainless steel, is made from 60–65% iron, 17–20% chromium, 12–14% nickel, and <0.030% carbon. Though it has strong corrosion resistance, it can corrode at crevices making itself easier to mimic the round shapes of inner joints. It is mainly forged at 1050 °C and continuously modified through forging and reannealing until reaching 23 °C or room temperature.

Duplex stainless steel has equal proportions of ferrite and austenite. Like 316L, duplex has a carbon composition of less than 0.030%, with the majority of duplex steel being 20–30% chromium and 5–8% nickel. Duplex stainless steels can be ductile and tough. Currently, they are not used for any medical implants.

Ferritic stainless steel has a body-centered cubic with 11–30% chromium. Sulfur and selenium can be mixed to increase machinability. It does not strengthen in heat. It is sensitive in cold as it can be reduced in ductility. Ferritic steel is mainly used in handles and pins.

Martensitic stainless steel has a microstructure of tetragonal crystal in the body center formed in heat treatments of 10.5–18% chromium and >1.2% carbon. Martensitic has moderate corrosion resistance but can gain increased toughness through alloying tungsten, silicon, and niobium. Martensitic stainless steel can increase in corrosion resistance and toughness through nickel. Alloying sulfur will increase its machinability.

The most popular usage of stainless steels in biomedical devices is in orthopedic implants for fixing fractures and replacing joints. Examples of biomedical devices that utilize steels include leg spacers, femur shafts, intramedullary pins, bone plates, and hip, ankle, and knee joints. Active research in stainless steel's applications for heart valves and mandibular bone plates are also ongoing.

Surface texture defined by surface roughness, lay, and waviness lead to corrosion resistance. Out of common mechanically finishing and other metallurgical techniques, electroplating is the most optimal treatment at increasing corrosion resistance in steels.

Cobalt-Chromium Alloys

As the name suggests, cobalt-chromium alloys are made of cobalt (Co) and chromium (Cr). Usually, a cobalt-chromium alloy is a solid mixture made from 65% cobalt and 35% chromium. Cobalt has a density of 8.85 g/cm^3 with a Young's modulus of 210 GN/m^2, whereas chromium has a density of 7.19 g/cm^3 with a Young's modulus of 279 GN/m^2. The elastic moduli of cobalt-chromium alloys usually range from 185 to 253 GN/m^2. This elastic modulus of cobalt-chromium alloys is approximately equal to that of *316L* austenitic stainless steel; and it is double that of titanium. Molybdenum can be added to enhance the fine grains of these alloys, which in turn increases the strength of them as well.

Four commonly used cobalt-chromium alloys for biomedical devices are ASTM F562, ASTM F75, ASTM F799, and ASTM F90. The following table (Table 3.1) outlines the different maximum percentage of chemical compositions of these four cobalt-chromium alloys.

ASTM F562 is a wrought alloy of Co-Ni-Cr-Mo also with the name MP-35N (multiphase). ASTM F562 has high ductility, corrosion resistance, and high strength. The cobalt has a face-centered cubic (FCC) crystal at 419 °C which changes to a hexagonal closed crystal below this temperature at equilibrium. These microstructures can be heat treated and be cold-worked to modify their strength. Cold working ASTM F562 can change it to a hexagonal close-packed configuration. MP-35N is multiphase because it produces a cobalt compound with molybdenum precipitates or Co_3Mo precipitates through precipitation hardening in a heat treatment at 430–650 °C. The process of precipitation hardening includes (1) solution heat treatment to a homogenous single-phase solid solution, (2) quenching to produce a supersaturated solid solution of alloying elements in the matrix, and (3) aging to form fine dispersion of precipitates. The strength is then enhanced by the precipitation of very fine, uniformly dispersed hard phase of the point dislocations within a softer matrix. In addition, cold treated ASTM F562 has one of the highest tensile strengths of most of metals in biomedical implant devices at 1795 MPa. It is mainly used for surgical implants.

ASTM F75 is a cast alloy of Co-28Cr-6Mo. It goes by the names of Vitallium™, Stellite 21™, Protasul-2™, and Zimaloy™. ASTM F75 has strong corrosion resistance from its high chromium composition which can form a chromium oxide layer on its surface. ASTM F75 is so strong that it can resist chloride. It can be molded through investment casting with ceramic coated wax as the beginning mold. Co-28Cr-6Mo melts at 1350–1450 °C, and the molten alloy is poured into the ceramic mold. If the grain size is large, then it will decrease its overall tensile strength and cause fatigue fracture. It is mostly used in orthopedic and dental implant devices.

ASTM F799 is a thermomechanical alloy of Co-Cr-Mo. It is hot forged after being cast in a mold and worked at temperatures of 800 °C or higher. The high temperature of forging also lessens the alloy's strength. Thus, high temperatures are used for deformation as an initial step and end with cold working. Regardless, ASTM F799 has twice the tensile strength of ASTM F75. It is mainly used in implants that deal with soft tissue and bones for long periods in time.

Table 3.1 Chemical compositions of the cobalt-chromium alloys

Elements	ASTM F562	ASTM F75	ASTM F799	ASTM F90
Cobalt	29.0–38.8	58.9–69.5	58.0–59.0	45.5–56.2
Chromium	19.0–21.0	27.0–30.0	26.0–30.0	19.0–21.0
Carbon	0.03	0.35	0.35	0.15
Iron	1.0	0.75	1.5	3.00
Molybdenum	9.0–10.5	5.0–7.0	5.0–7.0	0.0
Nickel	33.0–37.0	2.5	1.0	9.0–11.0
Silicon	0.15	1.0	1.0	0.40
Tungsten	0.0	0.2	0.0	14.0–16.0

ASTM F90 is a wrought alloy of Co-Cr-W-Ni. It goes by the name of Stellite 25™ and is made by the Cabot Corporation. It has a higher percent composition of tungsten than the other ASTM cobalt-chromium alloys. Tungsten and nickel make ASTM F90 easier to machine and manufacture. It has strong corrosion resistance but also the ductility to be used for sutures and cerclage in bone repair. It can also be used for stents for cardiovascular repair and eyelet markers.

Titanium-Based Alloys

Titanium is widely used because of its biocompatibility and light weight due to its density of 4.505 g/cm^3. Corrosion is not a major problem as titanium displays impressive corrosion resistance. However, titanium devices are subject to wear and fretting which results in the increase in the concentration of titanium near implant areas.

Alloying titanium with other metals allows for the creation of hip prostheses and bone fracture equipment. Alloys of aluminum, vanadium, and titanium have a density similar to the density of pure titanium since aluminum and vanadium are lighter elements with densities of 2.698 g/cm^3 and 6.110 g/cm^3, respectively. Titanium-6Aluminum-4Vanadium (Ti-6Al-4V) and pure titanium are the most used metals for biomedical applications.

Pure titanium has a melting temperature at 1665 °C. Its Young's modulus is around 110 GN/m^2 with a shear modulus of 38 GN/m^2. Titanium alloys have similarly low moduli and can be used for orthopedic devices. Pure titanium reacts easily, leading to material degradation. The four different grades of titanium range from Grade 1 to Grade 4 based on the percent impurities of other elements. Titanium can be treated in a couple different ways to enhance its biocompatibility, such as oxidizing the surface layer through anodizing, exposing the surface to a diffusion of interstitial atoms, flame spraying other metal to coat and alloy the surface, and electroplating metals onto the surface. The different grading system of impurities in titanium is shown in Table 3.2.

Notice that oxygen reacts with the most with the different grades of pure titanium. Increasing the oxygen composition increases the tensile strength, yield strength, and fatigue strength. It can also decrease ductility. Titanium alloys are used for maxillofacial, craniofacial, and dental implants as well as for total joint replacements and cases for pacemaker.

Table 3.2 Grade system of the impurities in titanium

Grades	Carbon percentage	Oxygen percentage	Nitrogen percentage	Hydrogen percentage	Iron percentage
Grade 1	0.10	0.18	0.03	0.015	0.20
Grade 2	0.10	0.25	0.03	0.015	0.30
Grade 3	0.10	0.35	0.05	0.015	0.30
Grade 4	0.10	0.40	0.05	0.015	0.50

Problems

Problem 3.1

An alloy bar has a strength coefficient of 0.50. The bar, which has an initial diameter of 1 cm and an initial gauge length of 3 cm, fails at an engineering stress of 120 MPa. After fracture, the gauge length is 3.5 cm and the diameter is 0.926 cm. No necking occurred. Calculate the true stress when the true strain is 0.05.

Problem 3.2

A 0.505 in. diameter metal bar with a 2 in. gauge length l_o is subjected to a tensile test. The following measurements are made in the plastic region:

Force (lb)	*Change in gauge length (Δl) (in.)*	*Diameter (in.)*
27,500	0.2103	0.4800
27,000	0.4428	0.4566
25,700	0.6997	0.4343

Determine the degree of strengthening and Poisson's ratio for the metal. Is the metal most likely to be FCC or BCC? Explain.

Problem 3.3

A 1.5-cm-diameter metal bar with a 3 cm gauge length (l0) is subjected to a tensile test. The following measurements are made:

Force (N)	*Change in gauge length (Δl) (cm)*	*Diameter (cm)*
16,240	0.6642	1.2028
19,066	1.4754	1.0884
19,273	2.4663	0.9848

Determine the degree of strengthening and Poisson's ratio for the metal. Is the metal most likely to be FCC or BCC? Explain.

Problem 3.4

An alloy material has the following properties:

Grain	*Diameter (mm)*	*Strength (MPa)*
0.015	1	70
0.025	1	158
0.035	1	151
0.050	1	145

Determine

(a) The constants in the Hall-Petch equation.
(b) The grain size required to obtain a strength of 200 MPa.

Problem 3.5

The strength of titanium is found to be 65,000 psi when the grain size is 17×10^{-6} m and 82,000 psi when the grain size is 0.8×10^{-6} m. Determine the following:

(a) The constants in the Hall-Petch equation
(b) The strength of the titanium when the grain size is reduced to 0.2×10^{-6} m

Problem 3.6

Please answer the following questions using Fig. 3.P1:

(a) We want a metallic bar to have a tensile strength of at least 70,000 psi and a final diameter of 0.375 in. Considering that we use cold work to compress and reduce the bar diameter, what is the minimum diameter of the original bar?
(b) We want a metallic sheet with a final thickness of 0.12 in. to have at least 50,000 psi yield strength and at least 10% elongation, after a cold work process called rolling. What range of original thickness must be used?
(c) We want to draw a 0.3 in. diameter metallic wire having a yield strength of 20,000 psi into a 0.25 in. diameter wire. (a) Find the draw force, assuming no friction. (b) Will the drawn wire break during the drawing process? Show why.

Problem 3.7

If the tensile strength of the base material for a steel alloy was 40,000 psi and the solid solution concentration constant for chromium was 60.000 psi, what would be the increase in strength if the chromium level increased (a) from 0.01% to 0.02%, (b) from 0.01% to 0.04%, and (c) from 0.01% to 0.1%?

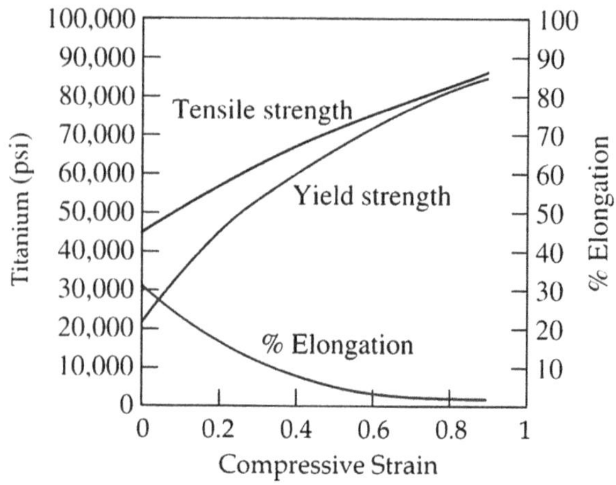

Fig. 3.P1 Tensile strength, yield strength and elongation of titanium against compressive strain

Problem 3.8

If the grain boundary strengthening constant is 400 psi/in. and the strength of a grain is 30,000 psi, what is the change in strength if the grain size goes from 0.001 in. to 0.003 in.?

Problem 3.9

A sheet titanium tensile specimen has the following original dimensions: 25 mm long, 10 mm wide, and 2 mm thick. After the load is applied, the final dimensions are 37 mm long and 8 mm wide. Considering the Poisson's ratio of titanium is 0.32, calculate the strain ratio along the load direction.

Problem 3.10

If the strain ratios of a specimen under load for the directions 0°, 45°, and 90° are 0.9, 1.2, and 1.1, respectively, what is the mean strain ratio (also called the mean anisotropic coefficient)?

References and Further Reading

1. Baboian, R.: Corrosion Tests and Standards: Application and Interpretation. ASTM International, West Conshohocken (2005)
2. Ong, J.L., Appleford, M.R., Mani, G.: Introduction to Biomaterials: Basic Theory with Engineering Applications. Cambridge University Press, Cambridge (2014)
3. Pilliar, R.M.: Biomedical Materials. Springer, New York (2009)
4. Askeland, D.R., Fulay, P.P., Wright, W.J.: The Science and Engineering of Materials. Cengate Learning Inc., Boston (2010)
5. Soboyejo, W.: Mechanical Properties of Engineered Materials. CRC Press, Boca Raton (2002)
6. Seetharaman, S.: Fundamentals of Metallurgy. Elsevier, Waltham, MA, USA (2005)
7. Dieter, G.E., Bacon, D.J.: Mechanical Metallurgy. McGraw-Hill, London (1986)

Chapter 4
Polymers

Abstract Polymeric materials are commonly used in many biomedical devices. The molecular structures of polymers determine their chemical and physical properties and, therefore, their utilization in various medical domains. The common polymers applied in biomedical applications include polyethylene, perfluorinated polymers, acrylics, hydrogels, polyurethanes, polyamides, silicones, and biodegradable synthetic polymers. This chapter provides basic polymer physics knowledge for better understanding of commonly used polymers in biomedical applications. This includes the molecular structures and the polymerization processes and the resultant physical properties. In particular, a few important mechanisms are discussed: melting point depression of polymeric mixtures, relationship between glass transition temperature and specific volumes, and related factors for the molecular weight. Materials properties of the commonly used polymers are further discussed.

4.1 Overview

Polymeric materials have long been used in many industries around the world such as plastics, rubbers, adhesives, and fibers. The first polymers were mostly made of natural products, including cotton, starch, proteins, and wool. However, previous scientists realized that they did not understand many of the relationships of the physical properties of polymers. Nowadays, biomedical devices have employed the usage of polymers in various aspects for many years. For example, polyimide (PI) is commonly used to encapsulate and insulate medical devices. Poly(methyl methacrylate) (PMMA) is commonly used to manufacture intraocular lens and bone cements in joint replacements as well as for surgical components such as rhinoplastics and cranioplastics. Polyethylene is used in bottles, polystyrene vials, rubber closures, plastic tubing for injection sets, and flexible plastic polyvinyl chloride bags to hold blood and other intravenous solutions.

The major consideration for implanting polymers is the toxicity of additives that may be released into the surrounding tissues. Residual monomers due to incomplete polymerization and catalysts used for polymerization may cause irritation. Another major concern is microbial contamination in which microorganisms adhere to the

R. H. W. Lam, W. Chen, *Biomedical Devices*,
https://doi.org/10.1007/978-3-030-24237-4_4

polymer's surface. In most cases, antimicrobial chemotherapy yields no results, and the danger can only be removed by the removal of the prosthesis or catheter. In addition to their usage in many medical devices and containers, synthetic polymers may also be added to drugs to act as excipients (substances added to aid in the delivery of a drug to its targeted location).

This chapter develops on basic polymer physics knowledge for better understanding of the material in the later chapters of this book. We will also discuss various synthetic polymers, their properties, and their applications in medicine.

4.2 Basic Characteristics

Polymer molecules are huge macromolecules formed by the linkage of smaller units. The smaller subunits are linked together by covalent bonds. The backbone of a polymer is usually a string of single covalently bonded carbon atoms. Polymers are composed of basic structures called mer units. A molecule with just one mer is a monomer. Different polymers are formed based on the arrangement of these molecules, which then dictates the polymer's physical and chemical properties. Monomers combine with each other to form polymeric macromolecules, and their related reactions are called polymerization. All macromolecules consist of at least one chain of atoms bonded together which run along the whole molecule. This chain is called the backbone of the macromolecules. It can contain carbon-carbon bonds, carbon-oxygen bonds, carbon-nitrogen bonds, or non-carbon-carbon bonds such as silicon-oxygen bonds. Figure 4.1 shows the conceptual illustrations of various polymeric structures. Molecular structures of polymers are very complex.

Polymeric structures have the following characteristics compared with small molecules:

- A polymer is simply a very large molecule formed by joining many small molecules known as monomers. (They can have a molecular weight of 10^3–10^5 g/mol). The repeating units are the same as the individual molecules which can polymerize into uniform polymers (when repeating units are the same) or copolymers (when the repeating units are different). The products can be linear molecules, branched molecules, or crosslinking network molecules.
- The backbone of polymeric macromolecules can rotate, bend, and be flexible. The shape of the backbone can change due to the movement of the segments.
- The process of polymerization is random, and therefore, some molecules grow much larger than others. A polymer is a mixture of molecules with different sizes. Molecular size is expressed as average molecular weight.
- The interactions between polymers can influence the condensed state structure of polymers significantly. After crosslinking between molecules, the mechanical properties of polymers change significantly. Crosslinked polymers do not melt on heating and do not dissolve in any solution.

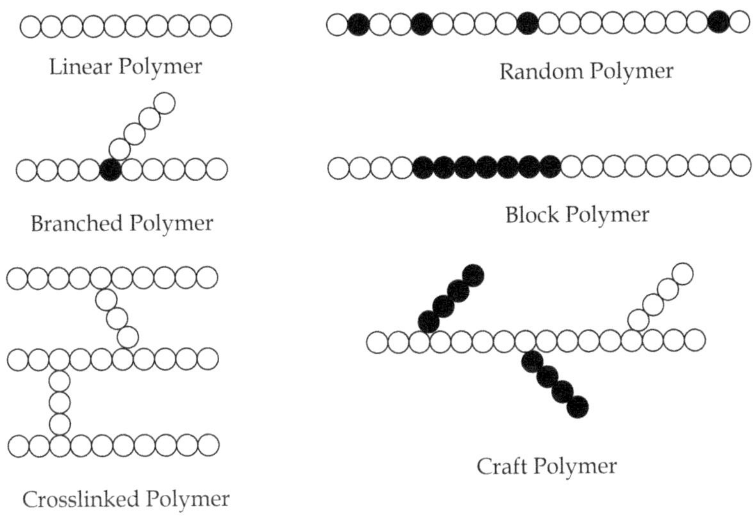

Fig. 4.1 The left side of the figure shows different possible polymeric macromolecular structures. The right side of the figure shows copolymers, which are polymers that are composed of two or more different monomers

- The integrated structure of polymers can be classified into crystalline and noncrystalline state structures. It should be pointed out that the crystallinity of polymers is less than that of small molecules.
- The mixture structures of a polymer have a significant influence on properties of polymer materials.

Crosslinking is a chemical process of bonding one polymer chain to another. Crosslinks can be deformed by the mixing of unpolymerized monomers with specific chemicals called crosslinking reagents. Postpolymerization of crosslinked polymers is also possible. Natural rubber, for example, consists of mostly linear molecules which can be crosslinked to a loss network with 1–3% sulfur (vulcanization) or to a hard rubber with 40–50% sulfur. The properties of the resulting crosslinked polymers depend strongly on the crosslink density. Furthermore, the structure of a polymer determines its properties as subunits interact with one another differently depending on their location. Generally, the factors that influence the mechanical properties of polymers include:

- Composition
- Molecular weight
- Amount of unreacted monomer in the polymer
- Morphology
- Crystallinity
- Configurational structure
- Additives

Long molecules can give rise to small crystallites. They lead to the chain-folded polymer crystalline. These polymers include crystallites with the chain axes normal to the plane of the polymeric film. The chains fold back and forth upon themselves, such that adjacent segments are parallel and in crystal register. A further type of crystallite, the chain-extended crystal, can also occur when samples are prepared in special ways. Extended chain crystallites contain straight chains at least 200 nm long and have been obtained for only a few polymers, such as polytetrafluoroethylene, polyethylene, and polychlorotrifluoroethylene, by using special crystallization techniques. Extended-chain polytetrafluoroethylene can be obtained by slow crystallization after melting; the other two are obtained by crystallization after melting under elevated pressure. Solution crystallization has so far not been shown to give rise to extended-chain crystals. These materials tend to be very brittle because although they are highly crystalline, there are few intercrystalline linking molecules. The crystalline layer is about 1–2 μm thick, and there appears to be no limit on the lateral dimensions of the crystallites except space and availability of the material.

The noncrystalline regions of polymers are usually considered amorphous, but they may sometimes have some shared organizations and orientations. They are the component of polymer morphology that is the least well understood and are often represented by a tangled mass of chains. Such regions can exist in molten polymers, totally amorphous polymers in the glassy or rubbery states, and as a component of semi-crystalline polymers.

In principle, almost any property that is different in the crystalline and noncrystalline regions could be used as the basis for a method of determining the degree of crystallinity or simply as the crystallinity of a polymer sample. However, it is important to distinguish between two slightly different measures: the volume crystallinity, X_V, and the mass crystallinity, X_M. If V_c is the volume of crystalline material and V_a is the volume of noncrystalline (amorphous) material within a sample, then $X_V = V_c/(V_a + V_c)$, and, in a similar way, $X_M = M_c/(M_a + M_c)$, where M_a and M_c are the masses of crystalline and noncrystalline material within the sample. From these definitions it is easy to show that

$$\rho_s X_M = \rho_c X_V \tag{4.1}$$

where ρ_c and ρ_s are the densities of the crystalline region and of the entire sample, respectively. It is easy to further show that, if ρ_a is the density of the amorphous material, then

$$X_V = \frac{\rho_s - \rho_a}{\rho_c - \rho_a} \tag{4.2}$$

with the assumption that the densities of crystalline and amorphous material are known, determination of the density of the sample easily provides a value of the crystallinity.

In particular, the elastic properties of polymeric materials as it undergoes elongation can behave very differently according to the polymeric crystalline structures as shown in Fig. 4.2. Information about the thicknesses of the crystalline layers can

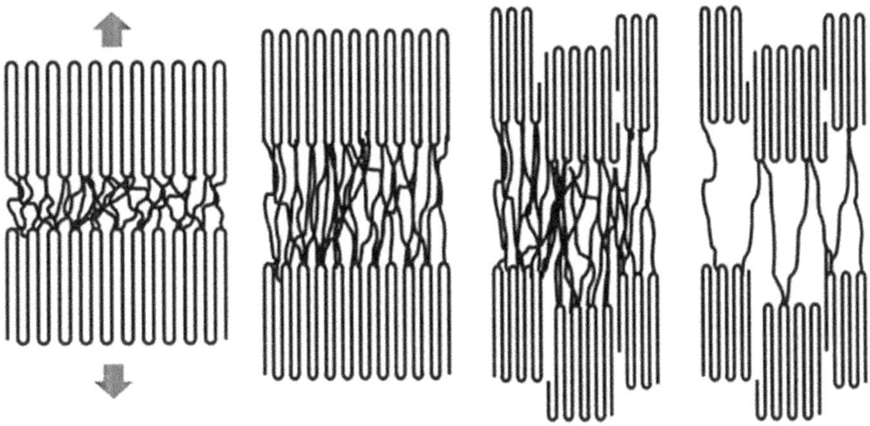

Fig. 4.2 The deformation of a semi-crystalline polymer

be obtained from Raman spectroscopy. During the measurement, certain low-frequency Raman-active modes of vibration of the chains within a monolayer of crystallite are measured like the longitudinal vibrations of elastic rods. These modes are called the longitudinal acoustic modes. The mode with the lowest frequency corresponds to a rod having a displacement node at its center and antinodes at its ends, so that the length of the rod is half a wavelength. If the modulus of elasticity and density of the rod are E and ρ, respectively, the frequency of this mode is given by

$$f = \frac{1}{2l}\sqrt{\frac{E}{\rho}} \tag{4.3}$$

where l is the length of the rod, which in the simplest case is equal to the crystalline plane thickness. There are complications, such as those based on how the frequency is affected by interactions of the chains in the crystallites with those in the noncrystalline regions, but it is generally accepted if l is expressed in nanometers and the frequency is expressed in cm^{-1}, for example, within about 20% absolute accuracy $l = 1000/f$ for polyethylene.

4.3 Polymeric Synthesis

4.3.1 Overview

Synthetic polymers are produced by the repetition of a "polymerization" reaction. By applying varying conditions of heat, pressure, and catalysts, monomers can be forced into reacting with each other to form a chain. The methods of polymeric synthesis are divided in two major groups: addition and condensation polymerization.

Since synthetic polymers are not naturally produced in living organisms, they are not as biologically compatible as natural polymers. However, they are popular because their syntheses are well understood and their properties can be modified for various uses. They can be made to have specific absorption times, scaled up for reproducibility, and do not have the potential to transmit diseases since living organism are not used in their production. They can also be fabricated readily into complex shapes and structures.

An inflammation response is initiated with the implantation of all polymers. In general, the implantation of any synthetic material initiates a wound healing mechanism, which causes an inflammatory response. Inflammatory cells can be triggered either through direct adherence to the polymer surface or when phagocytes in the local environment are activated through the immune system. However, the duration of the implant is an important factor that contributes to the biocompatibility of a polymer. A material intended to be used for a short period of time does not need to have the same biocompatibility characteristics as a material intended to be used for a long period of time. A systemic toxic effect may occur acutely or chronically from the diffusion of small molecular weight materials that were either formed or absorbed during the synthesis, fabrication, or sterilization process. Chronic toxicity may occur as these small molecular weight materials are released over a period of time or as the material undergoes degradation with usage.

Wear between two components of a prosthetic device may also lead to failure of materials. This frequently happens in hip and knee prostheses since they involve the abrasion of metals surrounded by a polymeric surface. This mismatch of the material properties frequently leads to wear on the polymeric component. Moreover, the generation and accumulation of micrometer and submicrometer size wear particles of some polymers have been found to cause local inflammation and osteolysis.

4.3.2 Addition

Addition polymers are obtained by subjecting olefins to polymerization. In an addition polymerization reaction, small molecules are generated during rearrangement in a chain reaction to create a polymer. Addition polymerization describes the method where unsaturated monomers are added one by one to an active site on a growing chain. The backbones of addition polymers consist of carbon-carbon bonds. There are three significant reactions that take place in addition polymerization: initiation, propagation, and termination.

During initiation, the initiators can be free radicals, cations, anions, or stereospecific catalysts. The initiator opens the double bond of the monomer, presenting another "initiation" site on the opposite side of the monomer bond for continuing growth. Light or heat can also act as an initiator to break down a molecule and create reactive free radicals. The formed free radical attaches itself to another molecule, forming yet another different free radical.

In the propagation step, the newly formed free radical attacks and attaches itself to the double bond of another monomeric molecule. This process repeats, increasing the length of the polymer. Rapid chain growth ensues until the reaction is terminated by reaction with another radical, a solvent molecule, another polymer molecule, an initiator, or an added chain transfer agent. In theory, the propagation reaction could continue until the supply of monomers is exhausted. However, most often, the growth of a polymer chain is halted by the termination reaction.

Termination typically occurs in two ways: combination and disproportionation. Combination occurs when the polymer's growth is stopped by free electrons from two growing chains that join and form a single chain. The polymerization reaction terminates when two reactive monomers combine together in a process known as disproportionation. Disproportionation halts the propagation reaction when a free radical strips a hydrogen atom from an active chain. A carbon-carbon double bond takes the place of the missing hydrogen. Disproportionation can also occur when the free radical reacts with an impurity. This is why it is extremely important that polymerization should be carried out under very clean conditions.

The properties of the polymer can be changed by changing the reaction time, reactants, and reaction conditions. Polymers produced by addition polymerization can be homopolymers containing only one type of repeat unit or copolymers with two or more types of repeat units. Depending on the reaction conditions and the reactivity of each monomer type, copolymers can be random, alternating, graft, or block copolymers. Some examples of the addition polymerization are shown in Fig. 4.3. Molecular weights of polymer chains are difficult to control with precision in free radical polymerization. Added chain transfer agents are used to control the average molecular weights, but molecular weight distributions are usually broad. In addition, chain transfer reactions with other polymer molecules may produce undesirable branched products that affect the ultimate properties of the polymeric material.

4.3.3 Condensation

Condensation polymerization is completely analogous to condensation reactions of low-molecular-weight molecules. Two monomers react to form a covalent bond, usually with elimination of a small molecule, such as water, hydrochloric acid, methanol, or carbon dioxide. Condensation polymers are typically formed from reactions of either alcohols with acids to form polyesters, acids, or esters with amines to form polyamides, alcohols, or amines with isocyanides to form polyurethanes or polyuria. Condensation polymers contain carbon-heteroatom bonds in the main chain. Silicone polymers have silicon-oxygen bonds as their backbone. Copolymer chains containing two or more types of monomers can be synthesized using copolymerization or block polymerization reactions. The polymer chains can be arranged in a linear, branched, crosslinked, or three-dimensional network form. Some examples of the condensation polymerization are shown in Fig. 4.4.

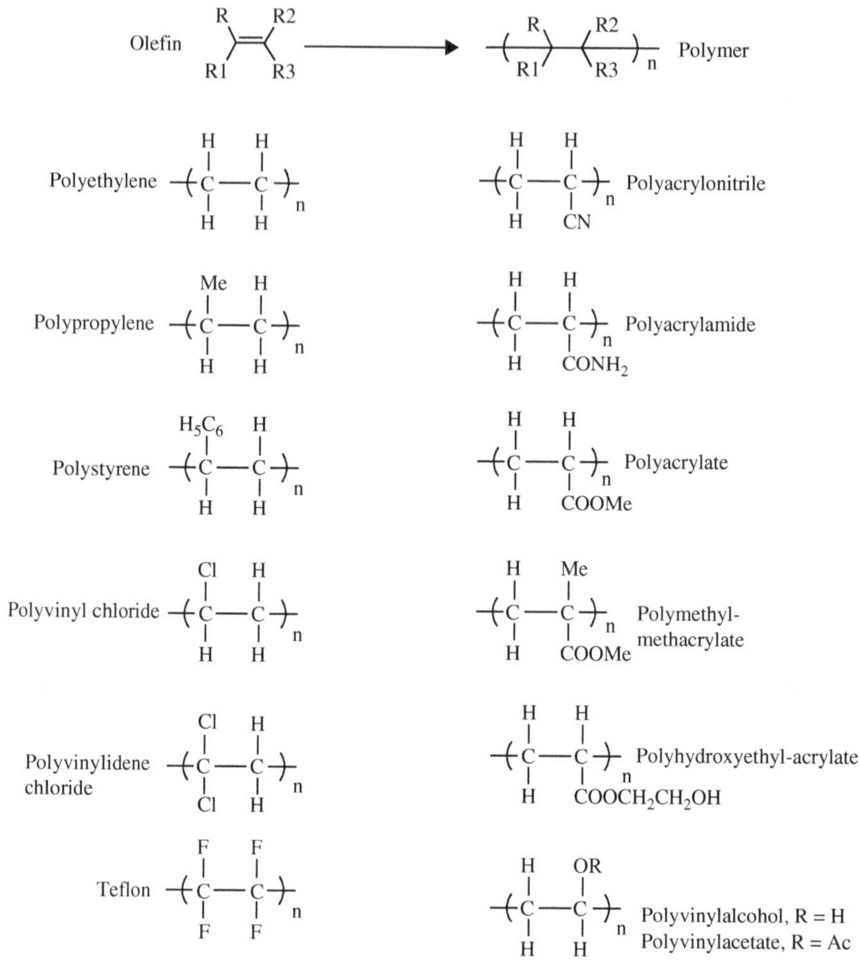

Fig. 4.3 The chemical structure of different addition polymers. Polymers are composed of multiple repeats of the subunit shown in parentheses. The n represents the number of times the subunit is repeated. (Bhat [5]. Adapted with permission from Springer)

The reaction continues until almost all of one reactant is used up. There are also polymerizations that resemble the stepwise growth of condensation polymers, although no small molecule is eliminated. Polyurethane is a polymer whose synthesis bears these characteristics. In condensation polymerization, the molecular weight of the polymer product can be controlled by the ratio of one reactant to another and by the time of polymerization. The use of functional monomers gives rise to linear polymers, while multifunctional monomers may be used to form covalently crosslinked networks.

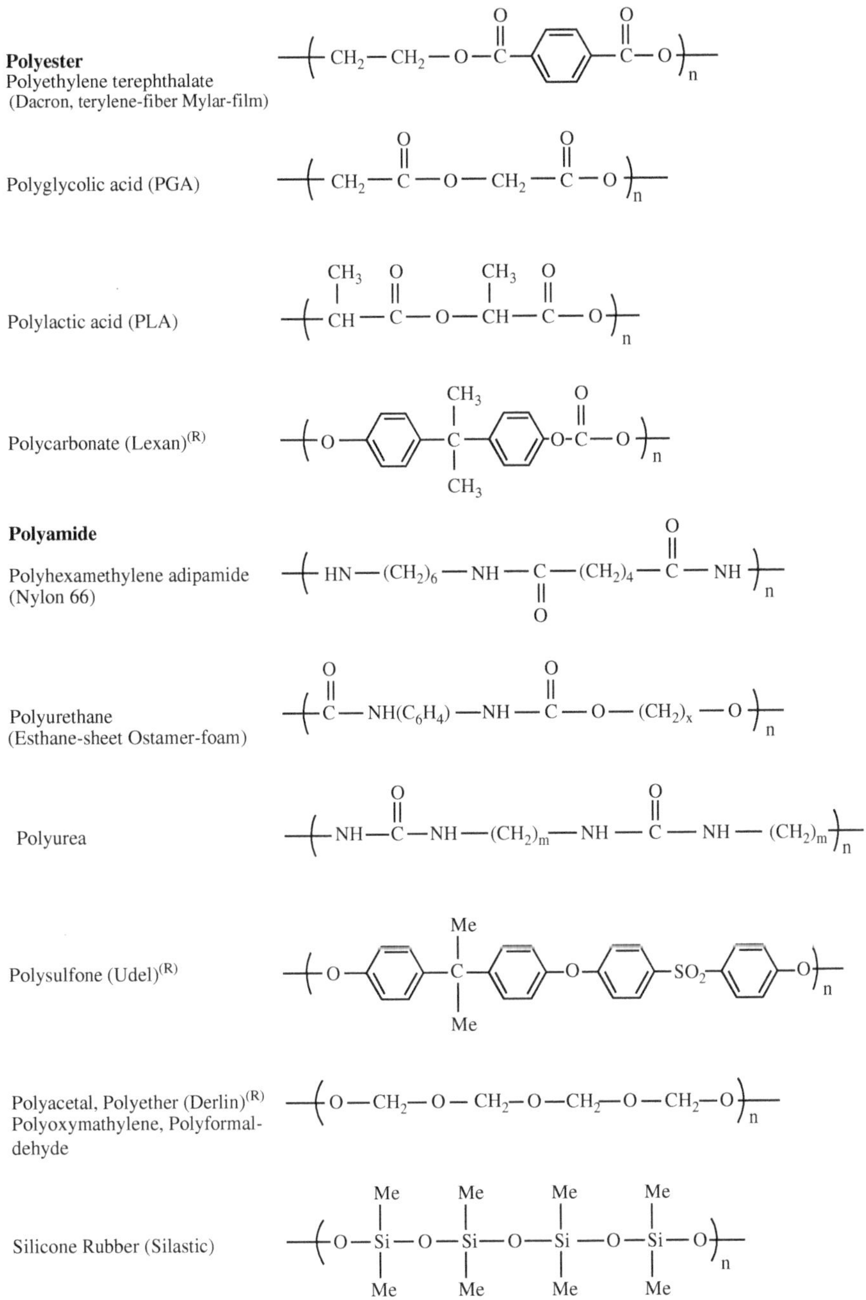

Fig. 4.4 The chemical structure of different condensation polymers. Polymers are composed of multiple repeats of the subunit shown in parentheses. The *n* represents the number of times the subunit is repeated. (Bhat [5]. Adapted with permission from Springer)

4.3.4 Number- and Weight-Average Molecular Weights

The degree of polymerization (DP) is one of the most important parameters in determining physical properties. It is defined as the number of monomers per chain. DP can be expressed by molecular weight alternatively. However, unlike a small molecule with a unique molecular weight, a polymer is usually produced with a distribution of molecular weights. Depending upon the DP, the polymer material may vary from thousands to millions of monomer units per chain.

To compare the molecular weights of two different batches of polymers, it is useful to define an average molecular weight. Two of the most widely used methods are number average molecular weight (M_n) and weight average molecular weight (M_w). M_n is the first moment of the molecular weight distribution and is an average over the number of molecules. M_w is the second moment of the molecular weight distribution and is an average over the weight of each polymer chain.

The number average molecular weight (M_n) is calculated by

$$M_n = \sum \frac{x_i M_i}{\Sigma x_i} \tag{4.4}$$

where M_i is the molecular weight of each species fraction and x_i is the number of molecules in each species fraction.

The weight average molecular weight (M_w) is calculated by

$$M_w = \sum \frac{w_i M_i}{\Sigma w_i} \tag{4.5}$$

where w_i is the weight fraction of the species i.

Linear polymers used for biomedical applications generally have M_n in the range of 2.5×10^4 to 1×10^5 and M_w from 5×10^4 to 3×10^5. In exceptional cases, for example, the M_w of polyethylene used in a hip joint may range up to a million. Figure 4.5 shows a typical distribution in molecular weights of polymers including

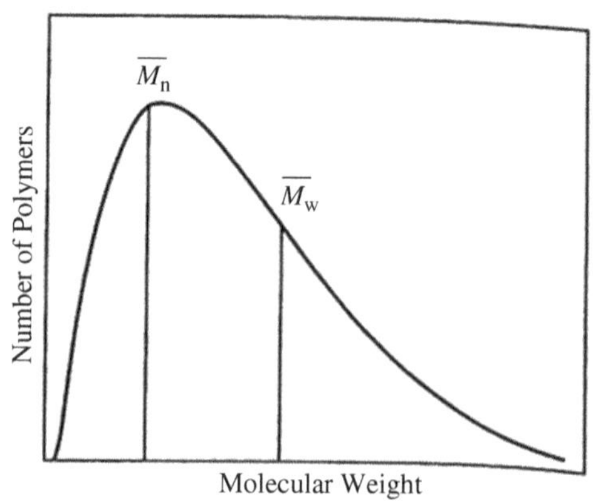

Fig. 4.5 Schematic of a simple molecular weight distribution showing the various averages

the weight and number average molecular weights. The ratio of M_w to M_n is known as the polydispersity index (PI) and is used as a measure of the breadth of the molecular weight distribution. Typical commercial polymers have PIs of 1.5–5.0, although polymers with PIs of less than 1.1 can be synthesized with special techniques, including "living polymerization" as mentioned previously.

In general, increasing molecular weight of polymers corresponds to increasing physical properties, such as strength and hardness. However, since melt viscosity also increases with molecular weight, the ability of a polymer to be processed will decrease, and an upper limit of useful molecular weights can usually be reached. Also, molecular weight may influence the ability of the polymer chains to crystallize and to exhibit secondary interactions, such as hydrogen bonding. Crystallinity and secondary interactions can give polymers additional strength.

4.4 Physical Properties

4.4.1 Relaxation, Transition, and Melt Viscosity

Polymers show a wide range of mechanical properties. Network polymers are often quite strong and have high yield strength and modulus of elasticity but poor ductility. On the other hand, linear polymers have lower strength and are easier to deform. Elastomers show very high ductility but variable modulus of elasticity. In general, the mechanical properties of polymers are dependent upon their molecular structures. Basically, long-chained polymers are usually stiffer than short-chained polymers. This is primarily due to the fact that longer chains result in larger entanglement which results in more rigidity in the bulk aspect. Furthermore, longer polymers have larger intermolecular forces because there are more bonds between the atoms. The regularity in arrangement of polymer fibers is defined by the percentage of crystallinity of the polymers. Apparently, crystallinity also influences the physical properties of substances. In general, crystallinity is directly proportional to the strength of the polymer and inversely proportional to the flexibility.

Apart from this, the presence of polar side groups makes a polymer stiffer due to attractive forces between the side groups. Branching of polymers can cause the links to be less tightly packed, resulting in a reduction in strength of the polymers. Crosslinking polymer chains, on the other hand, increases the rigidity of polymers and makes them more difficult to melt. In particular, transition refers to the change in state of a material due to alterations in temperature or pressure. Relaxation refers to the time required for a material to change in state due to a change in temperature or pressure.

A polymer undergoes multiple phases when subjected to different temperatures as shown in Fig. 4.6. At the low-temperature region in the glassy region (1), the polymer is glassy and brittle. The Young's modulus is roughly constant. In the glass transition region (2) below the glass transition temperature T_g, the modulus decreases by about a thousand times. Minute changes of degrees in temperature will also change the stiffness. In the rubbery plateau region (3) beyond the melting

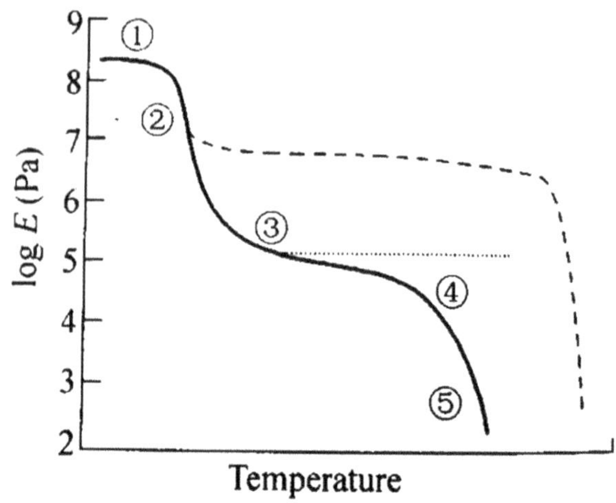

Fig. 4.6 Change of elasticity upon temperature for polymers

temperature T_m, the modulus becomes roughly constant. Polymers will also exhibit long-range rubber elasticity, i.e., the elastomer can be stretched up to several hundred times of its length and snap back to the original length. In the rubbery flow region (4), the polymer is marked by having both rubbery elasticity and flow properties, depending on the time scale of the deformation. When a polymer is subjected to a short deformation time, the physical molecular entanglements are not able to relax, causing the material to remain in a rubbery state. For longer times, the increased molecular motion imparted by the increased temperature permits the assemblies of chains to move in a coordinated manner (depending on the molecular weight) and flow. In the liquid flow region (5), the polymer flows readily. The increased thermal energy permits the polymeric chain to break apart, allowing the individual molecules to flow.

Flow occurs when polymer molecules slide past each other. The ease of flow depends on the mobility of the molecular chains and the forces or entanglements holding the molecules together. That an increase in temperature reduces viscosity is widely known and, in general, is an easily understandable property. The influence of environmental pressure and shearing history of the polymer melt are much less known, and it is only in recent years that their importance has begun to be appreciated. It is well established that for Newtonian liquids, the logarithm of viscosity μ is linearly proportional to the reciprocal of temperature T, i.e., $\ln(\mu) = £/T$, where £ is a constant. Yet, this does not apply to polymer melts. A number of attempts have been made to obtain a fundamental explanation of the differences in temperature dependence of viscosity between different polymers. To date, perhaps the most successful is the concept of "free volume." The polymer chains in the free volume can move with only a relatively low level of physical constraints, and therefore, their probability of movements P_m (e.g., rotation, or cooperative motion) follows the Maxwell-Boltzmann distribution as a function of temperature, i.e., $P_m = \exp.(-E_f/kT)$, where E_f is the free energy of activation and k is the Boltzmann's constant. Next, a large time scale t allows for greater probability of the required molecular motion. The free volume theory further assumes that Pt must reach a constant threshold value for the onset of the motion, and therefore

$$\ln(Pt) = -\frac{E_f}{kT} + \ln t \tag{4.6}$$

Furthermore, the quantity E_f is associated with free volume and qualitatively would be expected to decrease as the fraction free volume increases. It is assumed that $E_f = \Lambda kT/f$, where Λ is a constant and f is the fractional free volume. Taking the differential yields,

$$\Delta \ln t = \Delta\left(\frac{E_f}{kT}\right) = \Lambda\Delta\left(\frac{1}{f}\right) \tag{4.7}$$

This theory suggests that at a reference temperature T_0 (typically chosen as ~50 °C below the measured glass transition temperature T_g), there is a relatively very small fractional "free volume" of V_0 and fractional free volume V_{f0} between the molecules. This fractional free volume, f, is postulated to increase linearly with the temperature so that at T_g the fractional free volume has a value f_g. The expansion coefficient α_f is defined by the equation

$$f = f_g + \alpha_f(T - T_g) = f_0 + \alpha_f(T - T_0). \tag{4.8}$$

It has been proposed that V_{fg} has a universal value of 0.025 and α_f has a universal value of 4.8×10^{-4}. We can then reconsider Eq. 4.7 as

$$\ln\left(\frac{t}{t_0}\right) = \Lambda\left(\frac{1}{f} - \frac{1}{f_0}\right) \tag{4.9}$$

where t_0 is the time of molecular movement activation scale at the reference temperature. In fact t/t_0 is called the reduced variables shift factor A_T. Now, we can substitute Eq. 4.9 into Eq. 4.8 to obtain the theoretical form of the Williams-Landel-Ferry (WLF) equation:

$$\ln A_T = -\frac{(\Lambda/f_0)(T - T_0)}{f_0/\alpha_f + (T - T_0)} \tag{4.10}$$

Furthermore, for polymer melts, the viscosity μ is a time (shear-rate)-dependent quantity related to the reference viscosity μ_0 with the same reduced variables shift factor A_T as for time, i.e., $\ln(\mu/\mu_0) = \ln A_T$. Thus,

$$\ln\left(\frac{\mu}{\mu_0}\right) = -\frac{(\Lambda/f_0)(T - T_0)}{f_0/\alpha_f + (T - T_0)} \tag{4.11}$$

If the reference temperature is taken as T_g and the base of logarithm function is taken as 10, then the above equation can be rewritten as

$$\log\left(\frac{\mu}{\mu_0}\right) = -\frac{C_1^g (T - T_g)}{C_2^g + T - T_g} \tag{4.12}$$

where both C_1^g (=17.44) and C_2^g (=51.6) are universal constants. It can be easily proven that the value of the right hand side in the above equation is negative, increases with temperature T for $T \geq T_g$, and approaches to zero when T is very high. In other words, the viscosity μ decreases with T, with a reference value of μ_0 at the temperature T_g.

4.4.2 Theory of Melting Point Depression

Crystalline polymers refer to those monomers in repeating patterns that are harder, more rigid, and more durable than amorphous polymers. They are also more durable against solvent penetration but have lower impact resistance than amorphous polymers. Some polymers crystallize at temperatures below their crystalline melting points, T_m (the temperature in which they melt). At the crystalline melting point and above, the crystalline polymers melt and become viscous. In some applications for polymeric materials which offer physical support, the properties of crystalline polymers are highly desirable. By using inherently suitable polymer chains, crystalline melting points can be raised to higher temperatures. Some engineering plastics are capable of competing with metals and ceramics in engineering applications. Examples of crystalline polymers include nylon clothing, polyethylene, packing supplies, and seat covers.

The melting of polymers may be observed by any of several means. For linear or branched polymers, the sample becomes liquid and flows. However, their interpretation can be difficult. For example, simple liquid behavior may not be immediately apparent because of the polymer's high viscosity. If the polymer is crosslinked, it may not flow at all. It must also be noted that amorphous polymers soften at their glass transition temperature T_g, which is not the melting temperature. Even worse, the polymer always includes impurities. For polymers, the melting temperature is usually taken where the last trace of crystallinity disappears, and therefore, the melting temperature can be determined from the thermodynamics point of view. The depressed melting point T_f (unit, K) in crystalline substances from the pure state T_f^0 (unit, K) is given by the general equation

$$\frac{1}{T_f} - \frac{1}{T_f^0} = -\frac{R}{\Delta H_f} \ln a \tag{4.13}$$

where a represents the activity of the crystal in the presence of the impurity, ΔH_f is the heat of fusion per mole of crystalline mers, and R is the Boltzmann constant (= 0.0083 kJ/mol/K).

As a first approximation, the melting point depression depends on the mole fraction of impurity X_B and the mole fraction of crystallizable polymer X_A. For small values of X_B, $-\ln X_A = -\ln(1 - X_B) \approx X_B$. If we let M_0 be the molecular weight of the end mer of the polymer (and assuming that both ends are identical), the mole fraction of the chain ends is given approximately as $X_B = 2M_0/M_n$, where the number average molecular weight (M_n). Thus,

$$\frac{1}{T_f} - \frac{1}{T_f^0} = -\frac{R}{\Delta H_f}\frac{2M_0}{M_n} \tag{4.14}$$

If a solvent or plasticizer is added, the case is slightly more complicated. Here, the molar volume of the solvent, V_1, and the molar volume of the repeating polymer unit V_u cannot be assumed to be equal. Interactions between the polymer and the solvent should be taken into account. The results may be revised as

$$\frac{1}{T_f} - \frac{1}{T_f^0} = \frac{R}{\Delta H_f}\frac{V_u}{V_1}\nu_1, \tag{4.15}$$

where ν_1 represents the volume fraction of diluent. Apparently, T_f decreases with ν_1.

4.4.3 Glass Transition as an Iso-Free-Volume State

An amorphous polymer is a polymer that has monomers randomly polymerized resulting in a large twisted structure with no defined organization. The glass transition temperature, T_g, is the temperature at which amorphous polymers will become brittle, stiff, and rigid. Below their glass transition temperature, polymers become hard, stiff, and transparent, like ordinary inorganic glasses. As the temperature is lowered past a polymer's melting point, the glass transition temperature is reached, where polymeric materials undergo a change in properties associated with the virtual cessation of molecular motion. Examples of amorphous polymers include polystyrene eye glass lenses, tires, and rubber bands.

Molecular motion in the bulk state depends on the presence of holes or places where there are vacancies or voids. Different from other materials such as metals, a material volume constructed from polymer chains may require more "holes" in the same locality in order to allow the polymer macromolecules to move. Thus, for a polymeric segment to move from its present position to an adjacent site, a critical volume must first exist before the segment can jump. The important point is that molecular motion cannot take place without the presence of holes. These holes are called "free volume," collectively. For infinite molecular weight, the specific free volume, V_f, could be expressed at the T_g as a function of temperature T, the cubic (volume) expansion coefficients in the rubbery and glassy states, denoted as α_R and α_G, respectively. The same specific volume-temperature relationships of polymers are independent of molecular weight at a temperature below T_g. In short, below T_g, the local configurational arrangement of the polymer segments is independent of both molecular weight and temperature, and the glass transition temperature is an iso-free-volume state. The free volume at the temperature T_g (V_g) can be described as Fig. 4.7.

The volume at temperature T_g can be expressed as

$$V_g = V_f + \left(V_{R0} + \alpha_G T_g\right) \tag{4.16}$$

and

$$V_g = V_{R0} + \alpha_R T_g, \tag{4.17}$$

leading to the relation

$$V_f = (\alpha_R - \alpha_G) T_g \tag{4.18}$$

It has been found from testing of a wide range of polymers that $V_f \approx 0.113$. We can also examine α_G in Eq. 4.16 that expansion in the glassy state occurs at nearly constant free volume; hence, α_G is proportional to the occupied volume. As an approximated value, we may consider $\alpha_G T_g \approx 0.164$. Hence, we may estimate the material specific volume at a temperature, based on the value of T_g.

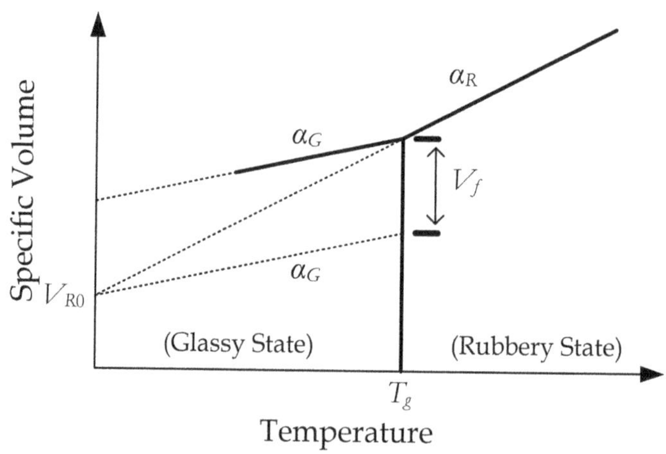

Fig. 4.7 A schematic diagram illustrating free volume and specific volume under different temperature conditions

4.4.4 Rubbery Elasticity

In this section, we discuss further on the mechanical elastic properties of polymers at a temperature above T_g. Rubbery polymers have both rubbery elasticity and viscosity. To discuss only the key rubbery elastic properties, we assume that a rubber is isotropic in the undeformed state, i.e., it has the same properties in all directions. Also, we assume the changes of volume on deformation are very small and can be neglected, i.e., the rubber is incompressible and the Poisson's ratio $\nu \approx 0.5$. For finite strain in isotropic media, only the state of homogenous pure strain without shear stress is considered here. In fact, a small shear strain is exactly equivalent to compressive and extensional strains applied at 90° to the original axes along which the shear is applied. In other words, a shear stain can be transformed to a state of homogeneous pure stain via an angle of rotation. Locally, all states of homogeneous strain can be regarded as "pure" if proper major axes are chosen.

We may now consider an extension ratio along a major axis "*i*" to be denoted as λ_i. Clearly, $\lambda_i = 1 + \varepsilon_i$, where ε_i is the pure strain along the major axis, and $\lambda_i^2 = 1 + 2\varepsilon_i + 2\varepsilon_i^2$. Recall Eq. 2.17, $\varepsilon_i \approx [\sigma_i + ч]/E$, where $ч$ is a variable dependent on Poisson's ration and stresses. We further let $ч = \sigma_i/2 + p$, where p is another variable. Then, we have

$$\lambda_i^2 = \frac{3}{E}[\sigma_i + p*] \tag{4.19}$$

where $p* = p + E/3$. For larger strains, the cross-sectional area changes significantly with stress. $p*$ does not correspond exactly to a pressure. Now, we further assume the direction '*i*' is along the *z*-direction. Then, $\sigma_x = \sigma_y = 0$ and $\lambda_x = \lambda_y$ by symmetry. From now on we simply consider λ_z as λ and $\sigma_z = \sigma$. Hence,

$$\lambda_x^2 = \lambda_y^2 = 3p * /E \tag{4.20}$$

Further, rubbers are nearly incompressible and $\lambda_x\lambda_y\lambda_z = 1$, then

$$\lambda = \frac{1}{\lambda_x\lambda_y} = \frac{E}{3p*}, \quad \text{thus } p* = \frac{E}{3\lambda} \tag{4.21}$$

Substituting the above relation into Eq. 4.19 leads to

$$\lambda^2 = (3/E)[\sigma + E/(3\lambda)] \tag{4.22}$$

In fact, σ is a local stress value, which is the true stress. We further consider the nominal stress along the force direction (*z*), and the relative cross-sectional area has a variation ratio of $\lambda_x\lambda_y = 1/\lambda$. We can define the nominal stress $\sigma_n = \sigma/\lambda$, and thus

$$\sigma_{\mathrm{n}} = (E/3)\left(\lambda - 1/\lambda^{-2}\right) \tag{4.23}$$

Notably, for very small ε, $\sigma_{\mathrm{n}} \approx E\varepsilon$, which is Hooke's law. The above relationship is compatible with the isotropy and incompressibility of a rubber and agrees with Hooke's law at small strains. Materials that obey these relationships are called neo-Hookean or hyperelastic solids. The model is applicable for plastics and rubber-like substances. A neo-Hookean material will initially be linear, but at a certain point, the stress-strain curve will plateau due to the release of energy as heat while straining the material. Then, at a higher strain point, the elastic modulus of the material will increase again. Although Eq. 4.23 is only a simple generalization of small-strain elastic behavior, it describes the behavior of a real rubber to a first approximation. In particular, it describes qualitatively the initial reduction of σ_{n} once the extension ratio λ rises even above a rather low level. However, it still fails to describe the material properties for a very high value of λ (>6).

Here, we can have an extended discussion on neo-Hookean solids. The nominal stress σ_{n} is indeed the rate of change of the "strain energy" U_{n} upon an extension, $\mathrm{d}U_{\mathrm{n}}/\mathrm{d}\lambda = \sigma_{\mathrm{n}}$. The strain energy is defined as the work done on unit volume of material. Therefore, we may find the expression of U_{n} by integrating Eq. 4.23 with respect to λ and considering the condition $U_{\mathrm{n}} = 0$ at $\lambda = 1$:

$$U_{\mathrm{n}} = (E/6)\left(\lambda^2 + 2/\lambda - 3\right) \tag{4.24}$$

From the other point of view to interpret the meaning of strain energy, U_{n} is proportional to the total strain square, i.e., $\varepsilon_x^{\ 2} + \varepsilon_y^{\ 2} + \varepsilon_z^{\ 2}$, and hence, for fairly small deformation $\lambda \approx 1$,

$$U_{\mathrm{n}} \approx (E/6)\left(\lambda_x^{\ 2} + \lambda_y^{\ 2} + \lambda_z^{\ 2} - 3\right) \tag{4.25}$$

A quick check on the validity of Eq. 4.25 can be performed by substituting the extension ratios for the case of uniaxial strain ($\lambda_x^{\ 2} = 1/\lambda$, $\lambda_y^{\ 2} = 1/\lambda$ and $\lambda_z = \lambda$), and the corresponding strain energy agrees with Eq. 4.24.

4.4.5 *Relationships of* T$_m$ *and* T$_g$ *with Molecular Weight*

Generally, adding chemical functional groups can change the glass transition temperature. There is a direct relationship as the side group molecular weight increases, the higher the glass transition temperature, and the lower the amount of side group, the lower the glass transition temperature (Fig. 4.8).

Both the glass transition temperature and melting temperature can be determined by different measurement methods such as differential scanning calorimetry. This procedure usually involves simultaneously heating a polymer sample on an aluminum pan and heating an empty aluminum pan. The temperature in the pans increases

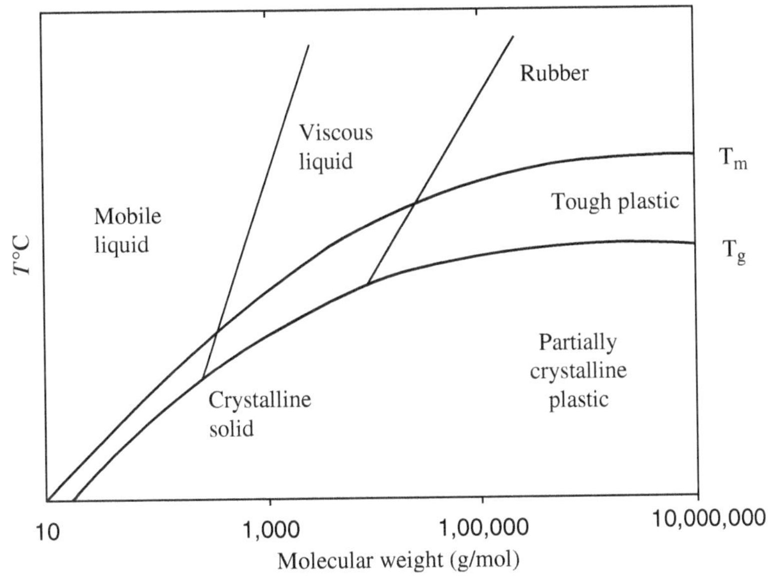

Fig. 4.8 The melting temperature, T_m, and glass transition temperature, T_g, of different types of polymers

10 °C per minute. The empty aluminum pan is the control, and the temperature increase of the polymer that changes the mechanical characteristics is the glass transition temperature (amorphous) or the melting temperature (crystalline).

4.5 Common Polymeric Biomaterials

The molecular structures of polymers determine their chemical and physical properties and, therefore, their utilization in various medical domains. For example, elastomers are able to withstand large deformations, returning to their original dimensions after being stressed. Plastics are more rigid materials and can be classified into two types: (1) thermoplastic polymers, which can be melted, reshaped, and reformed, and (2) thermosetting plastics, which cannot be remelted and reused, meaning any chemical reactions occurred are irreversible. Components with a smaller molecular weight in a polymer have a tendency to leach into the physiological environment of humans. This leaching may lead to an increase of the elastic modulus and decrease in yield strength of the component, leading to chemical, physical, or mechanical alterations that eventually lead to failure. Polymers may also undergo degradation through a variety of mechanisms such as absorption, leaching, hydration, oxidation, chain scission, and hydrolysis. In the following sections, we discuss some commonly used polymers in biomedical devices.

Polyethylene

Polyethylene is formed by polymerization of ethylene (C_2H_4), a gas at room temperature and pressure. The density of polyethylene (0.9–0.96 g/ml) varies, with its tensile strength, hardness, and chemical resistance increasing with density and molecular weight. High-density polyethylene (5×10^5 g/mol) does not contain branches, while low-density polyethylene does (3.4×10^3 g/mol). High-density polyethylene possesses better chain packing, which induce crystallinity of 80–95% and elasticity of 410–1240 MPa, compared to the low-density polyethylene with crystallinity of 40–70% and elasticity of 96–260 MPa. The longer-chain structures of high-density polyethylene also enables for higher tensile strength (23–40 MPa) and elongation (400–500%) than low-density polyethylene (strength 7.6 MPa and elongation 150%).

On the other hand, the type of polyethylene often used in surgery has a molecular weight of approximately $2–4 \times 10^6$ g/mol and a relatively high elastic modulus of 1.1–2 GPa. This kind of polyethylene is referred to as ultra-high molecular weight polyethylene (UHMW-PE). UHMW-PE is often used for fabrication of acetabular cups in artificial hips, the bearing surface of some knee prostheses, and blood contacting tubes.

Ultra-high molecular weight polyethylene (UHMWPE) is a common material used in total joint replacements because it has an extremely low coefficient of friction and is highly resistant to abrasion. However, since this material is frequently used in joints, abrasion and the subsequent creation of wear particles are high concerns. Wear particles of UHMWPE can cause cellulitis, giant cell reaction, and tissue death. Wear can also cause bone resorption due to causing an abnormal distribution of forces at the implant-bone interface. As shown in Fig. 4.9, wear particles of UHMWPE from the femoral stem implant in the human body can reduce the patient's bone mineral density (indicated by arrows) over 14 years after the femoral implant.

Perfluorinated Polymers

The unique chemical and thermal stability of perfluorocarbon polymers has caused early and sustained interest in their implant potential. The unique intrinsic stability of these materials rests on the extreme inertness and strength of the covalent bonds between elements in the polymer chain and the fluoride bonds. The linear homopolymers of tetrafluoroethylene (PTFE) are a widely used material type in clinical implantation, while others are rarely used in implant fabrications. PTFE has high density (2.15–2.2 g/cm^3), low tensile strength (17–28 MPa), and low modulus of elasticity. It has a very low surface tension (18.5 mN/m) and friction coefficient ($\mu_f = 0.1$). PTFE does not cause acute inflammatory changes and tissue reaction; therefore, PTFE is widely used within the cardiovascular circulation for cardiovascular and other sutures.

The perfluorinated polymer Teflon is another popular polymer used in devices for replacing joints. Like other polymers used in joints, wear is an important factor that affects the material. Wear particles of Teflon can cause osteolysis and death to other tissues. Also, extensive wear of the polymer may lead to failure of the prosthesis or metal on metal abrasion, leading to release of metal particles and illnesses associated with the release of metals.

Fig. 4.9 X-ray image of a 51-year-old male patient immediately (left) and 14 years (right) after the femoral stem implant. It shows the relatively low resection level at the femoral neck just above the minor trochanter. Calcar rounding and loss of medial cortical density are visible indicating second-degree stress shielding after 14 years of implantation. (Streit [6], International orthopedics

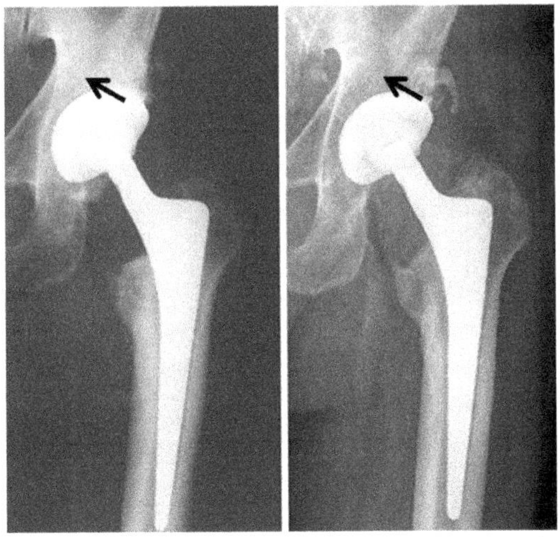

Acrylic Polymers

Simple acrylates have relatively high toughness and strength. These are obtained through addition polymerization of acrylic acid derivatives. The most widely used polyacrylate is poly-(methyl methacrylate, PMMA). It is relatively brittle compared to other polymers. It has an excellent light transparency (>92% transmission) and a high index of refraction (1.49). Therefore, it is sometimes referred as "organic glass." It has excellent chemical resistivity and is highly biocompatible in the pure form. PMMA is used extensively in medical and surgical applications for contact lenses, implantable ocular lenses, bone cement for joint fixation, dentures, and maxillofacial prostheses (Fig. 4.10). One issue against the biocompatibility of acrylic resins lies in the fact that they are capable of causing allergic reactions. Studies show that injection of PMMA (usually used for implants) for bone cement can trigger a granulomatous inflammatory reaction. Initially, a patient may experience vascular compromise, but later, he or she may see hardening and swelling in the implanted area, causing the cement to deteriorate. Some individuals who received acrylic dentures have had complaints of a sore mouth and burning sensation accompanied by symptoms such as swollen oral mucosa.

Hydrogels

Hydrogels derive their name from their affinity for water and incorporation of water into their structure (Fig. 4.11). The concentration of water in a hydrogel can significantly affect the interfacial free energy of the hydrogel and its biocompatibility. Hydrogels are a three-dimensional network of hydrophilic polymers held together by association bonds such as covalent bonds, weaker cohesive forces such as hydrogen and ionic bonds, and intermolecular hydrophobic association. These networks are able to retain a large quantity of water within their structures without dissolving. Therefore, hydrogels are mostly water by weight. While they behave like solids, hydrogels have inherently weak mechanical properties.

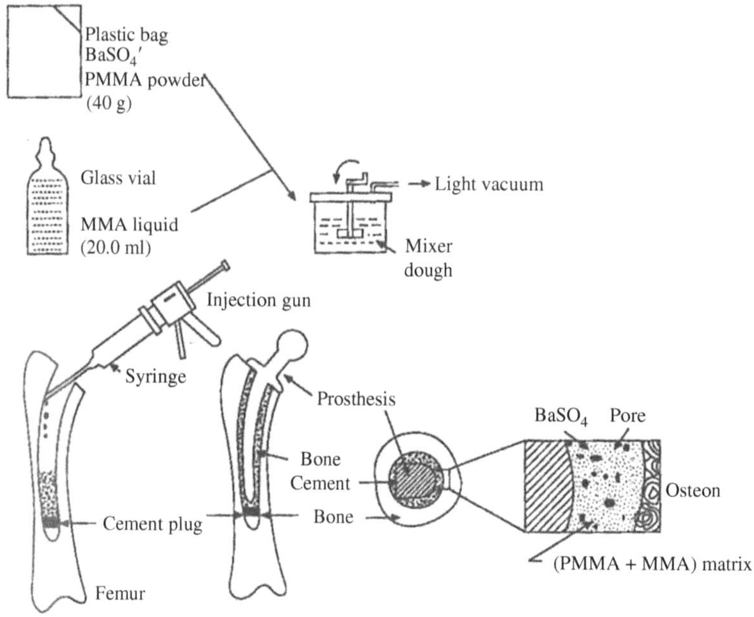

Fig. 4.10 The process of implanting a femoral stem. The cement must be freshly mixed and injected into the femur. The prosthesis is then wedged into the cement. (Bhat [5]. Adapted with permission from Springer)

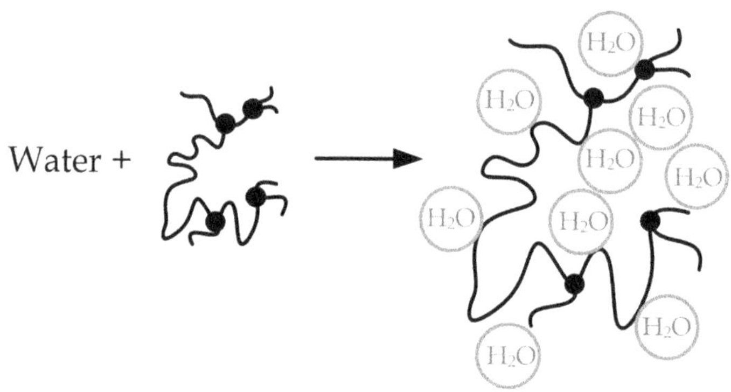

Fig. 4.11 Hydrogels have a high affinity for water as shown in the above image

Hydrogels exhibit ranges of chemical and physical properties, depending on their structures and compositions. For example, the soft rubbery nature of hydrogels minimizes mechanical and frictional irritation to surrounding tissues. These polymers may have low or zero interfacial tension with surrounding biological fluids and tissues, thereby minimizing the potential of absorbing proteins or attaching to cells. These characteristics endow hydrogels with good biocompatibility. Hydrogels, like living tissues, allow for the permeation and diffusion of low molecular weight metabolites, waste products, and salts.

Hydrogels swell distinctly in water due to the affinity between hydrophilic polymer networks and water. The polymer chains of hydrogels interact with the solvent molecules (water) and tend to expand to their fully solvated state. On the other hand, the crosslink structure works as the retractile force to pull back the polymer chain inside. The counterbalance between expanding and retracting forces attains equilibrium in a particular solvent at a particular temperature. The swelling ratio is expressed by the ratio of the weight of swollen sample over that of the dry sample. The water content of a hydrogel is expressed in terms of percentage of water weight in the hydrogel. For instance, most hydrogel contact lenses have water content between 38% and 75%.

In particular, hydrogels are attractive for a variety of pharmaceutical applications because of their good tissue compatibility, easy manipulation under swelling conditions, and solute permeability. Their hydrophilicities can impart desirable release characteristics to controlled and sustained release formulations of drugs and proteins. Hydrogels hold tremendous promise as protein delivery system since they do not denature their incorporated protein. However, one disadvantage of these materials is that they cannot be steam-sterilized.

Polyurethanes

Polyurethanes have found increasing applications in the field of medical devices due to their biocompatibility. Among the polyurethanes, the polyether urethanes are principally favored because of their hydrolytic stability and good mechanical properties. Vascular tubes made of these are used as aortic patch grafts. Polyurethane copolymers are the preferred choice for long implants because of their high hydrolytic stability. These polymers also have good blood compatibility. They have been successfully used in fabrication of artificial heart assist devices with minimal inflammatory reaction. Furthermore, there is no change in the concentration of plasma proteins or increase in the risk of thrombosis. They are also non-cytotoxic and do not cause adverse tissue reactions.

Polyamides

Polyamides are obtained through condensation of diamine and diacid derivatives or through polymerization of cyclic lactams. These polymers are known as nylons and are designated by the number of carbon atoms in the parent monomers. Nylons have density of ~1.1 g/cm^3, modulus of elasticity of ~2 MPa, and tensile strength of ~60–80 MPa with an elongation of at least 100%.

Silicone Rubber

Silicone rubbers are polymers with alternating atoms of silicon and oxygen in the main chain and organic side groups attached to the silicon atoms. Silicone rubbers typically have a density of ~1–1.2 g/cm^3 and an elongation of >300%. However, these materials have a limited elastic modulus (<10 MPa) and tensile strength (7–30 MPa). For medical applications, the most widely used polymer is polydimethylsiloxane (PDMS). Silicone rubbers intended for medical purposes must not contain a wide variety of additives. Their outstanding characteristics include thermal and oxidative stability at high temperatures of up to 150 °C, retention of flexibility and elasticity at low temperatures, ability to produce mutually nonadhesive and water-repellent surfaces, extreme inertness, resistance to weathering and sunlight, and good electrical insulation.

Due to its superior blood compatibility, silicone rubber has been extensively used for cardiovascular applications. Catheters made from silicone rubber are preferred for long-term parenteral nutrition (when a patient is fed intravenously). Silicone rubber is also a popular material in the replacement of destroyed or diseased interphalangeal (finger) joints. Other applications include artificial bladders, sphincters, and testicles.

Biodegradable Synthetic Polymers

Essentially all biopolymers are susceptible to enzymatic degradation because the enzymatic polymerization reactions responsible for their synthesis also have closely related depolymerization counterparts within organisms. The factors which affect biodegradability are the molecular structures (hydrophilicity and functional groups, crystallinity, and amorphous) and physical and biochemical environmental conditions (temperature, pH, humidity, and oxygen availability). Examples of the synthetic polymers based on polyester are poly(lactic acid) and poly(lactic-co-glycolic acid) copolymers, polycaprolactone, and poly(β-hydroxybutyrate). Biodegradable materials have the following major biomedical applications:

- Adhesives.
- Temporary scaffolding. This application has received the most attention and includes the absorbable (or soluble) suture. The suture is used to hold both sides of a wound together until sufficient collagen synthesis has occurred to hold the wound together unassisted.
- Temporary barrier. In this application, a layer of the polymer is placed in a blood vessel to stop fibrin from adhering together (forming a blood clot) and remains in situ until all the fibrin undergoes phagocytosis. This is often used after tendon, spinal, or open-heart surgery.
- Drug delivery matrix. These can be loaded with different drug concentrations. The matrix degrades at a predictable rate, releasing the drug into the tissues of the target organ in a controlled rate.

Problems

Problem 4.1

The following data (Table 4.P1) were obtained for a polymethylacrylate [monomer ($H_2C{=}CHCOOCH_3$)]

(a) Calculate M_n and M_w of this polymer.
(b) What is the polydispersity index ($=M_w/M_n$)?
(c) What is the degree of polymerization (DP)? The molecular weight of methyl acrylate is 86 g/mol.

Table 4.P1 Molecular data of polymethylacrylate

	Mean molar weight (g/mol)	Weight (g)	Number fraction
Group A	2×10^4	2.0	0.5
Group B	4×10^4	1.0	0.4
Group C	3×10^4	1.0	0.1

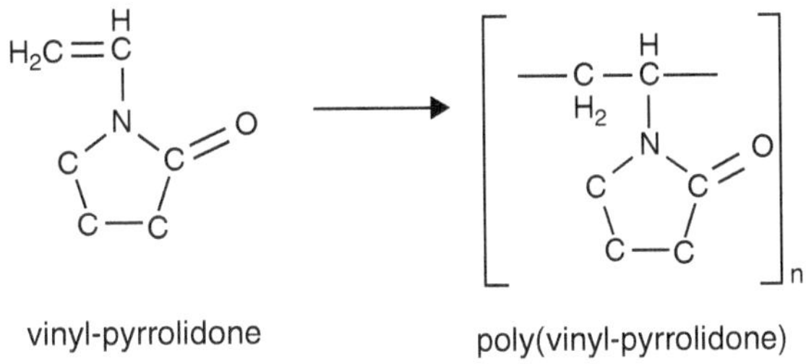

Fig. 4.P1 Polymerization of poly(vinyl-pyrrolidone)

Problem 4.2

The polymerization process of PVP can be described as Fig. 4.P1. What type of polymerization should this process be, and why?

Problem 4.3

Suppose that we swell poly(ethylene oxide) ($\Delta H_f = 8.29$ kJ/mol; $T_f^0 = 66$ °C; and molar volume, 36.6 cm^3/mol) with 10% of benzene (molar volume, 88 cm^3/mol). What will be the new melting temperature if it melted at 66 °C when dry? Will the addition of benzene change phase of the polymer to liquid at room temperature (~25 °C)?

Problem 4.4

Figure 4.P2 shows the bonding process of a widely used polymeric biomaterial called polydimethylsiloxane (PDMS). Please identify the type of this polymeric synthesis and explain for your answer.

Problem 4.5

Poly(decamethylene adipate) with a density of 0.99 g/cm^3 is mixed with various quantities of dimethylformamide with a density of 0.9445 g/cm^3, and the

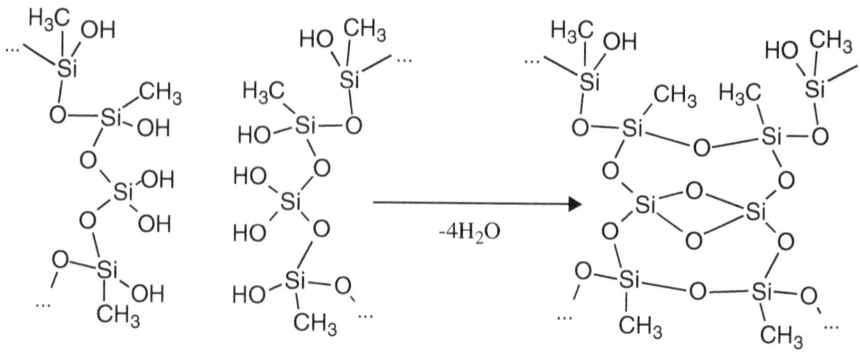

Fig. 4.P2 Polymerization of polydimethylsiloxane

Table 4.P2 Melting temperature of poly(decamethylene adipate)-dimethylformamide mixtures with different volumetric ratios

Volumetric fraction of the polymer	0.078	0.202	0.422	0.603
Melting temperature, T_m (°C)	72.5	66.5	61.5	57.5

Table 4.P3 Molecular data of poly(vinyl chloride)

Weight fraction of the polymer	0.04	0.23	0.31	0.25	0.13	0.04
Mean Mol. Weight ($M_i \times 10^{-3}$)	7	11	16	23	31	39

corresponding melting temperatures are observed as Table 4.P2. Please estimate the melting temperature of the pure polymer.

Problem 4.6

A sample (3.0 g) of poly(vinyl chloride) is composed according to the following fractional distribution (Table 4.P3):

(a) Calculate M_n and M_w of this polymer.
(b) How many molecules per gram are there in the polymer?

Problem 4.7

Show that the mass crystallinity X_m of a polymer sample is given by the expression $A(1 - \rho_a/\rho_s)$, where A depends on the polymer but not on the degree of crystallinity and ρ_a and ρ_s are the densities of the amorphous component and of the sample, respectively.

The values of X_m for two samples of a polymer with densities 1346 kg/m^3 and 1392 kg/m^3 are 10% and 50%, respectively. Now, please calculate the densities of the amorphous and crystalline material and the mass crystallinity of a sample of density 1357 kg/m^3.

Problem 4.8

A particular type of rubber behaves like a neo-Hookean solid with a value of $E = 1.2$ MPa. Calculate:

(c) The force required to extend a long piece of this rubber with a relaxed cross-sectional area of 1 cm^2 to twice its original length
(d) The force required to compress a thin sheet with an area of 1 cm^2 to half its original thickness if this could be done in such a way that it could expend freely in the lateral directions
(e) The true stress for (a) and (b)

Problem 4.9

A cylindrical piece of rubber with 10 cm in length and 2 mm in diameter is extended by a simple force F to a length of 20 cm. If the rubber behaves as a neo-Hookean solid with a Young's modulus of 1.2 MPa, calculate:

(a) The diameter of the stretched cylinder
(b) The value of the nominal stress
(c) The value of the true stress
(d) The value of F

Problem 4.10

The average molecular mass of a sample of commercial grade polypropylene (which is known to contain some polyethylene) is 33,400, and its degree of polymerization is 850. Given the molar masses of ethylene monomer (C_2H_4) and propylene monomer (C_3H_6) are 29.06 g and 42.09 g, respectively, calculate the relative proportion of ethylene monomers in the material.

References and Further Reading

1. Bower, D.I.: An Introduction to Polymer Physics. Cambridge University Press, Cambridge (2002)
2. Shalaby, W.: Polymers as Biomaterials. Springer Science & Business Media, Berlin (2012)
3. Thomas, S., Balakrishnan, P., Sreekala, M.S.: Fundamental Biomaterials: Polymers. Woodhead Publishing, Duxford (2018)
4. Rubinstein, M., Ralph, H.C.: Polymer Physics. Oxford University Press, Oxford (2003)
5. Bhat, S.V.: Biomaterials. Springer, London (2002)
6. Streit, M.R., Schröder, K., Körber, M., Merle, C., Gotterbarm, T., Ewerbeck, V., Aldinger, P.R.: Int. Orthop. **36**(6), 1129–1136 (2012)

Chapter 5
Ceramics

Abstract Ceramics are defined as nonmetallic, inorganic, solid materials. Due to their structures consisting of strong ionic bonds and/or covalent bonds, ceramics tend to be relatively strong and stiff when compared to metals and polymers. However, ceramics are brittle and susceptible to failure by the propagation of pre-existing cracks. For this reason, there are relatively few applications of ceramic matrix composites (compared to those of polymer matrix composites). The important ceramic characteristics include porosity, defects, fracture, and bridging contributions. Common bioceramics include carbons, bioactive glasses, alumina, hydroxyapatite, tricalcium phosphate, and calcium sulfate.

5.1 Overview

The word "ceramic" is derived from the Greek word for clay, *Keramos*. Ceramics are defined as nonmetallic, inorganic, solid materials. They are characterized by their brittleness, good compressive strength, and chemical inertness. The history of the use of ceramics as biomaterials goes back to the start of the nineteenth century, when porcelain was used as an implant in dental crowns. In 1892, Dressman, at Trendelenburg's Clinic, Bonn, West Germany, reported on the use of plaster of Paris, a ceramic, for filling defects in bone. The use of ceramics in human surgery began in the 1980s when the first commercially available hydroxyapatite-coated implants were used in surgeries. Since then, ceramics have been regularly used in orthodontic and orthopedic applications. They are highly sought after for their bioactivity, controlled biodegradation rate, and biocompatibility.

Due to their strong ionically and/or covalently bonded structures, ceramics tend to be relatively strong and stiff when compared to metals and polymers. However, ceramics are brittle and susceptible to failure by the propagation of pre-existing cracks. For this reason, there are relatively few applications of ceramic matrix composites (compared to those of polymer matrix composites). Most of the applications of ceramic matrix composites take advantage of their excellent high-temperature properties. Their moduli are relatively high due to the strong/stiff nature of their ionically or covalently bonded structures. However, the tensile strengths of

R. H. W. Lam, W. Chen, *Biomedical Devices*,
https://doi.org/10.1007/978-3-030-24237-4_5

these composites are moderate due to their inherent susceptibility to internal/inherent microcracks within ceramic matrices. On top of this, different bioceramics display different level of reactivity when implanted inside a host tissue. Bioceramics can be classified into three categories according to their interaction with the host tissue: (1) completely resorbable, (2) surface reactive, and (3) nearly inert. Some examples are the bioinert alumina dental implant, bioactive hydroxyapatite $[Ca_{10}(PO_4)_6(OH)_2]$ coating on a metallic dental implant, and the surface-active bioresorbable bio-glass known as tricalcium phosphate ($[Ca_3(PO_4)_2]$ found in implants.

Bioactive Ceramics
Bioactive materials interact with the surrounding bone and tissues in the body after implantation. Chemical reactions between the bioactive implant and the surrounding body fluids result in the formation of a biologically active layer that is typically chemically equivalent to the mineral phase in bone. They usually have low values of mechanical strength and fracture toughness and are not used as load-bearing materials. They are commonly used as temporary scaffolds which are gradually replaced by tissue, orthopedic implants, and dental implant coatings.

Bioactive glass-ceramics have the ability to bond to the bone in vivo through the formation of a hydroxyapatite surface layer. The mechanism involved in the formation of the hydroxyapatite layer mainly involves surface reactions on the glass surface; this bond is typically stronger than that of the host bone.

Bioresorbable Ceramics
Bioresorbable materials begin to dissolve once placed in the human body and are eventually replaced by advancing tissue. The chemicals produced during the resorption process can be processed through the normal metabolic pathways of the body without any deleterious effect. These types of bioresorbable ceramic materials have been commonly used in implants for the past three decades for the fixation of implants in the skeletal system and make direct and strong chemical bonds with tissue.

Bioinert Ceramics
Bioinert refers to a material that has minimal interactions with its surrounding tissue once implanted. For relatively inert bioceramics, these traits include resistance to corrosion and wear. Bioinert ceramics can also maintain their physical and mechanical properties while in the host. These materials are usually chosen for dental implants to serve as safe and effective long-term replacements for missing teeth.

Bioinert ceramics are highly biocompatible while maintaining high compressive strengths and toughness levels, ranging between three to five times higher than that of compact bone. This allows them to be easily applied in orthodontic and bone situations, ranging from load-bearing parts to surface-coating powders. Ceramics are often used for dental fillings, dental crowns, and dentures due to their relative inertness to body fluids and aesthetically natural appearance. In addition to dental applications, bioinert ceramics are frequently found in femoral heads or the balls of hip implants. They can also be used for bone plates and screws due to their strong and durable mechanical properties. However, ceramic biomaterials also have limitations including low ductility and brittleness. Nonetheless, ceramic biomaterials continue to be sought after as new development cycles allow for the creation of these useful and valuable materials.

5.2 General Characteristics

5.2.1 *Basic Physical Properties*

Ceramics are materials containing a mixture of metals or metalloids bonded with nonmetallic materials through ionic, covalent, and, in some instances, metallic bonds. However, the most common chemical bonds found in ceramic materials are covalent and ionic bonds. Covalent and ionic bonds are strong bonds, giving ceramics high compressive strength and chemical inertness while resulting in low ductility and low tensile strength. Unfortunately, this results in ceramics being extremely brittle under elastic deformation due to the development of cracks, causing quick failure once they develop. This trait is compounded during tensile deformation as ceramics are about 15 times weaker during this type of stress compared to during compressive deformation. Flexural strength levels are at intermediate levels between these two types of stress due to the effects of structural defects such as cracks.

Ceramics are largely made of repeating, ordered, crystalline structures (with the occasional amorphous structure depending on processing conditions). Each crystalline atom has the same orientation, and regularly repeated units of these atoms (known as unit cells) create the overall crystalline lattice. Each unit cell usually reflects the symmetry of the crystalline structure, but ceramics generally have more complex structures with lower amounts of symmetry as displayed by Bravais lattices without cubic or hexagonal crystal structures found in ceramics. This loss in symmetry is a major reason for the brittle nature of many ceramics.

The low levels of plastic deformation in ceramic materials can also be attributed to the reduced mobility of dislocation in their structures due to large Burgers (slip) vectors and ionically/covalently bonded structures which do not respond well to plastic deformation. Therefore, plastic deformation in ceramics is quite poor with very small strain levels between <0.1 and 1%, except for at very high temperatures where superplasticity can occur in fine-grained ceramics with strain levels up to 1000% due to creep. At normal temperatures, ceramics perform better under compression than under tension, probably owing to the fact that the compression tends to close up microcracks that may be present compared to the tension, which on the other hand opens up new cracks, causing small amounts of permanent strain which will affect the overall modulus of elasticity. We can use this fact as a global/scalar measure of change in a ceramic sample by assuming that the original "undamaged" sample has a damage of zero and the catastrophic failure has a damage state of 1. Therefore, to test for an intermediate damage state based on an initial elastic modulus E_o, the damage variable D is equal to $D = 1 - E/E_o$. Ceramics can also undergo plastic deformations by stress-induced phase transformations as seen in partially stabilized zirconia, ZrO_2, when alloyed with CaO, Y_2O_3, or CeO in order to stabilize the high-temperature tetragonal phase down to room temperature. While under monotonic loading, the metastable tetragonal phase of the zirconia undergoes stress-induced phase transformations from the tetragonal to the monoclinic phase, increasing its volume by 4% and toughening the ceramic via transformation toughening. These phase transformations, induced by stress, happen slowly in partially stabilized zirconia, gradually transitioning from the first linear elastic region to the second nonlinear region of the stress-strain curve. Afterward, the second stage ends as the stress-induced transformation spreads to the gauge sections of the specimen, and the final stage results in rapid hardening until failure. The amount of strain until failure during this process is quite small, and the transformation may feature increasing or decreasing stress-strain curves.

The structures of ceramics vary from relatively simple to very complex. The microstructure can be glassy, crystalline, or a combination of both. Ceramics differ from ionic compounds and polymers due to the fact that they do not form discrete molecules or polymeric chains. Due to versatility in bonding and crystalline structures, ceramics display variability in material properties. One example of this variation in properties is found in the expression of bioresorbability in different ceramics. Bioresorbability depends to a large extent on the crystalline structure of ceramics. Highly crystalline ceramics are generally uniformly arranged and take longer to degrade inside the human body. However, they are also stronger, harder, and more resistant to corrosion. Hence, bioresorbability of a ceramic is inversely proportional to its crystalline nature.

Even though ceramics display a wide range of properties, they have certain characteristics that differentiate them from metals and polymers. Next, we will take a brief look at these general characteristics.

- Brittleness: The abundance of ionic and covalent bonds in ceramics tends to make them brittle at room temperature. However, once temperature increases, some ceramics, like glass, lose their brittle nature and can be reshaped into desired shapes.
- Strength: Ceramics display high compressive strength. Hence, when creating implants from ceramics, it is important that the implants are designed to be under compressive stress. The same is not true for their tensile strength, which is relatively lower when compared to the compressive strength. This differentiates ceramics from metals, whose tensile and compressive strengths do not display as much variation.
- Inertness: Most ceramics are chemically inert, making them ideal for use in environments that are highly corrosive to other materials. This is the reason why most laboratory containers are made of glass.

The properties listed in Table 5.1 show the general mechanical characteristics of some representative ceramics. It is possible to find some ceramics which may show different properties from more typical ceramics. For example, some ceramics like hydroxyapatite have low toughness due to their covalent-ionic bonding. However,

Table 5.1 Typical room temperature properties of bioceramics and cortical bone

Property	Magnesium-PSZ	Y-TZP	Al_2O_3	Hydroxyapatite	Cortical bone
Density (g/cm^3)	5.7	6.0	>3.90	3.15	1.7–2.0
Grain size (um)	50	0.5	<7	1	–
Flexural strength (MPa)	600	1200	>400	100	60–160
Fracture toughness ($MPa{\cdot}m^{0.5}$)	9	10	5–6	0.5	2–12
Vickers microhardness (GPa)	12	10	23	4	0.2
Young's modulus (GPa)	200	200	380	100	3–30

such a low toughness property can be improved using hot pressing techniques. Next, we will take a look at various aspects related to the biocompatibility of ceramics.

5.2.2 Porosity

Pores are often the most important defects in polycrystalline ceramics. They act as locations for cracks to emerge, thereby compromising mechanical integrity, especially during tensile loading. The mechanical properties of ceramics can vary due to the variability in the size of pores and the overall level of porosity. On the other hand, interconnected pores are useful for increasing resistance to thermal shock in applications such as filters for hot materials.

Pores in a ceramic may be either interconnected or closed. Apparent porosity is used to measure interconnected pores and determine permeability, the ability of which gases and fluids to seep through a ceramic. The apparent porosity P_{apparent} is determined by weighing the dry ceramic (W_{d}), followed by reweighing the ceramic both when it is suspended in water (W_{s}) and after it is removed from the water (W_{w}). Using units of grams and cm^3

$$P_{\text{apparent}} = \frac{W_{\text{w}} - W_{\text{d}}}{W_{\text{w}} - W_{\text{s}}} \tag{5.1}$$

True porosity P_{true} includes both interconnected and closed pores, which is more reflective of the density ρ of the ceramic, and is

$$P_{\text{true}} = 1 - \frac{W_{\text{d}}}{\rho(W_{\text{w}} - W_{\text{s}})} \tag{5.2}$$

Considering that the porosity is actually the relative empty space in the material body, the elastic modulus reduces with the porosity. It has been widely adopted with the following empirical relationship between elastic modulus E and true porosity P_{true} for materials with a submicron size of the pores:

$$E = E_o\left(1 - 1.9P_{\text{true}} + 0.9{P_{\text{true}}}^2\right) \tag{5.3}$$

where E_o is the elastic modulus of a porosity-free solid or the so-called the intrinsic elastic modulus of the material.

Furthermore, most ceramic materials are made by compacting the ceramic materials in powder form and then heating them to cause fusion of these powders. This fusion process is known as sintering. Some porosity may remain after sintering. The resultant porosity acts as flaws, which produce stress concentration and result in failure at a lower applied stress according to the following relation:

$$\sigma_f = \sigma_S \exp(-n_f P_{\text{true}}) \tag{5.4}$$

where σ_f is the flexural strength, σ_S is the ceramic material strength without porosity, and n_f is an empirical constant.

5.2.3 *Fracture and Bridging Contribution*

Due to the ionic/covalent bonding and large lattice spacing of ceramics when compared to metals and alloys, slip is relatively difficult, contributing to their brittle modes of failure. The fracture surfaces of ceramics can be divided into four regions, as shown in Fig. 5.1: (1) the fracture source/initiation site, with a radius denoted as a_f; (2) a smooth mirror region with a highly reflective surface, (3) a misty region that

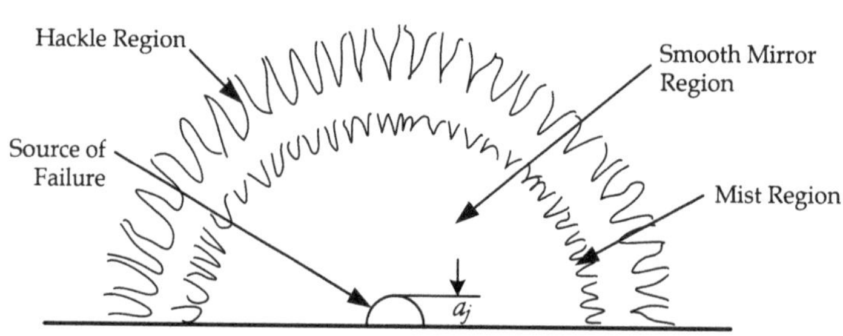

Fig. 5.1 Schematic of four regions on the fracture surfaces of glassy ceramics

contains small radical ridges and microcrack distributions, and (4) a hackle region that contains larger secondary cracks. The hackle region may also be bounded by branch cracks in some ceramic systems. The dimensions of the mirror, mist, and hackle regions have been shown to be related to the fracture stress.

In highly brittle solids with strong covalent/ionic bonds with little point movements of defects, any issues such as pores, inclusions, or gas bubble entrapments can act as areas for the development of cracks. In most brittle solids, residual stress at grain boundary facets creates microcracking during cooling after processing due to a thermal contraction mismatch between nearby phases or grains. These microcracks may become larger cracks, but usually, a variety of microcracks in different sizes will nucleate, ultimately determining behaviors of the solid. Microcracks may form in semibrittle solids, such as MgO, due to dislocation or microstructure interactions. Here, slip may occur when shear stress goes above some critical level on favorably oriented low-index planes. The sources of dislocation are activated, and as an effect, the glide of the dislocation continues. This moves the material away from these sources, but its progress may be hindered by obstacles such as grain boundaries, resulting in pileups and microcrack formation.

Crack growth due to fatigue is hard to monitor in brittle materials, especially when applying tension forces at room temperature. However, stable crack growth due to cyclic changes in load application can happen at these temperatures in single-phase ceramics, transformation-toughened ceramics, and ceramic composites. While the cause for macroscopic fatigue in ductile solids comes from cyclic slip, in ceramics, the main causes for crack growth involve degrading bridging zones behind the crack tip (Fig. 5.2), microcracking, martensitic transformations, and interfacial sliding.

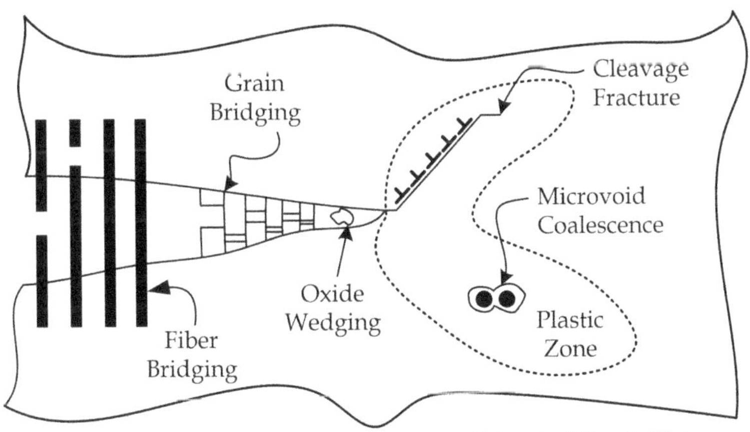

Fig. 5.2 Schematic illustration of intrinsic and extrinsic fatigue damage mechanics

Many ceramic materials show stable cracking under static or quasistatic stress. In these conditions, one can figure the time rate of crack growth, da_f/dt, where a_f is the radius of fracture source as shown in Fig. 5.1, as a function of the linear elastic stress intensity factor, K_E, with a unit of MPa·m$^{1/2}$. The factor K_E predicts the stress intensity near the tip of a crack caused by a remote load or residual stress, which has a unit of stress times the root of a distance. K_E can be approximated by

$$da_f/dt = A_c K_E^{p_c}, \tag{5.5}$$

where A_c and p_c are material constants that are obtained from crack growth experiments under static loading. In general, p_c is found to lie between 2 and 4. Equation 5.5 is known as the Paris-Erdogan law. Stable ceramic crack growth often induces toughening mechanisms such as crack bridging or transformation toughness by stress-induced martensitic phase transformations. The effective stress intensity factor of K_E due to such mechanisms is given by

$$K_E = K_\sigma - K_S, \tag{5.6}$$

where K_σ is the applied stress intensity factor and K_S is the shielding stress intensity factor. Stable crack growth in ceramics is largely due to the inelasticity of crack-tip shielding mechanisms during cyclic loading, possibly involving the cyclic breakdown of bridging ligaments due to wear. At high temperatures, glassy films added to most ceramics during processing often become viscous, leading to the in situ formation of viscous glassy films at grain bridges. This viscoelastic grain bridging can contribute to inelasticity during each fatigue cycle and to the stable growth of cracks.

Intermetallics are compounds made of mixed metals bonded through metallic and covalent bonds. While they are often similar to metals in their strength, they are also brittle due to their inconsistent bonding and nonsymmetric (noncubic) atomic structures. Regardless, some intermetallics are lightweight and heat-resistant, such as gamma-based titanium aluminides (TiAl) and niobium aluminides (Nb_3Al), which are of commercial interest for replacing heavier superalloys.

The fatigue damage characteristics of intermetallics lie somewhere between metals and ceramics. Fatigue crack growth occurs as a result of intrinsic crack-tip processes (partially reversible dislocation motion, deformation-induced twinning, and crack-tip/environmental interactions) and crack wake processes (degradation of bridging zones), as shown schematically in Fig. 5.3. Stable fatigue crack growth occurs in intermetallics due to stress-induced martensitic transformations. The slowest fatigue crack growth rates are typically observed in intermetallics in which crack-tip dislocation processes are predominant. However, most intermetallics crack more quickly than metals and alloys.

Instead of considering the time with loading, a more precise quantification is the number of loading cycles, N_{lc}. The relative contributions from each of these variables may be estimated from the modified two-parameter Paris equation, which is given by

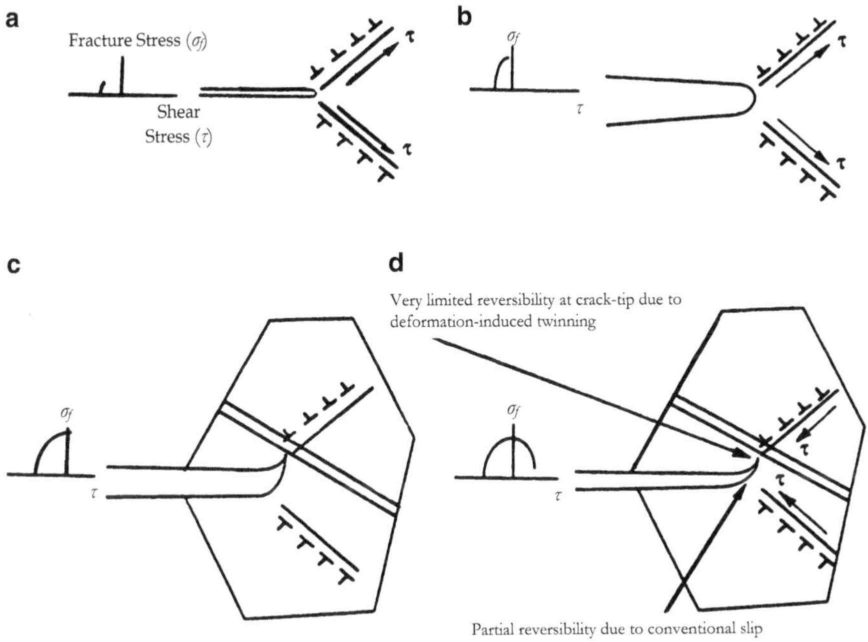

Fig. 5.3 (**a**) An intrinsic crack-tip in material, (**b**) partially reversible dislocation motion, (**c**) deformation-induced twinning, and (**d**) degradation of bridging zones

$$da_f/dN_k = CK_E{}^m T_{max}{}^n \tag{5.7}$$

where n is the exponent of T_{max}, which has a value typically close to 1. T_{max} is approximately equal to the fracture toughness of the material T_f; and m is the exponent of the effective stress intensity factor K_E. Fatigue crack growth in both brittle and ductile solids is affected by both K_E and T_{max}. For ductile metals and their alloys, the exponent m is usually much greater than n. However, brittle ceramics are the opposite with higher values of n than m, and intermetallics often have comparable values of m and n. The relative values of m and n can give useful information about the effects of K_E and T_{max} on the fatigue crack growth process. The above arguments are also consistent with fractographic data that show clear evidence of K_E-controlled ductile (striations) fatigue fracture modes in metals, T_{max}-controlled static fracture modes (cleavage-like fracture) in ceramics, and K_E- and T_{max}-controlled mixed ductile plus static (mostly cleavage-like plus striations in some ductile intermetallics) fracture modes in intermetallics.

Fracture-toughness tests can be made on ceramic specimens to determine the fracture toughness T_f. T_f values for ceramic materials are usually obtained as the following equation:

$$T_{\mathrm{f}} = Y\sigma_{\mathrm{f}}\sqrt{\pi a} \tag{5.8}$$

where T_f is the fracture toughness with a unit of MPa·m$^{1/2}$, σ_f is the fracture stress in MPa, a is the radius of the largest internal flaw in m, and Y is a dimensionless constant with a value about 1. For example, this relation can be used to determine the largest-size flaw that a particular engineering ceramic of a known fracture toughness and strength can tolerate without fracture.

Grain and fiber bridging (Fig. 5.2) of the crack surface behind the crack-tip by a strong discontinuous reinforcing phase which imposes a closure force on the crack is, at times, accompanied by pullout of the reinforcement. The extent of the pullout length of brittle, discontinuous reinforcing phases is generally quite limited due both to the short length of such phases and the fact that bonding and clamping stresses often discourage pullout. However, pullout cannot be ignored, as even short pullout lengths contribute to the toughness achieved, which is the ability of a material to absorb energy and plastically deform without fracturing with a unit of absorbed energy per unit volume.

Crack deflection due to reinforcements has been suggested to contribute to fracture resistance. Often, shear-related crack deflections are limited in length and angle and are best considered as means of breaking the reinforcement-matrix interface. Such interfacial breaking helps achieve frictional bridging (bridging by elastic ligaments which are partially debonded from the matrix) and pullout processes. Frictional bridging elastic ligaments can contribute significantly to fracture toughness and the stress intensity increment. Here, we will focus on toughening due to crack bridging by various brittle reinforcing phases, where the reinforcement simply bridges the crack surfaces and effectively reduces the crack driving force. The bridging contribution to the fracture stress intensity factor is given by

$$K_{\mathrm{cb}} = \sqrt{E_{\mathrm{c}}(J_{\mathrm{m}} + J_{\mathrm{cb}})}, \tag{5.9}$$

where K_{cb} is the overall fracture stress intensity factor of the composite, which can be considered as the shielding stress intensity K_S as previously mentioned in Eq. 5.7; E_c is the modulus of elasticity at the crack-tip, J_m is the matrix fracture energy per unit area; and the term J_{cb} corresponds to the energy change per unit area due to the bridging process. Such energy change associated with the bridging process is a function of the bridging stress divided by traction, σ_b, and the crack opening displacement, d_c, and is defined as

$$J_{\mathrm{cb}} = \int_0^{d_{\max}} \sigma_{\mathrm{b}} d d_{\mathrm{c}}, \tag{5.10}$$

where d_{max} is the maximum displacement at the end of the zone. One can equate the maximum crack opening displacement at the end of the bridging zone, d_{max}, to the tensile displacement in the bridging brittle ligament at the point of failure:

$$d_{\max} = \varepsilon_f^l l_{db}, \tag{5.11}$$

where ε_f^l represents the strain to failure of the whisker and l_{db} is the length of the debonded matrix-whisker interface (Fig. 5.4). The strain to failure of the whisker can be defined as

$$\varepsilon_f^l = \sigma_f^l / E^l, \tag{5.12}$$

where E^l is the elastic modulus of the reinforcing phase and σ_f^l is the fracture strength of the whisker.

On the other hand, the interfacial debonded length depends on the fracture criteria for the reinforcing phase versus that of the interface and can be defined in terms of fracture stress or fracture energy:

$$l_{db} = k_{db} r^l \gamma^l / \gamma^i, \tag{5.13}$$

where γ^l represents the fracture energy (unit, J/m^2) of the bridging ligament; γ^i represents the fracture energy of the reinforcement-matrix interface; and k_{db} is a constant. The critical debonded length just ahead of the main crack tip is defined as $k_{db} = 1/6$.

In Eq. 5.11, it can be noted that tensile strain displacement occurs in the bridging reinforcement, so the maximum crack opening displacement will occur at the end of the bridging zone and increase as the debonded length/gauge length of the reinforcing ligament increases. Equations 5.12 and 5.13 also show that increasing the reinforcing phase strength and/or enhancing the interface debonding will contribute to greater tensile displacement within the reinforcing ligament. Increases in the crack opening displacement supported by the bridging zone will enhance the toughening achieved by such reinforcements. Therefore, debonding of the matrix-reinforcement interface can be a key factor in the attainment of increased fracture toughness in these elastic systems. Indeed, in ceramics reinforced by strong ceramic whiskers, debonding is observed only in those systems which exhibit substantial toughening.

If we assume a simple case of a bridging stress which increases linearly from zero (at the crack-tip) to a maximum at the end of the bridging zone and immediately decreases to zero, Eq. 5.10 can be reduced to $J_{cb} = \sigma_{\max} d_{\max}/2$. The maximum

Fig. 5.4 Schematic of debonding whisker

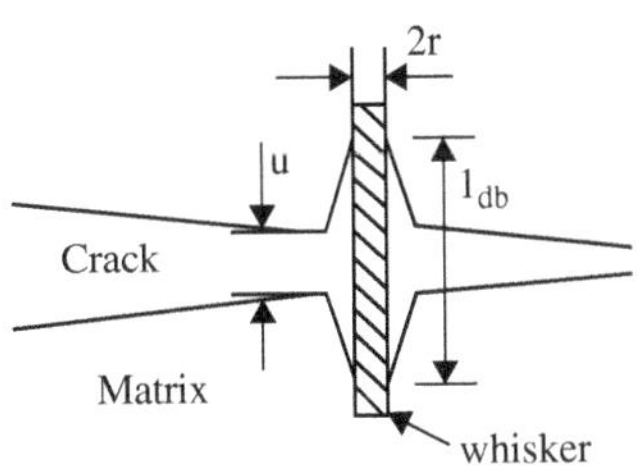

closure stress, σ_{max}, imposed by the reinforcing ligaments in the crack-tip wake is the product of the fracture strength of the ligaments σ_f^{l} and the real fraction of ligaments intercepting the crack plane:

$$\sigma_{max} \approx \sigma_f^{l} l_{db}/r^{l} \tag{5.14}$$

where r^{l} is the radius of the ligaments, which have large aspect ratios (e.g., $l_{db}/r^{l} > 30$ for whiskers). Recalling Eq. 5.11, reinforcement by frictional bridging introduces a change in energy per unit area equal to

$$J_{cb} \approx \frac{r^{l}}{2E^{l}}\left(\frac{\sigma_f^{l}\gamma^{l}}{6\gamma^{i}}\right)^2. \tag{5.15}$$

From these results, the stress intensity contribution from frictional bridging by the reinforcing phase in the crack-tip wake, K_{cb}, can be further estimated by recalling Eq. 5.9. For the case $J_{cb} \gg J_m$

$$K_{cb} \approx \sqrt{E_c J_{cb}} \approx \frac{\sigma_f^{l}}{6}\frac{\gamma^{l}}{\gamma^{i}}\sqrt{\frac{r^{l}}{2}\frac{E_c}{E^{l}}} \approx K_S. \tag{5.16}$$

To reveal the role of key factors during the crack propagation process, we further impose some large approximations, by considering that the material has homogenous material properties and stress distributions, we let $\gamma^{l} \sim \gamma^{i}$, $E_c \sim E^{l}$, and σ_f^{l} is equal to the ceramic material strength σ_S. Hence, combining Eqs. 5.5, 5.6, and 5.16 can then induce a highly simplified relation:

$$\frac{da_f}{dN_k} \approx CT_f\left(K_\sigma - \frac{\sigma_S r^{l}}{6\sqrt{2}}\right)^m \tag{5.17}$$

Furthermore, if we further consider that r^{l} is roughly equal to the average separating distance of pores within the material with a porosity of P_{true} and an average pore size close to $(4/3)\pi a^3$, this means the cracks in a material body would not propagate when the applied stress intensity factor is not exceeding a threshold level:

$$K_\sigma \ll \frac{\sigma_S a}{6\sqrt{2}}\left(\frac{4\pi}{3(1-P_{true})}\right)^{1/3}. \tag{5.18}$$

This relation also shows that the resistance to fracture could be increased by reducing the material porosity P_{true}. Notably, this relation also applies to metals and alloys.

5.2.4 Slip Dislocation

Ceramics and metals share many of the dislocation modes. In particular, the common slip systems appear in single-crystalline materials, e.g., BCC and FCC. There is no fundamental difference between the slip systems in ceramics (or other crystalline materials) and metals, since all the aforementioned structures appear in both materials.

Yet, materials with ionic bonding, including many ceramics such as MgO, also are resistant to slip. It is because movement of a dislocation must disrupt the charge balance around the anions and cations, requiring that bonds between anions and cations be broken altogether. Additionally, during slip, ions with a like charge must also pass close together, causing repulsion. Finally, the repeat distance along the slip direction, or the Burgers vector, is larger than that in metals and alloys. Again, brittle failure of ceramic materials typically occurs due to the presence of flaws such as small pores before the applied level of stress is sufficient to cause dislocations to move.

An edge dislocation (Fig. 5.5) is subjected to a shear stress τ that acts parallel to the Burgers vector and perpendicular to the dislocation line. A component of the shear stress must act parallel to the Burgers vector in order for the dislocation to move. During slip, a dislocation moves from one set of surroundings to an identical set of surroundings. The "line" stress, also well-known as the Peierls-Nabarro stress, is required to move the dislocation from one equilibrium location to another, with the following relation:

$$\tau = c\exp(-kdlb). \tag{5.19}$$

where τ is the shear stress required to move the dislocation; d is the interplanar spacing between adjacent slip planes; b is the magnitude of the Burgers vector, and both c and k are constants for the material.

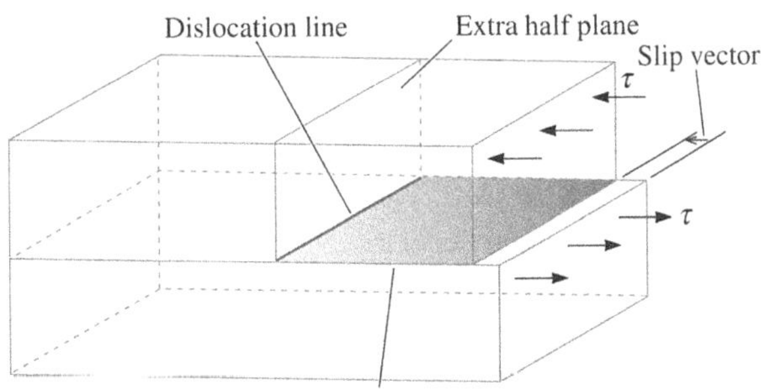

Fig. 5.5 Schematic of the dislocation line, slip plane, and slip (Burgers) vector for an edge dislocation

5.2.5 Biocompatibility

Biocompatibility is the ability of an implanted device to perform its desired action inside the host without any adverse reaction. Biocompatibility inside a human body requires that the implant be nontoxic, non-carcinogenic, non-allergenic, and noninflammatory. Ceramics are relatively inert to various bodily fluids and, hence, have excellent resistance to biodegradation. Biocompatibility of ceramics can be analyzed by looking at the corrosion, wear, and tear of an implant and the immunological responses that may arise in the presence of the biomedical device once inside the body.

Corrosion involves the deterioration of a material as it interacts with its environment. Corrosion generally involves oxidation and reduction. Since ceramics are already in their oxidative states, they are already fairly resistant to corrosion. However, bio-erosion, the degradation of bioceramics, can still happen due to chemical and mechanical processes in the microenvironment.

Biodegradation is caused by the chemical breakdown of a material mediated by its surrounding environment. Degradation could be either controlled or uncontrolled. Uncontrolled degradation is dependent on two factors: the mechanical environment and the porosity of the ceramic. Factors relating to the mechanical environment could refer to stress-induced degradation, such as when the material is under constant tension. If a crack forms, tensile stress could result in further dissolution and eventual fracture. As for ceramic porosity, pores introduce more areas for stress to affect and could promote the propagation of cracks. On the other hand, controlled degradation is sometimes desirable for tissue engineering and drug delivery. Controlled degradation could be used for temporary implants that do not need to be removed using surgical means. The rate of degradation can be affected by the chemical susceptibility of the material, amount of crystallinity, the material's surface area to volume ratio, and implant site.

Bioactive ions released from bioceramics could trigger inflammatory responses within the human body. The ultimate functional success of an implant depends invariably on the host's tissue response. If a biomaterial induces an immune reaction, it might reduce healing or even result in rejection by the body. Hence, a careful evaluation of the material's biocompatibility is essential to prevent potential inflammatory responses, particularly for bioactive and bioresorbable materials.

It is desirable that the products of degradation be transported out of the body through metabolic pathways without causing deleterious effects on the body or accumulating within the body. For example, in the case of dental implants, zirconia, the ceramic alternative to titanium, has been vigorously studied for its associated inflammatory responses. It has been found that zirconia has exceptional biocompatibility, with inflammatory responses being less prominent for zirconia implants as compared to those of titanium.

5.3 Common Bioceramics

5.3.1 Basic Bioceramics

Carbons

All of the carbon materials currently of interest for use in medical devices (Fig. 5.6) have the quasicrystalline turbostratic structure, a crystal structure in which the basal planes have slipped out of alignment. They are made in many allotropic forms, specifically crystalline diamond, graphite, nanocrystalline glassy carbon, and quasi-crystalline pyrolytic carbon.

Carbons also fall under the category of inert materials and have generally good biocompatibility. Graphite is one of the widely used forms of carbon for biomedical applications and has two types of bonding: covalent bonding of atoms in hexagonal layers and Van der Waals interactions. Van der Waals interactions are weak compared to covalent bonds, and they usually exist between layers. Hence, these weak bonds allow the layers to slide across each other. Depending on the carbon type, the mechanical properties can vary. For instance, graphite has a density of ~1.5–1.9 g/ml, an elastic modulus of 24 GPa, and a compressive strength of 138 MPa. Isotropic pyrolytic carbon has a density of 1.5–2.0 g/ml, an elastic modulus of 28 GPa, and a compressive strength of 517 MPa and is widely used in vascular implants. Poly-lactic-acid-carbon-fiber composites are used in artificial tendons and ligaments. Glassy (vitreous) carbon has a density of 1.5 g/ml, an elastic modulus of 24 GPa, and a compressive strength of 172 MPa.

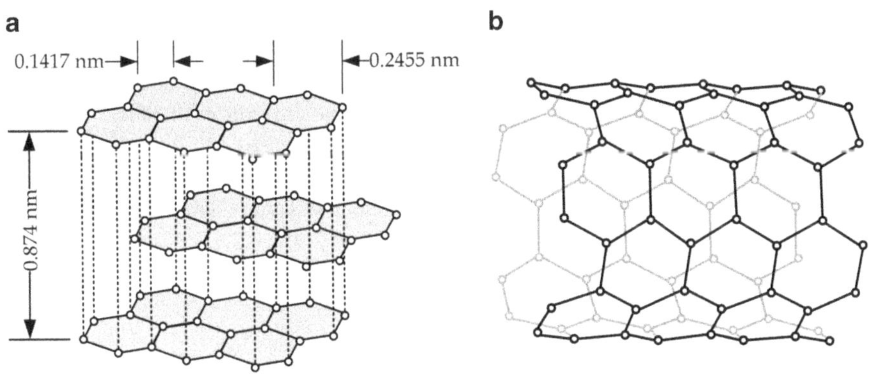

Fig. 5.6 Atomic structures of (a) graphite and (b) carbon nanotube

Bioactive Glasses and Glass-Ceramics

Glass-ceramics are crystalline materials obtained by controlled crystallization of an amorphous parent glass. This process requires very accurate and specific chemical compositions, heat treatment, and nucleation to grow a crystal of uniform size. A critical step in producing a glass-ceramic is to ensure that crystallization does not occur during the cooling process from the forming temperature. If a glass cools too slowly, uncontrolled nucleation and growth of the crystals will happen, resulting in devitrification. To prevent this, modifying oxides can be added to glass to achieve more uniform crystallization, even at slow cooling rates, similar to adding alloying elements to steel to improve metal properties. Glass-ceramics are generally very porous. This decreases the material's elastic modulus, strength, hardness, and thermal conductivity. However, high porosity can also be a desirable trait, particularly for non-loading implant applications, such as scaffolding for tissue engineering. It allows for vascularization which promotes tissue growth. Another benefit is that the material's porosity can be varied to obtain a large range of elastic moduli to match that of bone. There are two types of glass-ceramics commonly used in biomedical applications: bioglass and ceravital.

Bioglass is composed of SiO_2, CaO, Na_2O, and P_2O_5 and is formed through nucleation and crystallization processes which require acute temperature controls. Nucleation of bioglass in the crystalline phase is controlled in two ways. First, oxidizing agents, such as TiO_2, react with each other to form phases to act as nucleation sites. Next, the material is heated at low temperatures to first maximize the rate of nucleation and then at higher temperatures to maximize the rate of crystal growth. This two-step process can crystallize as much as 99% of the bioglass in a development cycle. The structure of bioglass allows for it to have high strength and toughness while having a low thermal expansion coefficient and high-temperature corrosion resistance.

Glasses behave as a viscous liquid above its glass transition temperature. Under stress at such temperature, groups of silicate atoms slide on each other, allowing permanent deformation. The viscosity of the glass decreases and facilitates viscous flow within the material. Recalling Eq. 4.12, the equation relating viscosity to temperature for viscous flow of bioglass above its glass transition temperature is

$$\mu = \mu_0 \exp\left(-\frac{C_1^g (T - T_g)}{C_2^g + T - T_g}\right) \tag{5.20}$$

where μ is the viscosity; μ_0 is the reference at the glass transition temperature; T is the temperature; T_g is the glass transition temperature; and $C_1{}^g$ and $C_2{}^g$ are constants. Alternatively, an approximation of Eq. 5.20 would be

$$\mu = \mu_0 e^{E_f/RT} \tag{5.21}$$

where E_f is the molar activation free energy for viscous flow and R is the universal molar gas constant. The approximation of the above equation is more applicable at very high temperature, i.e., T is significantly larger than T_g. In this case, the material viscosity is inversely proportional to the atomic diffusivity, given that a less "viscous" liquid should be less resistant to the diffusion process. During the atomic diffusion process, an atom should possess a kinetic energy level above a threshold level, which is the molar activation energy E_f, such that the atom can "jump" to a neighboring gap between other atoms. The occurrence probability and the total count of the atomic jumps both increase with the material temperature and so does the atomic diffusivity D. D has a relation similar to the Boltzmann energy distribution, i.e., $D = D_o \exp(-E_f/RT)$ with D_0 as the diffusivity at the absolute zero temperature.

Glass-ceramics can be produced with minimal porosity. Glass at a moderate temperature can be considered to be a supercooled liquid with an amorphous structure. For instance, glasses are often made of silica-alumina. By introducing an appropriate modifier and with careful heat treatment, glass-ceramics can be produced with 90% of the materials crystallized and with grain sizes of ~0.1–1 μm. The remaining 10% glass-ceramics tend to be stronger and have higher thermal conductivity, implying that they are less susceptible to thermal shock.

Glass-ceramics can be designed to directly bond with bone for up to 30 days. They accomplish this by forming a calcium phosphate and SiO_2-rich film layer on their surface. Glass-ceramics are generally used in orthopedic and dental implant coatings, facial reconstructions, bone graft substitute materials, and percutaneous access devices.

Alumina

High-purity alumina bioceramics provide an alternative to surgical metal alloys. Their high hardness, low friction coefficient, and excellent corrosion resistance offer very low wear rates at articulating surfaces in orthopedic applications. Medical-grade alumina has very small grain size (<7 μm) and a narrow grain size distribution capable of inhibiting static fatigue and slow crack growth while under load. However, unless its surface is modified or used directly in articulating areas, alumina has a fundamental limitation as an implant material in that, like other "inert" biomaterials, a nonadherent fibrous membrane may develop at the interface which can result in interfacial failure.

The only thermodynamically stable crystallographic modification of alumina is α-Al_2O_3 or corundum. Corundum has a hexagonal crystal lattice with the cell parameters $a = 4.754$ Å and $c = 12.99$ Å as shown in Fig. 5.7. On approaching equilibrium, the crystal lattice becomes more ordered until the stable modification is formed. All metastable modifications have a partially deformed, closely packed hexagonal O_2 sub-lattice with various configurations of interstitial aluminum atoms. The type of metastable polymorph influences the morphology of the formed α-Al_2O_3 particles.

Alumina is a composite material, meaning it is a mixture of two or more phases bonded together so that stress transfer occurs across the phase boundary. Typically, composite materials are designed to provide a combination of properties that cannot be achieved with single-phase materials. With these materials, there is the potential for producing a lightweight, high-strength structural component with anisotropic

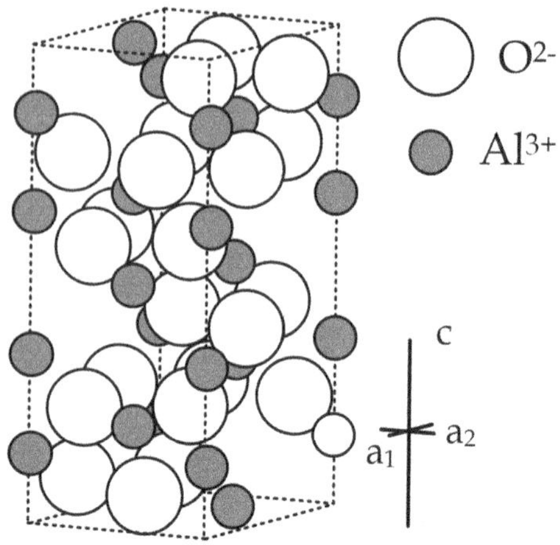

Fig. 5.7 Crystal structure of α-Al_2O_3

Table 5.2 Mechanical properties of medical-grade alumina

Property	ASTM F 603-83	Frialit-biolox-bioceramics
Density (g/cm^3)	3.9	3.98
Alumina content (%)	99.5	99.9
SiO_2 content (%)	0.1	0.05
Average grain size (μm)	7	2.5
Elastic modulus (GPa)	380	380

properties similar to those of natural bone. For example, two types of medical-grade alumina are provided in Table 5.2. Alumina-based materials can be divided roughly into three groups:

Solid-state sintered alumina: This type is a nanocrystalline material with superior mechanical properties (e.g., hardness and wear resistance) and potential transparency to visible light. The prerequisites for the successful preparation of sub-micrometer alumina with these desired properties are sufficiently fine-grained and reactive nano-powders of high purity. Their synthesis and characterization have, therefore, been intensively pursued during the past years.

Liquid-phase sintered alumina: Despite the fact that liquid-phase sintered alumina represents a substantial part of industrially produced alumina-based materials and the amount of research surrounding it, many unknowns about this material remain. Although sintering additives have a profound influence on mechanical properties (especially on hardness, creep, wear resistance, and to a certain extent bending strength and fracture toughness), there remains some confusion as to how individual additives or their combinations influence the microstructure and behavior of alumina-based materials. These sintering additives include mostly silica, alkali oxides, alkali earth metal oxides, and combinations thereof. Enhancing alumina with rare earth oxides continues to be studied with the aim of understanding and enhancing the creep resistance of polycrystalline alumina.

Alumina-based composites: These comprise especially zirconia-toughened alumina (ZTA) and alumina-based nanocomposites with non-oxide second phases such as SiC or TiC.

Alumina has low fracture toughness and tensile strength, but it also has very high compression strength. Its low wear resistance limits its applications as a bioceramic. Nonetheless, alumina is an ideal material for orthopedic load-bearing applications such as joint replacement components and dental implants. Alumina is also applied as coatings for tissue ingrowth (cardiovascular, orthopedic, dental, and maxillofacial prosthetics for maxillofacial reconstruction). Alveolar ridge augmentations and otolaryngological applications are other examples.

5.3.2 *Resorbable Ceramics*

Ceramics that have the property of being able to be absorbed in the body after their function is performed are called bioresorbable ceramics. These are materials that, like bones, are replaced and reabsorbed by the body after they have assisted in repair. They form a very important category of biomaterials since they are used to create devices that don't need surgery for removal. Some of the most important types of bioresorbable ceramics are discussed in the following subsections.

Hydroxyapatite
Hydroxyapatite, sometimes referred to as "HA," is a naturally occurring mineral that is highly biocompatible because of its similar mineral composition as bone. Synthetic HA $Ca_{10}(PO_4)_6(OH)_2$ is prepared through the reaction of $Ca(OH)_2$ and H_3PO_4 in an aqueous solution. It is a bioresorbable material; however, the resorption process takes up to decades. This could be attributed to its more uniform crystalline structure and homogeneous composition as compared to human bone. There is a wide variation in the reported mechanical properties of HA. However, HA generally has a higher elastic modulus than mineralized tissues. For example, $Ca_{10}(PO_4)_6(OH)_2$ HA has an elastic modulus of ~50–144 GPa which is stiffer than many human bone materials such as enamel (74 GPa) and cortical bone (~30 GPa).

HA can be used as a coating for chemical bonding with human bone or as dental implants. The composites of HA, like HA-autogenous bone composite and HA-PLA (polylactic acid), can be used for alveolar ridge augmentations. HA-PLA can also be applied for maxillofacial reconstruction. For example, by developing porous coatings of bone-like HA ceramic, it is possible to make titanium implants more bioactive, promoting the growth of bone around the implant sites.

Tricalcium Phosphate
Tricalcium phosphate (TCP) is another ceramic material with a composition very similar to hydroxyapatite. TCP is generally more soluble than hydroxyapatite because it has a less ordered crystalline structure. Bio-resorption can be up to 10–20 times faster in the case of TCP. Due to these characteristics, it is typically

used as an osteo-conductive scaffold where it supports bone growth on a surface during the healing process. TCP is also used for temporary bone space filling and periodontal pocket obliteration. One drawback of calcium phosphate ceramics is their rather complicated fabrication process and difficulty in being to mold into a desired shape.

Calcium Sulfate

Calcium sulfate is more commonly known as plaster or gypsum. It is extremely useful for eliminating dead space and is widely used as a bone void filler or as a mean to transport antibiotics. Its solubility in a host can be controlled accurately for weeks. Hence, it is only ideal for short-term scaffolding that does not require bone growth.

Problems

Problem 5.1

The specific gravity of Al_2O_3 is 3.96 g/cm^3. A ceramic part is produced by sintering alumina powder. It weighs 80 g when dry, 92 g after it has soaked in water, and 58 g when suspended in water. Calculate the apparent porosity and the true porosity.

Problem 5.2

A sintered zirconium oxide (ZrO_2) ceramic has a true porosity of 28%, and the volume ratio between the material part and the bulk porous ceramic volume is 0.5. If the weight after soaking in water is 760 g, what is the dry weight of the ceramic? The pure density of ZrO_2 is 5.68 g/cm^3.

Problem 5.3

The density of Al_2O_3 is 3.96 g/cm^3. A ceramic part is produced by sintering alumina powder. It weighs 80 g when dry, 92 g after it has soaked in water, and 58 g when suspended in water. Calculate the apparent porosity and the true porosity.

Problem 5.4

Silicon carbide (SiC) has a density of 3.1 g/cm^3. A sintered SiC part is produced, occupying a volume of 500 cm^3 and weighing 1200 g. After soaking in water, the part weighs 1250 g. Calculate the bulk ceramic density, the true porosity, and the volume fraction of the total porosity that consists of closed pores, which cannot contain water inside.

Problem 5.5

A 96% silica glass has a viscosity of 10^{12} kg/(ms) at its annealing point of 940 °C and a viscosity of 10^{7} kg/(ms) at its softening point of 1470 °C. Calculate the activation energy in kJ/mol for the viscous flow of this glass in this temperature range.

Problem 5.6

We can induce chemical resistance in a glass if we include more B_2O_3 into silica (SiO_2). To ensure the proper glass-forming process, the O:Si ratio in the resultant material must be no more than 2.5. Given that the molar mass of B_2O_2 and silica are 69.62 g/mol and 60.08 g/mol, respectively, design the glass in terms of the mole fraction of B_2O_3, denoted as *f*.

Problem 5.7

A fully stabilized, cubic polycrystalline ZrO_2 sample has a fracture toughness of $T_f = 3.8$ MPa·m$^{1/2}$. Assume $Y = \pi^{1/2}$. The sample is under a four-point bend test.

(a) If the sample fails at a stress of 450 MPa, what is the size of the largest surface flaw?

(b) The same test is performed with a partially stabilized ZrO_2 specimen. This material is transformation-toughened and has a $T_f = 12.5$ MPa·m$^{1/2}$. If this material has the same flaw distribution as the fully stabilized sample, what stress should be applied to cause sample failure?

Problem 5.8

(a) A sketch of a dislocation in magnesium oxide (MgO), which has a lattice parameter *a* of 0.396 nm, is shown in Fig. 5.P1. Determine the length of the Burgers vector (the arrow next to "b").

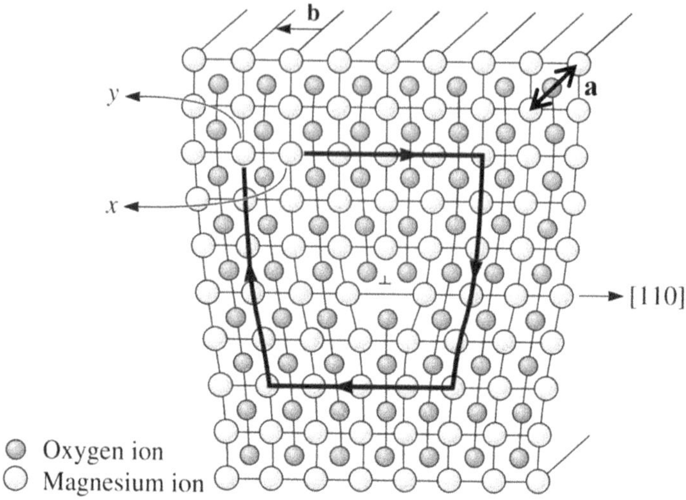

Fig. 5.P1 Atomic configuration of a line defect in magnesium oxide

(b) Based on the principles of the bonding of MgO and Mg and the Peierls-Nabarro stress described in Eq. 5.19, explain why slip in MgO is much more difficult than in Mg from the perspective of the Burgers vector.

References and Further Reading

1. Smith, W.F., Hashemi, J.: Foundations of Materials Science and Engineering. Mcgraw-Hill Publishing, Boston (2006)
2. Arturo, P., Cenni, E., Ciapetti, G., Granchi, D., Savarino, L., Stea, S.: Integrative Biomaterials Science. Springer, New York (2002)
3. Wunderlich, W.: Ceramic Materials. Sciyo, Rijeka (2010)
4. Riedel, R., Chen, I.-W.: Ceramics Science and Technology 2: Properties. Wiley-VCH, Weinheim (2010)
5. Bhat, S.V.: Biomaterials. Springer, London (2002)
6. Askeland, D.R., Wright, W.J.: Essentials of Materials Science & Engineering. Cengage Learning, Stamford (2013)
7. Dowling, N.E.: Mechanical Behavior of Materials: Engineering Methods for Deformation, Fracture, and Fatigue. Pearson, Boston (2012)
8. Pelleg, J.: Mechanical Properties of Ceramics. Springer, Cham (2014)
9. Soboyejo, W.: Mechanical Properties of Engineered Materials. Marcel Dekker, New York (2002)
10. Budiansky, B., Hutchinson, J.W., Evans, A.G.: Matrix fracture in fiber-reinforced ceramics. J. Mech. Phys. Solids. **34**, 167–189 (1986)

Part II
Manufacturing Processes

Chapter 6
Common Manufacturing Process

Abstract The main goal of the manufacturing process is to generate the defined geometry of a chosen material for construction. In order to achieve the successful development and manufacturing of a biomedical device, all the material properties, manufacturing processes, and shape features of the device should be considered altogether with optimized performance and compatibility in mind. Any factors that might contradict with each other should be eliminated or replaced before starting the manufacturing process. In this chapter, we go through the common manufacturing processes for metals, polymers, and ceramics, classified into primary, secondary, and tertiary processes. The process compatibility between different materials and manufacturing processes is also summarized.

6.1 Introduction

In order to achieve the successful development and manufacturing of a biomedical device, all the material properties, manufacturing processes, and shape features of the device should be considered altogether with optimized performance and compatibility in mind. Any factors that might contradict with each other should be eliminated or replaced before starting the manufacturing process. Otherwise, it will be costly to backtrack on the development process.

The main goal of the manufacturing process is to generate the defined geometry of a chosen material for construction. Based on the level of precision in shaping a material, manufacturing processes can be classified into three classes: primary, secondary, and tertiary processes. The primary processes shape raw materials to a desired form or shape close to the target geometry, but some detailed shape features can be missing in the processed materials. The secondary processes define detailed geometric features of the materials generated after the primary processes such that the material shape and dimensions match the specifications. The tertiary processes include surface treatments and assembly of different parts into the final functional product.

R. H. W. Lam, W. Chen, *Biomedical Devices*,
https://doi.org/10.1007/978-3-030-24237-4_6

6.2 Primary Processes for Metals and Alloys

6.2.1 *Rolling*

Metal rolling is one type of primary deformation process which reshapes a metallic material body. Generally, deformation refers to any process that changes the shape of a solid material without changing its phase. Permanent deformation of materials will occur during this process and transform the simple initial shape to the desired shape by applying sufficiently large stresses over the material body. Notably, excessive stresses generated at local regions over the material body may cause defects or even cracks in the processed body. Deformation processes are of utmost importance in today's industrial mass production operations as these processes enable efficient production and do not rely on lengthy metallurgical processes.

In rolling, a metal slab passes through a pair of constantly rotating rolls and is permanently deformed in the roll zone where the rolls compress the metal slab. The friction generated due to the spinning of the rollers moves the metal slab forward. The body thickness is reduced and equalized by the compressive forces during the rolling process (Fig. 6.1a), while the volume is largely maintained. As shown in the half block model (Fig. 6.1b), the metal block will be drawn in between the rolls with a normal force F_n and a friction force F_f with the relation $F_f = \mu_f F_n$, where μ is the friction coefficient, and hence, $\tan(\alpha) = \mu_f$. Furthermore, $\cos(\alpha) = 1 - T_R/(2R)$, where R is the radius of the roller and T_R is the thickness reduction of the block after rolling. Along with this, $\sin(\alpha) = [1 - \cos(\alpha)]^{1/2} \approx (T_R/R)^{1/2}$. Hence, the maximum feasible T_R is $\mu_f^2 R$. In addition, the rolling force (F_R) can be estimated by the forming pressure P_F length of contact L_C and width of the workpiece W: $F_R = P_F L_C W$. P_F can be set based on the material yield strength σ_{yield}. Here, we adopt the von Mises criterion based on the maximum distortion theory on all stresses in three dimensions. Yielding with perfect plasticity, which induces a Poisson's ratio of $\nu = 0.5$, is obtained if

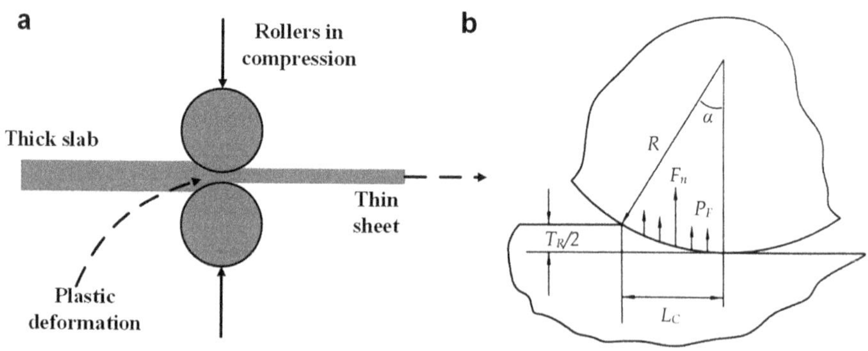

Fig. 6.1 (**a**) Basic shape rolling operation. (**b**) Force and geometry of fully engaged rolling

$$\left(\sigma_x - \sigma_y\right)^2 + \left(\sigma_y - \sigma_z\right)^2 + \left(\sigma_z - \sigma_x\right)^2 = 2\sigma_{\text{yield}}^2 \tag{6.1}$$

We may consider that the rolling process would induce a compressive stress on the plate surface along the y-direction, while the plate is fed along the x-direction. Then, if we wish to configure the rolling process such that there is no deformation along the z-direction, $\sigma_z = \nu(\sigma_x + \sigma_y)$ according to Eq. 2.17. Plugging this relation in Eq. 6.1, we can obtain that

$$\sigma_x - \sigma_y = 2\sigma_{\text{yield}}/\sqrt{3} \approx 1.155\sigma_{\text{yield}} \tag{6.2}$$

To ensure a sufficient F_{R}, we may set $P_{\text{F}} = 1.2\ \sigma_{\text{yield}}$. Further considering that $L_{\text{C}}^2 = RT_{\text{R}}$, $P_{\text{F}} = 1.2\ \sigma_{\text{S}}W(RT_{\text{R}})^{1/2}$, and the required torque τ_{R} of the roller is $\tau_{\text{R}} = F_{\text{R}}L_{\text{C}}/2$. Thus

$$\tau_{\text{R}} = 0.6\sigma_{\text{S}}WRT_{\text{R}} \tag{6.3}$$

Rolling can be generally classified into hot rolling and cold rolling depending on the operation temperature. Hot rolling operates at a temperature which is higher than the recrystallization temperature of the metals/alloys involved; there will also be dislocation, propagation, and softening during this process. The maximum temperature for hot rolling is typically limited to 50 °C below the melting temperature of the metal/alloy being used. An ultrahigh operating temperature can cause melting or excessive oxidation, affecting the processed material shape and characteristics. On the other hand, in cold rolling, the materials are first rolled at room temperature followed by an annealing process to temper the rolling process. It is useful for boosting the yield strength and desired grain orientation of the material, but the ductility will simultaneously decrease as well.

Comparatively, hot rolling machines can work with larger material bodies since heated material bodies with reduced stiffness require smaller rolling forces and corresponding stress generation. The production cost of hot rolling is lower since the process does not involve reheating of the rolled products. Ductility of the processed product is better with higher temperatures as the material body can deform to a larger extent, and hence, more complicated shapes can be achieved. On the other hand, cold rolling causes no oxidation, and the rolled body has better surface finish and smaller dimensional tolerance. The resultant strength is higher due to the heavier grain dislocation (grain hardening). If lubrication is applied during cold rolling, a thinner wall can be generated.

6.2.2 *Drawing and Extrusion*

Drawing is a primary deformation process involving squeezing out the product through a hole using a pulling force (tensile strength) as shown in Fig. 6.2a. Although the generated shape of the product from this drawing process is limited,

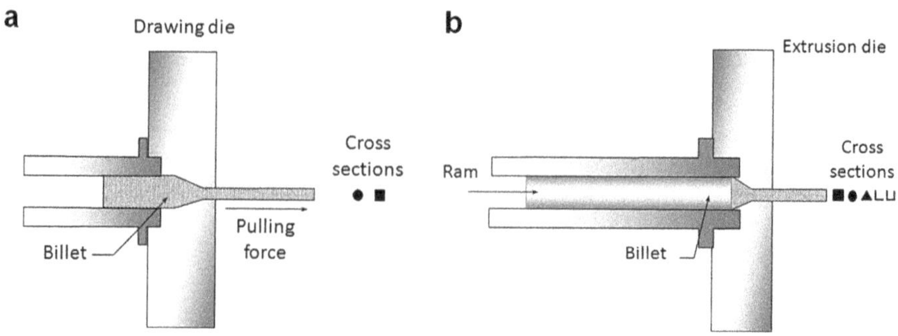

Fig. 6.2 (**a**) Drawing and (**b**) extrusion process

it is still very useful for producing wires from round metal bars. As an example, we may consider a wire being drawn through a conical dye with negligible friction between the material and the die. The material cross section is reduced from the initial diameter D_{in} to a final diameter D_{out}. We assume that such drawing operation does not induce noticeable volume variation in the drawn wire. The relationship between the initial material length L_{in} and final material length L_{out} can be related by the material volume. The true strain ε produced in this operation can be expressed by

$$\varepsilon = 2\ln\left(D_{in}/D_{out}\right) \tag{6.4}$$

Notably, this strain should be within the range for the material yielding, and the required drawing stress can be estimated accordingly based on the material stress-strain curve. In the manufacturing process, the exerted drawing force F_{draw} must be sufficient for supporting the tensile yielding strength σ_{yield}, such that $\sigma_{yield} = F_{draw}/(\pi D_{out}^2)$, according to the von Mises criterion (Eq. 6.1) mentioned in the previous section.

Similar to drawing, extrusion is another primary deformation process for manufacturing the product with a uniform cross section by squeezing the metal out through a hole by using a pushing force (Fig. 6.2a). Compared to other manufacturing processes, extrusion can produce more complicated cross sections of a product. The typical materials used in extrusion include ductile metals (Cu, Steel, Al, Mg), plastics, rubbers, and brittle materials, which are difficult to manipulate in other processes. Recalling the wire-drawing example above, extrusion of a wire has a similar relation, except that the compressive yielding strength now has to be considered.

6.2.3 *Forging*

Forging is a primary deformation process involving applying compressive forces to a metal in order to reshape the material body and create a desired geometric change defined by a forging die. Figure 6.3 shows one type of forging-impression die

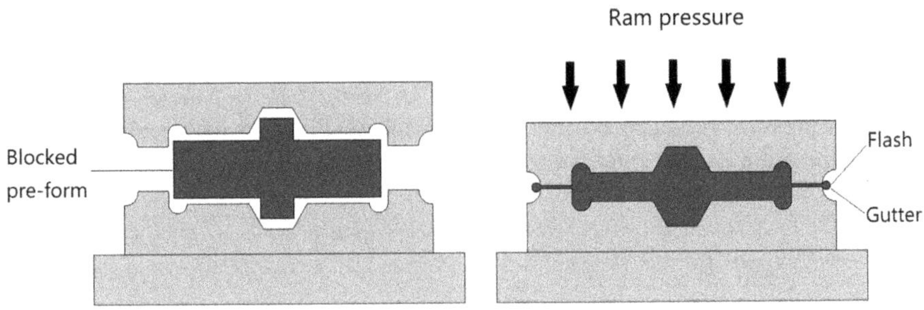

Fig. 6.3 Material in a die before (left) and after (right) forging

forging in which the workpiece will be enclosed in a mold and then compressed to fill up the space of the mold. The final metal piece will be reshaped to the geometric shape of the mold. During impression die forging, some excess metal pieces will flow out of the mold, forming flash, a thin layer of metal surrounding the forged material.

To further understand the more detailed considerations in the forging process, we may consider the cold forging of a cylindrical material body with a radius R and a thickness H to be another cylindrical body with a reduced thickness and a larger radius. Here, we would adopt cylindrical coordinates with radial position r, angular position θ, and height position h. Since the deformation is isotropic over the circular cross-sectional area, i.e., the shapes before and after deformation are both circles, the strains ε_r and ε_θ are equal, and thus, $\sigma_r = \sigma_\theta$. Thereby, according to Eq. 6.1, $\sigma_z - \sigma_r = \sigma_{\text{yield}}$. Ideally, we wish to exert a compressive pressure $-p_c(r)$, which may vary along the radial position r, such that the yielding occurs over the perfect plasticity region with rather consistent yield strength at $r \approx R$. The compressive pressure has a negative sign as it induces a compressive stress over the cylindrical material body. Hence, over the upper and lower material surfaces

$$d\sigma_r/dr = d\sigma_z/dr = -dp_c(r)/dr \tag{6.5}$$

During the forging under the ram pressure, the material cross section expands with the resistance of the frictional forces on both the upper and lower material surfaces along the negative radial direction:

$$\int_0^H (d\sigma_r/dr)dh = -\int_0^H (dp_c(r)/dr)dh = 2F_c \tag{6.6}$$

where F_c is the Coulomb static friction that $F_c = \mu_f F_n = \mu_f p_c$. And then,

$$\frac{dp_c}{p_c} = -\frac{2\mu_f}{H}dr. \tag{6.7}$$

Since $p_c(r) = -\sigma_{\text{yield}}$ at $r = R$,

$$p_c(r) = \sigma_{\text{yield}} \exp\left[\frac{2\mu_f(R-r)}{H}\right]. \tag{6.8}$$

The ram pressure provided by the forging machine P_{ram} can be calculated as the average pressure exerted over the upper/lower material surfaces:

$$P_{\text{ram}} = \frac{1}{2\pi R^2}\int_0^R 2\pi r p_c(r) dr = \frac{2\sigma_{\text{yield}}}{R^2}\exp\left(\frac{2\mu_f R}{H}\right)\int_0^R r\exp\left(-\frac{2\mu_f r}{H}\right) dr.$$

$$P_{\text{ram}} = \frac{\sigma_{\text{yield}}}{2}\left(\frac{H}{\mu_f R}\right)^2\left[\exp\left(\frac{2\mu_f R}{H}\right) - 1 - \frac{2\mu_f R}{H}\right] \tag{6.9}$$

6.2.4 *Casting*

Casting is a primary manufacturing process that can efficiently generate material shapes with rather complex geometry, like internal cavities and hollow parts. Casting is generally used for mass production of these complex parts, which is less economical, difficult, and slow to achieve by other manufacturing methods. The basic steps for casting are (1) refactoring a cast with a cavity of the desired shape, (2) pouring liquid metal into the cast, (3) allowing the metal/alloy to solidify inside the cast, and (4) removing the cast metal/alloy part from the cast. Casting can be applied for a large variety of materials, ranging from most metals to different cold setting materials. This process can be applied to a wide range of material sizes, ranging from small scales (~10 g) to very large scales (~1000 kg). The process induces homogeneous material properties regardless of directions or positions. Cast production is costly, but such production requires only one process to form a complex body shape. Hence, the overall process is still economical for mass production. Furthermore, as the excessive material around the casted body can be trimmed, remelted, and reused in the following casting processes, there is no material wastage, ideally.

Die casting (Fig. 6.4) is one type of casting process involving the usage of molds called die. In die casting, metal is usually first melted. Then, it is injected into the die cavity by a plunger under high pressure (~1000 to 20,000 psi). The cavity is formed by two hardened die with the desired geometry. Once the die cavity is filled, the molten metal is allowed to cool down and solidify inside the die. After a solid product is formed, it is removed from the die by being pushed out by ejection pins. Sometimes, there may be trimming following the removal to remove any sprue surrounding the final product to finish the whole process.

Investment casting (Fig. 6.5), also known as the lost wax process, is one of the oldest manufacturing processes. It was developed over 5000 years ago with its origins tracing back to ancient Egypt and China. The main characteristic of investment casting is the formation of a target metal shape involving the use of wax and ceramic molds. The basic steps for investment casting are:

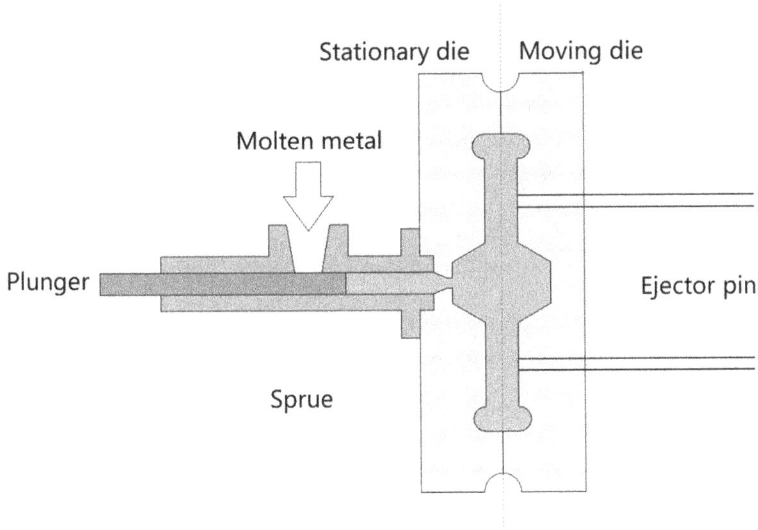

Fig. 6.4 Die casting process

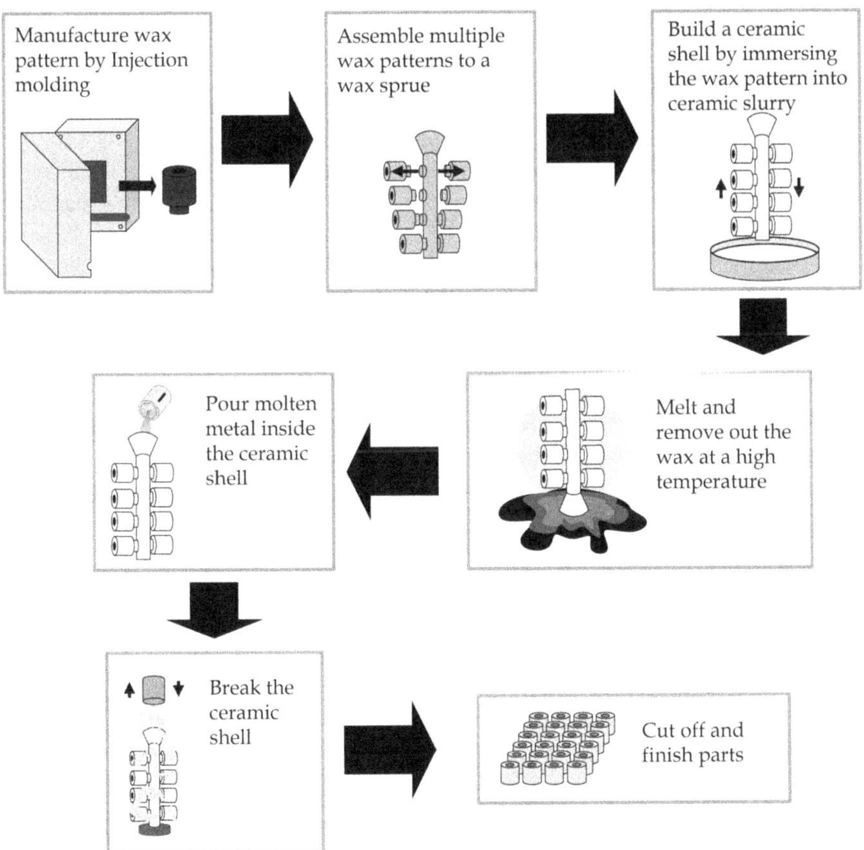

Fig. 6.5 Investment casting process

1. Pattern formation: Parts are first molded in wax, and multiple wax parts are attached to a wax runner.
2. Mold formation: The whole wax part is then soaked in a wet ceramic slurry repeatedly to form a thick ceramic layer. Once the ceramic layer is hardened, the ceramic mold is inverted and heated to flow out the inner wax, and the hollow cavity takes the shape for casting.
3. Pouring: Molten metal is poured into the cavity of the hot ceramic mold (~1000 °C).
4. Cooling: The metal inside the mold solidifies by cooling, and the ceramic mold layer is then broken out by vibration or water jets.

The applications of investment casting are very diverse, including the production of dental fixtures, hammers, cams, turbine blades with complex geometry, triggers, and machinery components of cooling system.

6.3 Secondary Processes for Metals and Alloys

6.3.1 *Sheet Metal Processing*

Sheet metal processing refers to any process that applies forces to metal sheets to form a desired geometry. Actually, metal can be extended or deflected by applying forces which are greater than the metal yielding strength and cause plastic deformation. Common sheet metal processes include punching, bending, drawing, shearing, shaving, etc.

Punching is a hole-making process in which a punch tool removes a slug from the metal sheet by pressing and cutting a hole (Fig. 6.6), cutting the sheet into the required shape. It is the cheapest and most efficient method for cutting holes in a metal sheet with controllable hole sizes. Because of the high production rate (~600 strokes per minute), it is appropriate for mass production of sheet metal components. The punching force F_{punch} for the boundary with a perimeter L_{p} and a plate with strength σ_{s} thickness H should have a punch force level:

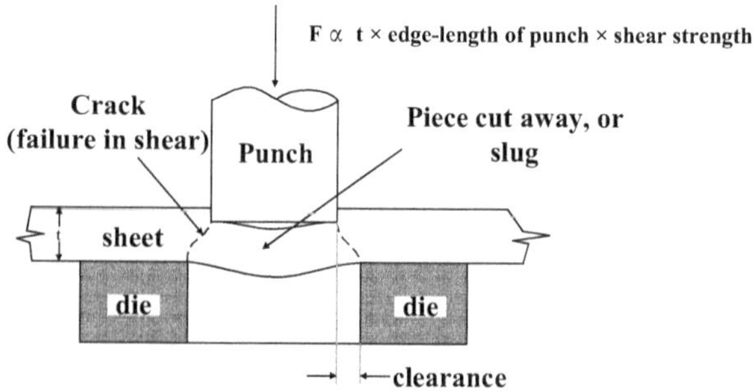

Fig. 6.6 Sheet metal punch process

$$F_{\text{punch}} = \sigma_S L_p H \tag{6.10}$$

Bending is a metal forming process that bends a sheet metal to a certain angle and transforms the metal to a required geometry with plastic deformation. During the bending process, bending forces are controlled between the metal's yield point and the tensile strength. There are many different types of bending method as shown in Fig. 6.7. Selecting the proper bending method under different situations should depend on the features of the desired bend shape and the workpieces.

Along the bending edge, the exerted force produces yielding across the thickness of the plate in those sections that should be bent. As a demonstration, we consider a material that yields in the perfect plasticity range both in tension and compression at a constant yield stress σ_{yield}. With the length of bending edge W, plate thickness H, and the bending force F_{bend} exerted with a distance L from the supporting point, the moment of the stresses M_{bend} should be equal to the moment from the external bending force F_{bend}:

$$M_{\text{bend}} = \sigma_{\text{yield}} \frac{WH}{2} \times \frac{H}{2} = \frac{F_{\text{bend}}}{2} \times \frac{L}{2} \tag{6.11}$$

And therefore,

$$F_{\text{bend}} = \sigma_{\text{yield}} WH^2/L \tag{6.12}$$

Sheet drawing is another sheet metal processing that is similar to the bending operation in that it also involves the curving of sheet metal. In a sheet drawing process, metal is pulled by a tensile force from a punch (Fig. 6.8). The sheet drawing tool looks similar to a punching machine. While no slug is removed in sheet drawing, the sheet metal is instead stretched into the shape of a die, usually concave shaped. For example, we may consider the sheet drawing of a cup-shaped product

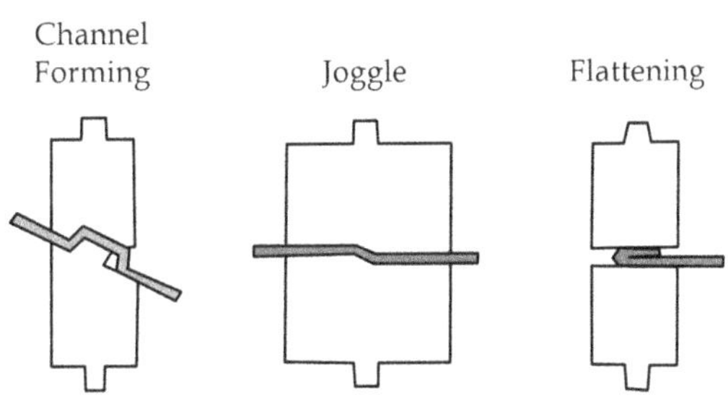

Fig. 6.7 Sheet metal bending process

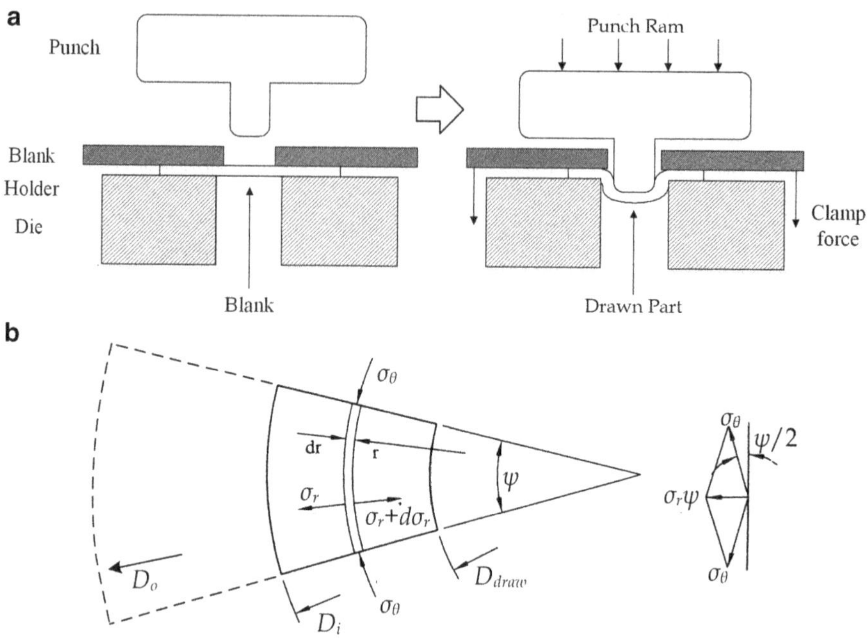

Fig. 6.8 (**a**) Sheet metal drawing process. (**b**) Balance of forces on an arc element

using a circular metal sheet with an initial diameter of D_o. We may define the resultant diameter of the can has a diameter of D_{draw}, which is defined by the diameter of the punch ram shape. The depth of drawing L_{draw} can be estimated by assuming that the sheet metal area is maintained after the process:

$$D_o^2 = D_{\text{draw}}^2 + 4D_{\text{draw}}L_{\text{draw}} \tag{6.13}$$

We adopt the cylindrical coordinates in radial (r), angular (θ), and thickness (z) directions. During the sheet drawing process, the outer sheet diameter reduces from D_o to D_i and finally to D_{draw}; and we assume the sheet thickness does not change, and therefore, $\varepsilon_z \approx 0$. Recalling Eq. 2.17, $\sigma_z = \nu(\sigma_r + \sigma_\theta)$, and the Poisson's ratio ν should be 0.5 if the process induces perfectly plastic deformation around the boundary of the drawing hole:

$$\sigma_r - \sigma_\theta = 2\sigma_{\text{yield}}/\sqrt{3} \approx 1.155\sigma_{\text{yield}} \tag{6.14}$$

where σ_{yield} is the shear yield strength.

For the outer ring-shaped region of the material prior to sheet drawing, according to the force balance described in Fig. 6.8b, $\sigma_r\psi(r + dr) - (\sigma_r + d\sigma_r)\psi r - 2dr\sigma_\theta\sin(\psi/2) = 0$ and $(\sigma_r - \sigma_\theta)\psi(dr) - \psi r(d\sigma_r) \approx 0$. Substituting Eq. 6.14, we obtain

$$d\sigma_{\rm r} = 1.155\sigma_{\rm yield}dr/r. \tag{6.15}$$

For the material region along the ram boundary,

$$\sigma_{\rm r} \approx \int_{D_{\rm o}/2}^{D_{\rm draw}/2} 1.155\sigma_{\rm yield}dr/r = 1.155\sigma_{\rm yield}\ln\left(\frac{D_{\rm o}}{D_{\rm draw}}\right). \tag{6.16}$$

Furthermore, in this boundary region, $\varepsilon_\theta \approx 0$, and thus, σ_z and $\sigma_{\rm r}$ should be in the same order of magnitude. The sheet drawing force $F_{\rm draw}$ must be large enough for supporting the sheet drawing process. The lower bound of $F_{\rm draw}$ can be estimated from the z-stress around the sheet drawing $F_{\rm draw} = \pi D_{\rm draw}\sigma_z H$:

$$F_{\rm draw} \geq 1.155\pi\sigma_{\rm yield}D_{\rm draw}H\ln\left(\frac{D_{\rm o}}{D_{\rm draw}}\right) \tag{6.17}$$

Typically, we may adopt a larger level for $F_{\rm draw}$:

$$F_{\rm draw} = 1.5\pi\sigma_{\rm yield}D_{\rm draw}H\ln\left(\frac{D_{\rm o}}{D_{\rm draw}}\right) \tag{6.18}$$

6.3.2 *Machining: Removal of Materials*

Machining is the process involving the removal of material by cutting the raw object into the expected product. Materials used in machining are widely covered, including metal, wood, plastic, and ceramic. There are three widely applied machining processes – drilling, turning, and milling.

Drilling
Drilling processes are machining operations for hole machining, which involve reaming and boring. Figure 6.9 shows the key features of drilling bits and gives a detailed description about them. Boring is used for enlarging an existing drilled hole; reaming is used for finishing an existing hole; and lapping is used for the surface finishing. Drilling processes can generate a significant material removal rate. They can achieve medium to high surface finish quality and good dimensional control of the holes they produce.

By applying different configurations of drilling parameters, holes with different sizes and depths can be created in a variety of materials. Different drilling bit features correspond to different operating situations. For example, the point angle is the angle at the tip of a drilling tool. It is determined by the drilling material: a larger point angle is chosen for harder materials, while a sharper point angle is chosen for softer

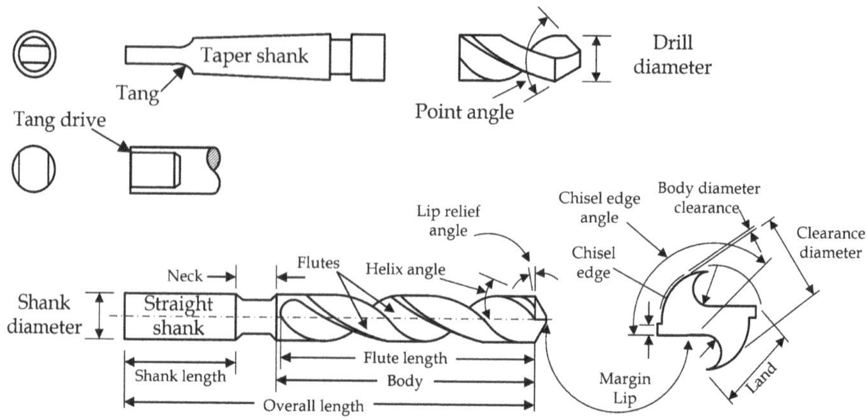

Fig. 6.9 Features of drilling cutter

materials. The drill diameter is meant to equal the hole diameter, but sometimes, it is chosen to be smaller than the hole due to vibrations and misalignments of the tools during the drilling process. Typically, the resultant hole diameter should not be smaller than 0.5 mm, or else there may be adverse effects like abrasion of the drill tip, drill breakage, and chipping. The shank length depends on how deep the hole is required to be. The lip angle depends on the level of physical support of the tool cutting edge. A larger lip angle supports cutting more aggressively under the same amount of point pressure, whereas a smaller lip angle cuts less aggressively.

Milling

In milling, the workpiece surface material is removed by the rotary cutting tool; and hence it can cut a product into arbitrary 3D shapes. The characteristic traits of milling are its high material removal rate, high surface finishing, and dimensional control. The orientation of milling can be either vertical or horizontal. For instance, for vertical milling, a vertical milling machine with a vertical spindle axis is used (Fig. 6.10a).

Some important features of milling can be modified via the following ways. By adjusting the power supply, the revolutions per minute of the milling cutter, which is held in the spindle, can be adjusted. Also, by applying different types of milling cutters (Fig. 6.10b), we can produce products with different milling properties. A selected milling cutter is mounted on the quill of a milling machine. The quill moves vertically in the head with an adjustable depth and contains the spindle installed with cutting tools. The quill feed handwheel moves the quill up and down within the head as does the quill feed lever. The knee moves up and down by sliding parallel to the column. The main function of the saddle on the knee is to allow for the translation of the worktable, and the column on the base holds the turret, which allows for the rotation of the milling head around the column's center. Also, the main function of the overarm (ram) on the turret is to allow for the milling head to be repositioned over the table. For the different dimensional movements of the

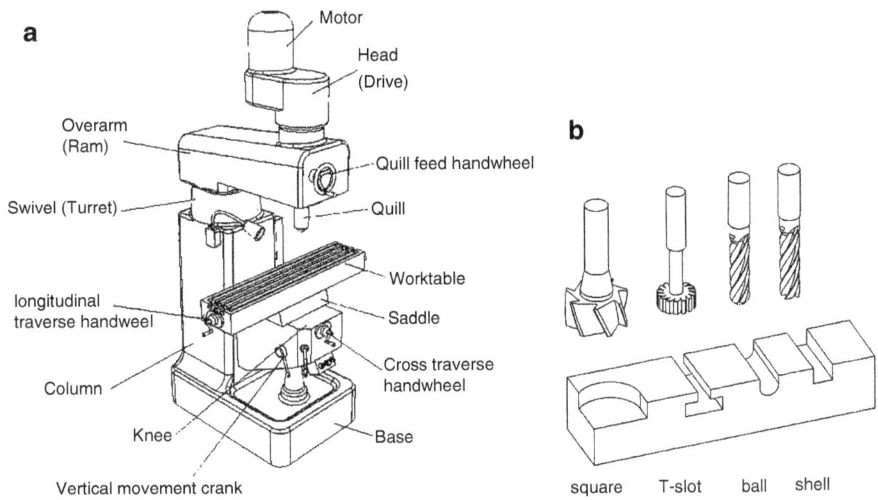

Fig. 6.10 (**a**) Milling machine. (**b**) Sample milling cutters and the resultant cut profiles

Fig. 6.11 Turning operation

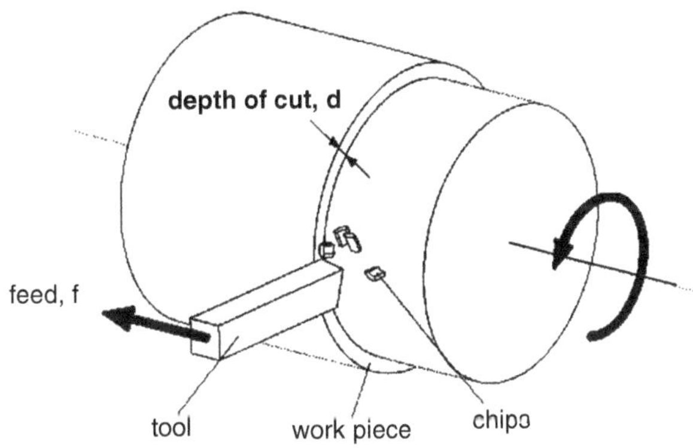

worktable, the longitudinal traverse handwheel moves it horizontally (left and right), the cross traverse handwheel moves the work in and out, and the vertical movement crank moves the worktable and also the knee and saddle vertically (up and down) in unison.

Turning

Turning is the machining process used to cut tubular stock into revolved shapes with precise diameters, which depends on the depth of cutting, by rotating the workpiece in a turning machine at high speed. The configuration of a turning process is shown in Fig. 6.11. The workpiece is then cut by the fixed cutting tool and creates waste material called chips. Products manufactured by turning are rotational, usually axisymmetric, with different features, including grooves, tapers, and holes. Turning

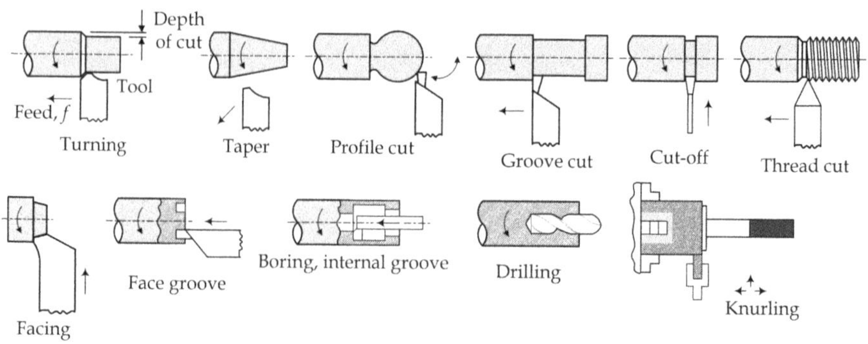

Fig. 6.12 Different turning operations

is also applied as a secondary process to add or remove features on parts that were pre-generated by other processes such as casting, forging, or drawing.

In the turning process, a variety of operations can be executed on the workpiece to produce various shapes as shown in Fig. 6.12. Operations can be done either outside of the workpiece or inside, classified as "external" or "internal," respectively. External operations modify the outer diameter of the workpiece, while internal operations modify the inner diameter.

6.4 Tertiary Processes for Metals and Alloys

6.4.1 Grinding and Abrasive Machining

Grinding is the most common abrasive machining process for removing metal from a surface in the form of chips by using a grinding wheel as the cutting tool. Therefore, it can be used for the finishing and smoothing of a surface as well as the final manufacturing process. Grinding can induce a resultant surface roughness down to <0.5 μm. A grinding wheel, the key component in grinding, is composed of a matrix of small abrasive grits held together with bonding materials. Each grinding particle acts as a cutting tool to cut the workpiece into waste chips and create a fine finishing surface. Grinding consists of some distinctive characteristics, such as a very low material removal rate, very high surface finish, and good dimensional control. Grinding is typically applied for finishing the surface of manufactured parts, improving the dimensional tolerance of manufactured parts, slitting and partings, and removing unwanted materials of a cutting process.

Surface roughness is defined as the high-frequency, short-wavelength component of a measured surface, also expressed as the surface profile roughness. Surface roughness affects different material characteristics, but it is most remarkable at the movable components of an implant.

Asperity contact, which describes the unevenness of surfaces, occurs between opposing surfaces when the roughness of both surfaces is high. The hardness of each surface also influences rubbing greatly when the two surfaces come into contact. If both surfaces are hard, significant friction is induced owing to the asperity contact. If one surface is harder than the other, the softer one will experience wear and be damaged. Other than surface friction, surface roughness also causes surface wear, reduces surface life span, and more importantly, sheds wear particles.

6.4.2 Joining

Joining is a collective term which includes many processes that combine separate components into a whole, such as fusion welding, solid-state welding, brazing, soldering, gluing, and mechanical joining. Each joining technique is suitable for a specific situation requirement or joint requirement.

Fusion welding is a joining technique particularly applicable for metal workpieces. It joins metals by heating them to their melting points, which is followed by solidification. As technology develops, modern welding is not limited to only using heat to melt metal. For example, plasma arc welding applies the heat obtained from an electrical arc setup formed between electrodes and the workpiece. It is a deep and narrow welding technique often used for medical device production. Laser beam welding generates rapid heat from radiation emitted from a concentrated beam of coherent light of a single wavelength and power around 100 kW. Electron beam welding involves concentrated high-velocity electron beams which can be used for deep and narrow welds while also being comparatively expensive.

Solid-state welding is a collection of welding techniques that join workpieces together at temperatures below the melting point of the materials (usually 0.3–0.5 T_m, where T_m is the melting point). Therefore, no melting is involved, and no heat-sensitive properties are affected. A possible configuration of solid-state welding is shown in Fig. 6.13a. The bond at the coalescence is formed by the diffusion of the interface atoms. The temperature can be set even lower for some materials. For example, cold welding is a solid-state welding process in which joining occurs at room temperature without heating; instead, pressure is exerted on the interface of two pieces to be joined.

Furthermore, resistance welding is a welding technique for metals which joins them using the heat generated from a strong current passing through the metal body (Fig. 6.13b). Force is applied to clamp the components together during the welding process. This technique includes the spot welding, which is defined as the welding of overlapping metal workpieces at the small points by application of opposing forces and electric current. Resistance welding can also be applied with a traveling heat source, a process called arc welding. Though there are many other welding techniques which involve generating localized heat at the welding site, we will only discuss the physical aspects for resistance welding.

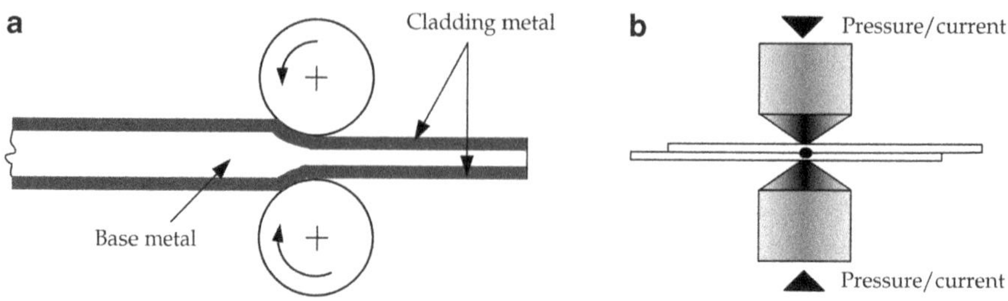

Fig. 6.13 (**a**) Cold welding. (**b**) Spot welding

The heat generated by current flow in resistance welding is driven by Joule heating. The heating site is mainly at the contact interface between the workpieces because such tiny interfacial gaps should have the relatively highest resistance. In the case of arc welding, the heat energy input (H_r) is defined as the quantity of energy introduced per unit length of weld from a traveling heat source. H_f can be estimated as the ratio of total input power to the traveling velocity v_r. The electrical input power is the product of applied voltage V_r and current I_r. Neglecting the heat loss to the environment, a brief estimation of H_r is

$$H_r = V_r I_r / v_r \tag{6.19}$$

We further assume that the cross-sectional area of the weld metal is roughly a circular region with a radius r_r. The corresponding specific theoretical quantity of heat is Q_r for melting a given volume of metal from a room temperature environment. An estimation of Q_r is $Q_r = \Phi_r T_m{}^2$, where T_m is the melting temperature of the weld metal in Kelvin and Φ_r is roughly a constant with an empirical value of 3×10^5. Hence the width of welded metal along the heating track can be approximated as

$$r_r = \frac{1}{T_m}\sqrt{\frac{V_r I_r}{\pi \Phi_r v_r}} \tag{6.20}$$

6.5 Polymer Processes

The conversion of raw polymers into finished products involves a series of steps. The first step is mixing, which modifies polymer properties by adding specific additives. The second step is forming, in which polymers are given desired shapes. It can be either a simple two-dimensional forming or complex three-dimensional forming. The last step is the finishing of the formed product into the final product.

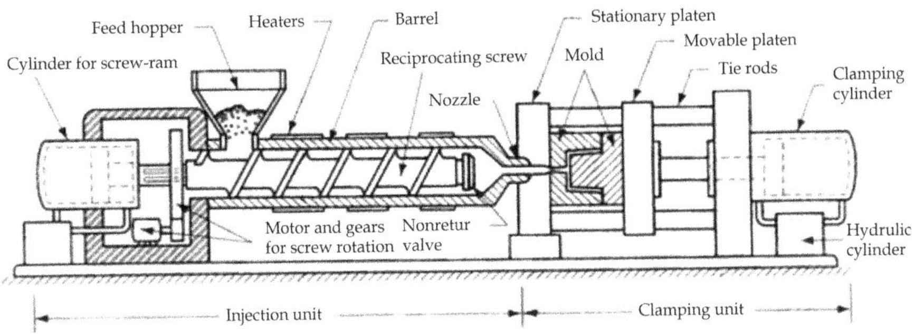

Fig. 6.14 Configuration of an injection molding machine

6.5.1 Injection Molding

Injection molding is a widely used manufacturing process mainly used for the mass fabrication of plastics as well as biopolymers. High versatility of the process makes it highly adaptable to different factors, including size, shape, complexity, and components of the products. Other advantages of this process include its high versatility, rapid production rate, low production cost, and low scrap rate. An injection molding machine (Fig. 6.14) can produce tens of thousands of parts successively in a short period, leading to an effective production rate. Even though the initial cost of the process is very expensive, especially the molding dies, the per-part cost is extremely low since production cost tends to drop greatly as more parts are created. Besides, injection molding produces lower scrap amounts than older manufacturing processes like machining.

There are four basic components of the mold that are important in injection modeling: sprue, gates, runners, and vents. Typically, an injection molding cycle consists of four stages (Fig. 6.15):

1. Clamping: The movable plates move toward the stationary plates and are put together. The clamping unit generates great force and exerts large pressure to the mold to ensure the molds are securely closed with no leakage.
2. Injection: The raw plastic is first melted by heat and pressure. Then, the molten plastic is forced into the cavity of the mold quickly to ensure no solidification occurs during injection. Two types of machines are used to melt and inject the resin into the mold: a plunger injection molding machine and a reciprocating screw injection molding machine. (Basic design of a reciprocating screw is asked and discussed in Problem 2.7.)
3. Cooling: The molten plastic inside is allowed to cool down and solidify into the desired final shape. Note that shrinkage may occur due to the improper cooling of plastic, so a holding pressure is maintained to compensate for it.
4. Ejection: After the plastic part is sufficiently cooled, the moving plate retracts and opens the mold. Natural shrinkage during cooling may cause the plastic part to

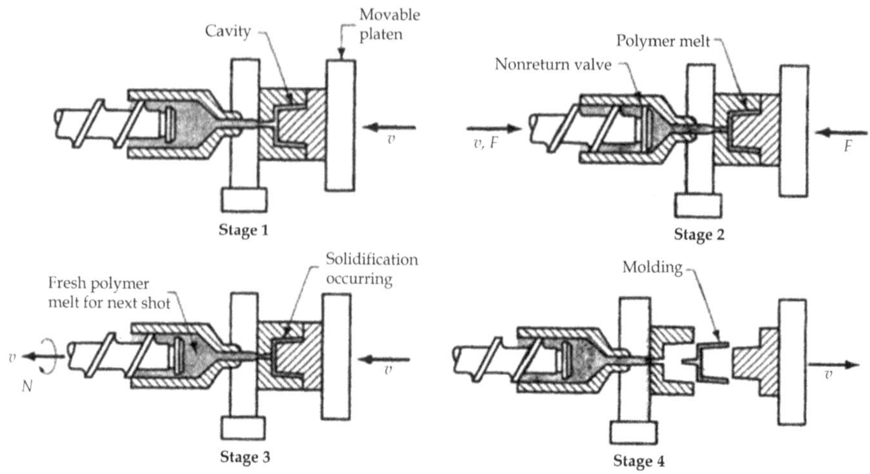

Fig. 6.15 Injection molding cycle

adhere to the moving plate. Thus, ejector pins are often designed, usually on the stationary plate, to help safely remove parts from molds after they have been made. These let technicians apply a force to eject a part from the mold. After ejection, the product is now released and ready for further modification.

During injection, the polymeric liquid has a significant level of viscosity with non-Newtonian behaviors. The relationship between the zero shear rate viscosity (μ_0) and the weight average molecular weight (M_w) is in the form

$$\mu_0 = K_\mu (M_w)^{\alpha_\mu}. \tag{6.21}$$

where K_μ and α_μ are constants. Typically, below a critical value of M_c of the molecular weight, the relationship is linear ($\alpha_\mu = 1$). Above M_c the dependence of μ_0 on M_w is much greater, and α_μ is found to be in the range of 3.4–3.5.

A narrower molecular weight distribution (i.e. the smaller polydispersity index M_w/M_n) reflects a more Newtonian behavior of the liquid as well as a greater dependence of viscosity on temperature. Experiments have shown that polymers with long side chains have lower zero shear rate viscosities μ_0 than linear polymers with the same molecular weight. Intuitively, this seems reasonable if it is considered that the linear polymer is more likely to become entangled and resist deformation under a low shear stress. On the other hand, a branched polymer with a similar molecular weight should include comparatively multiple, shorter molecular side chains than the linear chain polymer. The branched polymer molecules are less likely to entangle among each other, so they can be considered as macroparticles sliding around each other under a low shear stress. Hence, the corresponding μ_0 is lower than that of the linear polymer with an equivalent molecular weight. For instance, an unbranched polyethylene melt has a value of μ_0 50 times greater than the melt of a branched polyethylene with a similar molecular weight.

6.5.2 Blow Molding

Blow molding is a manufacturing process used to produce thin-walled hollow workpieces from thermoplastic materials as shown in Fig. 6.16. The main principle of the process comes from the idea of glassblowing, which involves inflating molten glass into a bubble (or parison). In general, there are three main types of blow molding: extrusion blow molding, injection blow molding, and injection stretch blow molding. The blow molding process begins with melting down plastic and forming it into a hollow tube (also known as a parison) or, in the case of injection and injection stretch blow molding, a preform. The parison is a tubelike piece of plastic with a hole in one end through which compressed air can pass. The parison is then clamped into a mold cavity, and compressed air is then blown into the parison, inflating the melted plastic into the shape of the mold cavity. Once the plastic has cooled and solidified, the mold is opened and the shaped part is ejected. Blow molding can be used to process many different types of polymers, such as polyethylene of various densities, polypropylene, polyvinyl chloride, thermoplastic elastomers, polystyrene, and more. Because lower pressure is needed, the mold and the machinery costs in extrusion blow molding are lower compared to those of injection molding. On the other hand, extrusion blow molding requires longer cooling times than injection molding and is not possible to form parts with holes except when making them after the process.

The three main phases of extrusion blow molding are (1) softening the resin by the use of heat, (2) forming the parison, and (3) blowing the parison in the mold. The first of these phases involves the use of an extruder to heat the plastic to a molten state and compress the material at the die head. This part of the process is the same as typical extrusion. This phase can create objects of a fixed cross-sectional profile. A

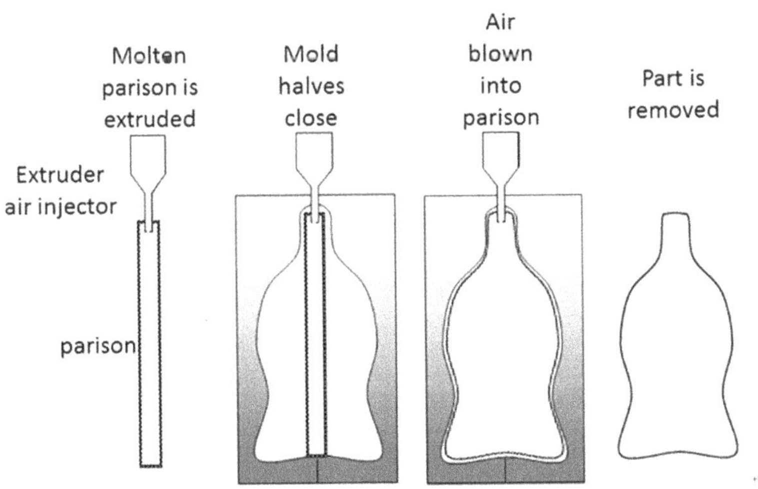

Fig. 6.16 Extrusion blow molding

material is pushed through a die of the desired cross section. The two main advantages of this process over other manufacturing processes are its ability to create very complex cross sections and form parts with excellent surface finishes. Second, the die or multiple dies form the diameter and wall thickness of the extruded parison which is ready to be clamped between the mold halves. The third phase involves the closing of the mold halves by hydraulic pressure and pinching of the parison. Air pressure of around 25 to 150 psi is used to expand the resin to conform to the mold cavity. Notably, as the diameter of the mold increases, the corners and areas with the thinnest walls will have the closest tolerance during this phase. A water cooling system is turned on to maintain the correct mold temperature. Once complete, the mold is reopened, and the plastic part is ejected until it is completely cooled. The product from extrusion blow molding will need to be trimmed afterward. In some operations, the container is conveyed to an automatic trimmer, while in other operations the trimming step is separate.

For a basic understanding about material deformation during blow molding, we shall discuss a simple case of a spherical shell expanding during the process. Consider a small sphere is first formed by extrusion following the pumping of a gas that exerts a gauge pressure, p_{blow}, on the inner surface of the shell. According to symmetry, the stress at any distance r (from the shell center) and any angular position on the inner shell surface is the same. The radial direction is a principal direction, and σ_r is the principal stress. All directions tangential to the radial direction will be equivalent, and hence, all angular directions have the same stress, i.e., $\sigma_\theta = \sigma_\psi = \sigma_{\text{blow}}$, which is another principal stress. In this spherical "pressure vessel," the radial stress is a compressive stress, and the angular stress is a tensile stress. If we further assume the radius of the shell is r_{shell} and the shell thickness is t_{cell}, by considering the force balance on a hemisphere as shown in Fig. 6.17, we may obtain a relation

$$\pi {r_{\text{shell}}}^2 p_{\text{blow}} = 2\pi r_{\text{shell}} t_{\text{shell}} \sigma_{\text{blow}}, \text{ or } r_{\text{shell}} p_{\text{blow}} = 2 t_{\text{shell}} \sigma_{\text{blow}} \tag{6.22}$$

Additionally, recalling Eq. 4.24, and by replacing x, y, and z by θ, ψ, and r

$$U_n \approx (E/6)\left({\lambda_\theta}^2 + {\lambda_\psi}^2 + {\lambda_r}^2 - 3\right) \tag{6.23}$$

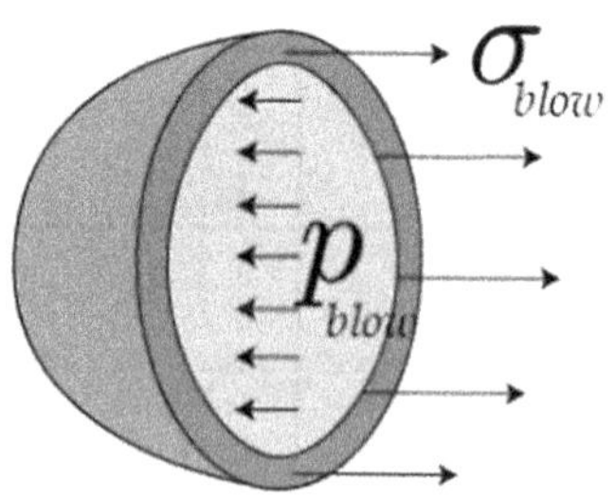

Fig. 6.17 Stress of a spherical shell upon an inner gauge pressure

For the case of inflation or equibiaxial stretch, there is equal stress along the two angular directions in the spherical coordinates (with radial r-direction and the two angular θ- and ψ-directions). This implies the stress relations $\sigma_\theta = \sigma_\psi = \sigma_{\text{blow}}$ and $\sigma_r = 0$ and the extension ratio relations $\lambda_\theta = \lambda_\psi = \lambda_{\text{shell}}$ and $\lambda_r = 1/\lambda_{\text{shell}}^2$. The strain energy (Eq. 6.23) of a unit length of shell area with the shell thickness t_{shell} becomes

$$U_{\text{shell}} = (E/6)\left(2\lambda_{\text{shell}}^2 + 1/\lambda_{\text{shell}}^4 - 3\right)t_{\text{shell}}. \tag{6.24}$$

If we define F_n to be the stretching force acting on the unit length of the strained material shell along either θ- or ψ-directions, which should be $2\sigma_{\text{blow}}\lambda_{\text{shell}}t_{\text{shell}}$, given that the cross-sectional area is $\lambda_{\text{shell}}t_{\text{shell}}$ and a pair of opposite stretching force is acting on both sides of the unit length shell segment, then the stretching stress along an angular direction is

$$\sigma_{\text{blow}} = (E/3)\left(1 - 1/\lambda_{\text{shell}}^6\right). \tag{6.25}$$

Substituting Eq. 6.22 into this relation,

$$\frac{r_{\text{shell}}p_{\text{blow}}}{2t_{\text{shell}}} = \left(\frac{E}{3}\right)\left(1 - \frac{1}{\lambda_{\text{shell}}^6}\right). \tag{6.26}$$

The angular extension ratio, $\lambda_{\text{shell}} = 1 + \varepsilon_{\text{blow}}$, where $\varepsilon_{\text{blow}}$ is the pure angular strain during the blowing process. $\varepsilon_{\text{blow}}$ can be approximated by $(r_{\text{shell}} - r_{\text{shell\_o}})/r_{\text{shell\_o}}$, where $r_{\text{shell\_o}}$ is the initial shell radius before blowing. If we further assume that the material volume of the polymeric shell (with a Poisson's ratio ≈ 0.5) can be maintained, t_{shell} can be expressed as $\lambda_{\text{shell}}^{-2}t_{\text{shell\_o}}$, where $t_{\text{shell\_o}}$ is the initial shell thickness. Hence, Eq. 6.26 can be expressed as

$$p_{\text{blow}} = \left(\frac{2E}{3}\right)\left(\frac{1}{\lambda_{\text{shell}}^3} - \frac{1}{\lambda_{\text{shell}}^9}\right)\left(\frac{t_{\text{shell\_o}}}{r_{\text{shell\_o}}}\right). \tag{6.27}$$

The term $\lambda_{\text{shell}}^{-3} - \lambda_{\text{shell}}^{-9}$ would have a maximum value of 3.85 for $\lambda_{\text{shell}} = 1.2$. In this case, if we can configure the blowing process to have a material extension ratio of $\lambda_{\text{shell}} > 1.2$, the required p_{blow} would then just need to be above a critical value determined using Eq. 6.27. Such critical value of p_{blow} would depend only on the elastic modulus and the extruded geometry of the material.

6.5.3 *Thermoforming*

Thermoforming is one of the major processing techniques used in the plastics industry. Thermoforming involves the heating of a plastic sheet up to a pliable temperature and then forcing it to form a shape around the contours of a mold while under pressure. The required pressure is usually a mechanical or air pressure, often assisted by localized heating and bending.

The thermoforming process involves three main steps: (1) heating the thermoplastic material to its softening point, (2) forming the sheet to a specific shape in a mold under a mechanical pressure, and (3) trimming the formed parts. The mold is normally water-cooled to control the mold temperatures. Matched mold forming requires the two mold halves to fit together perfectly, especially for products requiring excellent reproduction details. Since the mold is directly in contact with the surface of the softened sheet, it is required that the mold surfaces be highly polished or textured to the product requirements. The molds are usually made of aluminum or steel and mounted in a hydraulic or pneumatic press.

Vacuum forming is the most versatile mode of thermoforming. In vacuum forming, the mold is placed directly under the softened sheet, and a slight pressure is applied to seal the plastic to the upper mold edge. A vacuum is applied through small holes in the mold cavity, and the atmospheric pressure forces the softened sheet against the contour of the mold walls to form the desired shapes. Upon cooling, the solidified product is removed while retaining the shape of the mold.

Straight pressure forming is another thermoforming method that uses a female mold over which a sheet of thermoplastic material is clamped. A radiant heater softens the sheet before a cover is quickly placed over the hot sheet, and preheated compressed air is then blown through the cover opening. The sheet is thus pushed against the contours of the mold, and any air trapped below the sheet escapes through vent holes in the mold. After cooling, the formed part is removed from the mold and trimmed, similar to vacuum forming.

6.5.4 *Transfer Molding*

Transfer molding is a molding process for thermoset polymers. It has advantages from both injection and compression molding and is an ideal manufacturing method for many applications. In the transfer molding process as illustrated in Fig. 6.18, the required amount of uncured molding material is accurately measured and then inserted and preheated by the molding pot which heats the polymer beyond its melting temperature. The melted molding material fills the sprue by gravity. The plunger then exerts a force and pushes the melted molding material from the pot into the mold cavity. The mold remains closed and heated to ensure a faster flow rate and optimal curing of the molding material. After the molding material is cooled and hardened, the mold cavity is opened, and the molded part is removed from the mold. Any flash at the sprue and gate is trimmed and removed after the completion of the process.

Like injection molding, the geometric design is relatively flexible. Once the mold is established, the required setup time and production cycle for transfer molding are shorter than injection molding. The corresponding setup and maintenance costs are lower. The dimensional control of the final product is good, too. However, the tooling cost of mold generation is high. Air may be trapped in the molded material if the mold design is not optimized.

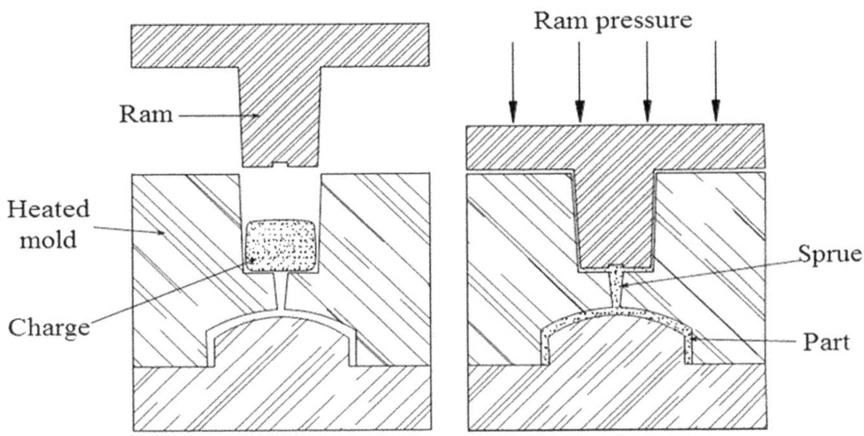

Fig. 6.18 Transfer molding

6.5.5 Spinning

Spinning is a specialized extrusion process used to produce continuous polymer fibers. The three major types of spinning are melt, dry, and wet. Melt spinning uses polymers that can be melted easily. Dry spinning involves dissolving a polymer into a solution that can be evaporated. Wet spinning is used when the solvent cannot be evaporated and must be removed by chemical means.

Figure 6.19 shows the general process of spinning. In the first stage, the polymer is either dissolved or melted into a liquid state. If the polymer is a thermoplastic, then it can be simply melted; otherwise, it is dissolved in a solvent or chemically treated to form soluble or thermoplastic derivatives. Dissolution is performed mostly in batches, while melting in the majority of contemporary processes is carried out continuously in screw melters. In both cases, the polymer solution or melt is transported under pressure to the spinning block where an exact metering pump such as a gear pump maintains a highly even issue of polymer. After passing the metering pump, the liquid polymer is forced through the final filter, usually called the filtering pack, which has more functions than just mechanical purification. Finally, the liquid raw material is forced through a plate with many small holes or capillaries, called the spinneret, and in this way the liquid moves in endless, fine streams. Tiny streams of polymer that emerge from these holes, called filaments, are wound together at speeds of up to 120 m/s as they solidify, forming a long fiber. Following the spinning process, the fibers are stretched 3–8 times their original lengths to create increased chain alignment and enhanced crystallinity in order to yield improved strength upon stretching.

The spinning blocks for melt and dry spinning have similar design principles, except the higher viscosities of melt polymers require higher strength and heavier construction. The spinnerets for wet spinning are usually mounted not directly in the spinning block but at the end of special transfer tubes attached to the block. Such a

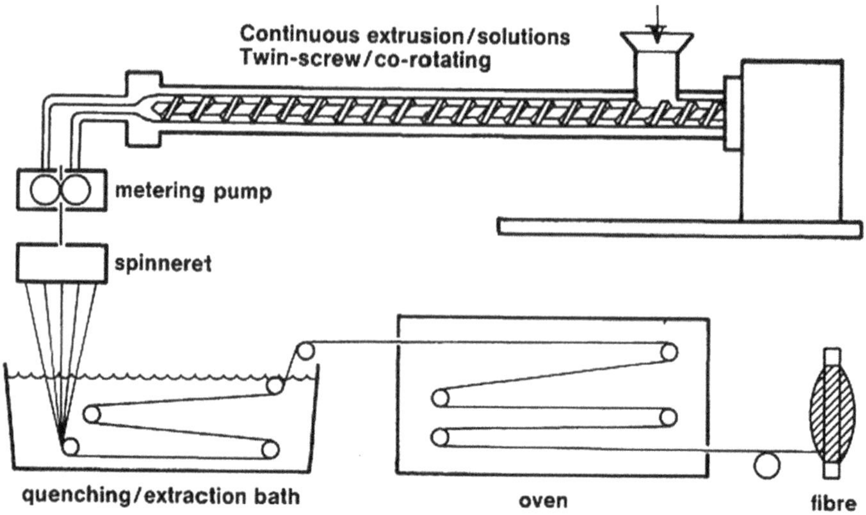

Fig. 6.19 Schematics of the gel-spinning process

design change, in comparison with melt and dry spinning, is more convenient, since the spinnerets must be submerged in a liquid.

6.5.6 Polymeric Scaffold Fabrication

Phase Separation

For designing scaffolds (refer to Chap. 11), phase separation techniques based on controlling temperature are used to separate polymeric solutions into two phases based on either low (polymer-lean phase) or high (polymer-rich phase) polymer concentrations. A polymer is first dissolved in phenol or naphthalene, followed by biologically active molecules being dispersed into these solutions. By lowering the temperature, a liquid-liquid phase is separated and quenched to form a two-phase solid. The solvent is removed by extraction, evaporation, and sublimation to integrate porous scaffolds with bioactive molecules. Proper liquid-liquid phase separation is crucial for preparing nanofibers, but it cannot happen in all types of solvents. As such, the selections of solvent and phase separation temperature are crucial for the formation of nanofibers. When conditions are favorable, liquid-liquid phase separation produces 3D fibrous structures with nanoscale architecture similar to that of type I collagen, which is used in numerous biomedical applications. This phase separation technique can be easily combined with other fabrication technologies (e.g., particulate leaching) to design 3D structures with controllable pore morphology. It can also be combined with rapid prototyping to create nano-fibrous scaffolds for tissue engineering applications.

Membrane Lamination

Membrane lamination is used for constructing 3D biodegradable, polymeric foam scaffolds. These scaffolds feature precise anatomical shapes and notable mechanical and structural properties. During the fabrication process, membranes with appropriate shapes are soaked with solvent. Afterward, they are stacked up layer by layer into 3D assemblies with defined pore structures and morphology. The properties of the final combined 3D scaffolds are identical to those of the individual constituent membranes. This method generates porous 3D polymer foams with defined anatomical shapes. They are also able to be designed using computer-aided modeling in order to create implant shapes with desired templates. However, layering porous sheets on top of each other can ultimately reduce the effectiveness of pore interconnectivity, and the creation process is time-consuming.

Fiber Mesh

Fabricating scaffolds using fiber mesh involves individual fibers being woven or interwoven into 3D patterns with variable pore sizes. Polyglycolide was the first biocompatible and biodegradable polymer spun into a fiber mesh and is used as a synthetic suture thread. It is prepared by the deposition of polymer solution over a nonwoven mesh of another polymer followed by subsequent evaporation. This technique provides a large surface area for cell attachment and rapid diffusion of nutrients favorable for cell survival and growth. However, one of the main drawbacks of this technique is the lack of structural stability which can be partly overcome by hot drying polylactic-acid fibers to improve the structural orientation and crystallinity of the mesh.

Self-Assembly

Self-assembly is the spontaneous organization of biomolecules via noncovalent interactions into an ordered structure with specific functions. Self-assembly naturally occurs in much of biochemistry, and its properties have been widely exploited for creating biomedical materials. For example, self-assembly of natural or synthetic molecules can produce nanoscale fibers known as nanofibers. Self-assembly of amphiphilic peptides is a common method used for the fabrication of 3D nanofibrous structures in tissue engineering and regenerative medicine applications. In an aqueous solution, the hydrophobic and hydrophilic domains within these peptides interact together with the help of weak noncovalent bonds (e.g., hydrogen bonds, Van der Waals interactions, ionic bonds, and hydrophobic interactions) to produce distinct, fast-recovering hydrogels. Along with peptides, synthetic polymer nanofibers can also be prepared through the self-assembly of di-block polymers with ordered morphologies at nanometer scales, making them ideal materials for various biomedical applications. Polymeric dendrimers can also be self-assembled into nanofibers. The di- and tri-block peptide ampholites (PAs) are designed to be self-assembled into rodlike architectures. A new technique for the self-assembly of PAs into nanofibers by controlling pH and engineering the peptide head group of PAs is also currently being developed. The salient features for synthesis of the PA involve (1) phosphoserine residue incorporation to increase hydroxyapatite mineralization, (2) Arg-Gly-Asp (RGD) peptide incorporation to enhance integrin-mediated cell adhesion, (3) four consecutive cystine residues incorporation to form intermolecular disulfide bonds that polymerize to provide improved structural stability, and (4) flexible linker region incorporation consisting of three glycine residues to provide flexibility to the head group. Due to the ease of modifying PA structures and reversing changes, this process allows for a variety of self-assemblies, including layered and lamellar structures, and provides system-wide flexibility. Therefore, self-assembly techniques show great potential for designing and manufacturing novel scaffolds for tissue engineering applications.

6.5.7 Polymer Orientation Generated by Processing

Polymers, especially crystalline polymers, can exhibit molecular orientation. There are many methods by which orientation can be produced in synthetic polymers, which generally fall into two groups depending on if the orientation is induced during or after the polymer solidification process. In the first group, the orientation is introduced after the polymer has first solidified in an unoriented state or a state of low orientation. There are essentially four types of processes that may be involved: drawing, extrusion, blow molding, and rolling. In these processes, the polymers may deform upon a directional external force with the following effects:

1. Yield is followed by a drop in a directional load. The load then remains constant for increasing extension, and the polymeric sample exhibits necking. When the neck has moved through all the material, the load increases rapidly for further extension, with breaking occurring after a little further extension. Then, drawn material is stable when the load is removed and is highly oriented.
2. Homogeneous deformation throughout, with an indistinct yield point. Some orientation is obtained once the strain is above the yield point.
3. Rubberlike behavior above the glass transition temperature, T_g. If the load is removed at the stretching temperature, the strain disappears, leaving no residual orientation, but fast cooling to below T_g in the stretched state results in the orientation being frozen in.

In the other group, the orientation is achieved by solidification of polymer from a fluid state in which the molecules are sometimes partially aligned. Further orientation is then often produced by subsequent drawing.

An example of the fluid-state polymer processes for producing fibers from polymer chains with defined molecular orientation is melt spinning. The principal requirement is that the polymer must not degrade before becoming molten. Examples of these polymers include nylon and polyester. The essential melt-spinning process is the production of polymer filaments by allowing molten polymer to pass downward through small holes in a spinneret. The precise mechanism by which orientation is produced in the fibers depends on the polymer and the details of processing and is not always easy to determine. It can take place through the crystallization or glass formation of molecules already oriented in the molten thread-line, by subsequent drawing after solidification, or by a mixture of the two processes. However, it has been shown that at a very high spinning speed, the melt can become orientated before solidification occurs, even with conventional polymers.

We may consider the deformation scheme of a polymer during these processes to predict the distribution of orientations. As a basic approximation, it is assumed that at the temperature of drawing or other deformation, the polymer is in the rubbery state, either as a true rubber containing chemical crosslinks or as a polymer containing tangled amorphous chains in which the entanglements can behave like crosslink points. Once deformation has been accomplished, the polymer is cooled down into the glassy state before the stress is removed so that the orientation is frozen at that moment.

The simplest version of the rubber model makes the assumption of a parallel deformation: when the polymer is stretched, the crosslink points move exactly as they would as if they were points in a completely homogeneous medium deformed to the same macroscopic deformation. As an approximation, we assume that each chain (i.e., the portion of a polymer molecule between two successive crosslink points) consists of the same number n of freely jointed random links, each of length l. We further assume that the expected length of the end-to-end vector r_0 of each chain in the unstrained state is equal to the root-mean-squared vector length of the free chains. If we consider the x-component of an end-to-end vector r of a polymer chain in the Cartesian coordinates, then $r_x = \Sigma_i l \cos\theta_i$, where θ_i is the angle between a link along the polymeric chain and the x-axis and i is the index of a polymeric link. Then

$$r_x^2 = l^2\left(\sum_{i=j} \cos^2\theta_i + \sum_{i \neq j} \cos\theta_i \cos\theta_j\right). \tag{6.28}$$

The second term in the above equation is approaching zero for long-chain polymers, i.e., $n >> 1$. If r_x^2 is averaged over a large number of randomly oriented polymer molecules, denoted as $<r_x^2>$, then $<r_x^2> = nl^2 <\cos^2\theta_i>$. Recall that

$$< \cos^2\theta_i >= \int_{\theta_i=0}^{\theta_i=\pi} \cos^2\theta_i d(\cos\theta_i) / \int_{\theta_i=0}^{\theta_i=\pi} d(\cos\theta_i) = \frac{1}{3}, \tag{6.29}$$

We can obtain $<r_x^2> = = nl^2/3$. We can also apply the geometric relations $<r^2> = <r_x^2> + <r_y^2> + <r_z^2>$ and $<r_x^2> = <r_y^2> = <r_z^2>$ and conclude that

$$r_0 = r_{\text{RMS}} = \sqrt{< r^2 >} = n^{1/2} l. \tag{6.30}$$

On the other hand, we neglect the change in volume due to deformation so that $\lambda_x \lambda_y \lambda_z = 1$ where the quantities λ_x, λ_y, and λ_z are the extension ratios defined along a Cartesian direction. Notably, the quantities have an obvious relation to the material strains: $\lambda_x = 1 + \varepsilon_x$, $\lambda_y = 1 + \varepsilon_y$, and $\lambda_z = 1 + \varepsilon_z$. For a basic consideration, only uniaxially oriented samples with a strain along z are considered here, which means that $\lambda_x = \lambda_y = (1/\lambda_z)^2$. In the following descriptions, we call λ_z as the draw ratio λ.

When the polymer is drawn, the end-to-end vector of each chain orients toward the draw direction; and the random links of each chain orient toward the end-to-end vector of the chain and, therefore, toward the draw direction. If we consider the angle θ_0 between the end-to-end vector r and the draw direction along the z-axis before drawing, it is trivial that $\cos^2\theta_0 = r_z^{\,2}/(r_x^{\,2} + r_y^{\,2} + r_z^{\,2})$ and $\tan\theta_0 = (r_x^{\,2} + r_y^{\,2})/r_z^{\,2}$. Accordingly, the angle θ_0 will change to θ_e after being drawn:

$$\frac{1}{\cos^2\theta_e} = \frac{r_x^2/\lambda + r_y^2/\lambda + \lambda^2 r_z^2}{\lambda^2 r_z^2} = 1 + \frac{\tan^2\theta_0}{\lambda^3} = \frac{\lambda^3\cos^2\theta_0 + (1 - \cos^2\theta_0)}{\lambda^3\cos^2\theta_0}; \text{ thus} \tag{6.31}$$

$$\cos\theta_e = \frac{\lambda^{3/2}\cos\theta_0}{\sqrt{1 + (\lambda^3 - 1)\cos^2\theta_0}}. \tag{6.32}$$

Here, we further introduce the concept of the distribution of orientations of the random links after drawing $N(\theta_e)$ with respect to the draw direction. To obtain the expression of $N(\theta_e)$, we first assume the polymer material initially has no orientation preference. In other words, θ_0 is randomly oriented, and the corresponding probability density function is $N_1(\theta_0) = 2/\pi$, given that $0 < \theta_0 < \pi/2$. We then apply the relation for changing the variable to find the probability density function $N_2(\cos\theta_0)$ for $\cos\theta_0$:

$$N_2(\cos\theta_0) = \left|\frac{d\theta_0}{d(\cos\theta_0)}\right| N_1(\theta_0) = \frac{2}{\pi\sqrt{1 - \cos^2\theta_0}}. \tag{6.33}$$

We can then use Eqs. 6.29 and 6.30 to calculate the probability density function $f_2(\cos\theta_e)$ for $\cos\theta_e$:

$$\begin{aligned} N_3(\cos\theta_e) &= \left|\frac{d(\cos\theta_0)}{d(\cos\theta_e)}\right| N_2(\cos\theta_0) \\ &= \frac{2\lambda^{3/2}}{\pi\left[\lambda^3 - (\lambda^3 - 1)\cos^2\theta_e\right]^{3/2}\sqrt{1 - \cos^2\theta_e}}. \end{aligned} \tag{6.34}$$

We apply again the change of variable technique to obtain $N(\theta_e)$ for $0 \le \theta_e \le \pi/2$:

$$N(\theta_e) = \left|\frac{d(\cos\theta_e)}{d\theta_e}\right| N_3(\cos\theta_e) = \frac{2\lambda^{3/2}}{\pi\left[\lambda^3 - (\lambda^3 - 1)\cos^2\theta_e\right]^{3/2}} \tag{6.35}$$

We can predict the distribution of θ_e changing with the draw ratio λ using the above ratio. In fact, the more useful prediction would be on the material properties (such as elastic modulus) changing with λ. As a basic understanding, we may continue the discussion with some "big" assumptions. We assume that the bulk elastic modulus E_θ along the draw force F_d is an integrated effect from all N_l links of the polymer chain, with each link having an effective elastic moduli along the link E_d and a shear modulus perpendicular to the link G_d. We may consider a link with an orientation angle θ_e from the draw direction deformed under an effective drawing force on each link f_d (with $f_d = F_d/N_l$). The occupied area of a link over a cross section is roughly proportional to $l\sin\theta_e$. Then, considering a cross-sectional layer of the polymeric links, we may obtain the relationship based on the average strain over a link layer:

$$\frac{F_d}{\sum_{N_l} l \sin\theta_e}\frac{1}{E_\theta}^{\theta} \approx \sum_{N_l}\left(\frac{f_d \cos\theta_e}{E_d}\cos\theta_e + \frac{f_d \sin\theta_e}{G_d}\sin\theta_e\right) \Big/ \sum_{N_l} l \sin\theta_e, \text{ and thus} \tag{6.36}$$

$$\frac{1}{E_\theta}^{\theta} \approx \frac{1}{E_d}\frac{\sum_{N_l}\cos^2\theta_e}{N_l} + \frac{1}{G_d}\frac{\sum_{N_l}\sin^2\theta_e}{N_l}. \tag{6.37}$$

By taking note of the anisotropy of the highly stretched polymeric molecules and making approximations by the orientation averages, it can be shown that for a sufficiently small θ_e, the resultant bulk elastic modulus E_d can have a reasonably good approximation as a function of the shear modulus G_d governing the torsional rigidities of both the fiber and the unit and the orientation of the polymer chains:

$$\frac{1}{E_\theta}^{\theta} \approx \frac{1}{E_d} + \frac{< \sin^2\theta_e >}{G_d} \approx \frac{1}{E_d} + \frac{\sin^2 N(\theta_e)}{G_d}. \tag{6.38}$$

The equation with fixed values of E_d and $< \sin^2\theta >$ has often been found to fit experimental data for the temperature dependences of E_θ and G_d. It has been well demonstrated that both the elastic modulus and strength of crystalline polymers can be enhanced with the draw ratio (Fig. 6.20). Extrusions and molding usually have high orientation in the surface layers due to there being high shearing forces at the walls of the extrusion die. The tensile modulus can be improved for extrudates by designing the dies so that the material can be drawn down, i.e., stretched, subsequent to extrusion to produce the final cross section required. This produces greater alignment of the polymer in the interior of the extrudate and thus improves its properties, but this cannot be done from moldings, for which the final shape is usually determined by the shape of the mold.

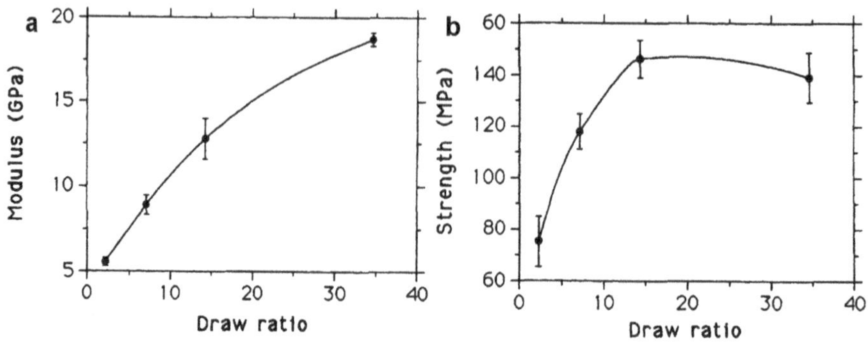

Fig. 6.20 Variations of (**a**) elastic modulus and (**b**) strength with draw ratio for crystalline copolyesteramide. (Lin [11]. Reproduced with permission of Wiley)

6.6 Ceramics Processes

6.6.1 Common Processes

Ceramic biomaterials are manufactured to cater to different needs with varying properties such as porosity, thickness, and pore size. Some other considerations during the fabrication process are surface cracks, inclusions, and excessive grain sizes. Inclusions refer to additives added to the materials to improve strength. Since melting, casting, and thermomechanical processing are generally not viable options for polycrystalline ceramics, ceramics are typically processed into useful shapes starting at ceramic powders. Ceramic powder is a collection of fine ceramic particles that are ready for shaping by crushing, grinding, separating impurities, blending different powders, drying, and spray drying to form soft agglomerates.

Compaction

Compaction is used to create what are called "green ceramics." These have respectable strengths and can be handled and machined. The compaction process can be completed within 1 minute for smaller parts; thus, uniaxial compaction can create large numbers of smaller, simpler components. Larger pieces with dimensions up to a few feet in diameter or length can be produced using a process called cold isostatic pressing, where pressure is applied using oil. Such large pieces then are sintered with or without pressure. Cold isostatic pressing is used to achieve higher green ceramic or complex shape densities.

In some cases, parts may be sintered using applied pressure in a process called hot sintering, used for refractory and covalently bonded ceramics that do not show good pressure-less sintering behavior. Hot isostatic pressing similarly sinters and compacts large pieces of metals and alloys using isostatic pressing by applying pressure against internal pore pressure, enhancing density without causing grain growth. Hot pressing or hot isostatic pressing also is used for making ceramics or metallic parts with almost no porosity. Some recent innovative processes that make use of microwaves have also been developed for drying and sintering ceramic materials.

Slip Casting

Slip casting typically uses an aqueous slurry of ceramic powder, known as the slip, which is poured into a plaster of Paris ($CaSO_4{:}2H_2O$) mold. As the water from the slurry begins to seep out by capillary action, a thick mass builds along the mold wall. When a sufficient thickness of ceramic builds up, the rest of the slurry is poured out (drain casting). More slurry can also be poured in to form a solid piece (solid casting). Pressure may also be used to inject the slurry into polymer molds. The green ceramic is then dried and "fired," or sintered, at a high temperature.

Extrusion and Injection Molding

Extrusion and injection molding are popular techniques to form relatively complex shapes with ceramics. The extrusion process uses a viscous, doughlike mixture of ceramic particles containing a binder and other additives. This mixture has a claylike consistency, which is then fed into an extruder where it is well-mixed in a machine known as a pug mill, sheared, deaerated, and then injected into a die where a continuous shape of green ceramic is produced by the extruder. This material is cut at appropriate lengths and then dried and sintered. Cordierite ceramics used for making catalyst honeycomb structures are made using this extrusion process.

Injection molding of ceramics is similar to injection molding of polymers. Ceramic powder is mixed with a thermoplastic plasticizer and other additives. The mixture is then extruded and injected into a die. Ceramic injection molding is better suited for complex shapes. The polymer contained in the injection-molded ceramic is burnt off, and the rest of the ceramic body is sintered at a high temperature.

6.6.2 Ceramic Scaffold Fabrication

Solvent Casting and Particulate Leaching

In solvent casting and particulate leaching as shown in Fig. 6.21, a polymer is dissolved in an organic solvent. Particles, usually salts, with specific dimensions are then added to the solution, and the mixture is shaped into its final geometry. For example, it can be cast onto a glass plate to produce a membrane or in a 3D mold to produce a scaffold. When the solvent evaporates, it creates a structure of composite material consisting of the particles together with the polymer. The composite material is then placed in a bath which dissolves the particles, leaving behind a porous structure.

Solvent casting is based upon evaporation of solvents in order to form scaffolds by either one of the two routes. One method is to dip the mold into polymeric solution and allow sufficient time to draw off the solution; as a result, a layer of polymeric membrane is created. The other method is to add the polymeric solution into a mold and provide sufficient time for the solvent to evaporate, creating a layer of polymeric membrane which adheres to the mold. One of the main drawbacks of this technique is that a toxic solvent may remain in the scaffolds. To overcome these problems, scaffolds need to be fully dried by vacuum in order to remove any remaining toxic solvent. However, this is very time-consuming, but some

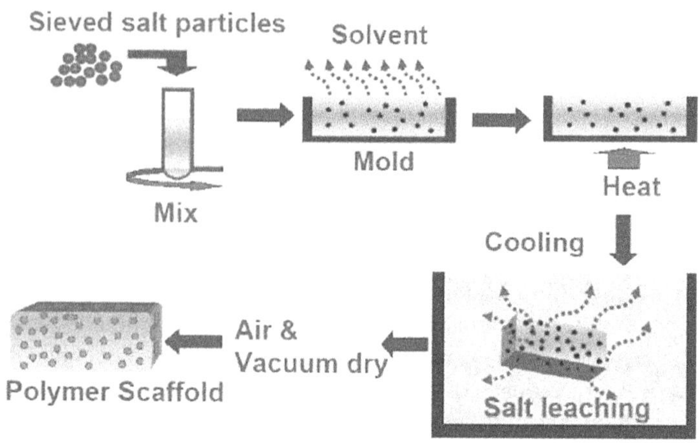

Fig. 6.21 Solvent casting and particulate leaching technique. (Adapted with permission from: Nanotechnology for the Regeneration of Hard and Soft Tissues, Huinan Liu and Thomas J. Webster, Copyright @ 2007 World Scientific Publishing Co.)

researchers have combined it with particulate leaching techniques with proteins optimized to create smart scaffolds for tissue regeneration.

Particulate leaching is a popular technique widely used to fabricate scaffolds for tissue engineering applications. Salt, wax, or sugars known as porogens are used to create pores or channels. Salt leaching produces an eventual porous scaffold, and this can be adjusted by using various diameters of salt particles and ratios of salt particles to polymer mixture. Here, salt is grounded into small particles, and those particles that have a desired size are poured into a mold and filled with the porogen. A polymer solution is then cast into the salt-filled mold. After the solvent evaporates away, the salt crystals are leached away using water to form the pores of the scaffold. This process is easy to carry out, and the pore size can be controlled by based on the amount, size, and shape of the added porogen. The particulate leached scaffold possesses pore size ~500 μm, a percentage of porosity ~94–95%, and the desired crystallinity. It is then casted in a mold and vacuum dried. The advantage of this method is the low amount of polymer it requires. However, certain critical variables, such as pore shape and inter-pore openings, are not well controlled. This method cannot be used to produce thick layers greater than 3 mm.

Melt Molding

Melt molding is a process of filling a mold with polymer powder and molding it into the desired shape either by injection or compression. Related fabrication techniques have been developed to control the pore interconnectivity and geometry, which are important for the exchange of nutrients and waste between pores in tissue engineering applications. Melt molding involves the filling of a Teflon mold with poly(lactic-co-glycolic acid) (PLGA) powder and gelatin microspheres of specific diameters. This is followed by heating the mold above the glass transition temperature of PLGA while applying pressure to the mixture, causing the PLGA particles to melt together.

Once the mold is removed, gelatin microspheres are dissolved by immersing the mixture in water. Scaffolds are then dried and assume the shape of the mold. The melt molding process has been modified to incorporate short fibers of hydroxyapatite as well. Uniform distribution of HA fiber throughout the PLGA scaffolds can only be achieved by using the solvent casting technique to prepare the composite material of hydroxyapatite fiber, PLGA matrix, and gelatin or salt porogen.

Gas Foaming

The gas foaming method (Fig. 6.22) combines two techniques: melt molding and particulate leaching. Air is incorporated into the mixture, where chemicals will bind to surround the air bubbles, hence trapping them in the polymer. As the mixture hardens in the mold, it forms a highly porous scaffold. This method allows for porosity as high as 90%.

Many of these fabrication techniques require the use of organic solvents and high temperature. The organic solvent residues that remain after completion of these processes can damage cells and nearby tissues and may also denature biologically active molecules incorporated within the scaffolds. To avoid this problem, the gas foaming scaffold fabrication technique does not require the utilization of organic solvents and high temperature. Instead, it uses high-pressure carbon dioxide gas for the fabrication of its highly porous scaffolds. The porosity and porous structure of the scaffolds depend upon the amount of gas dissolved in the polymer. This process involves exposing highly porous polymer with carbon dioxide at high pressure

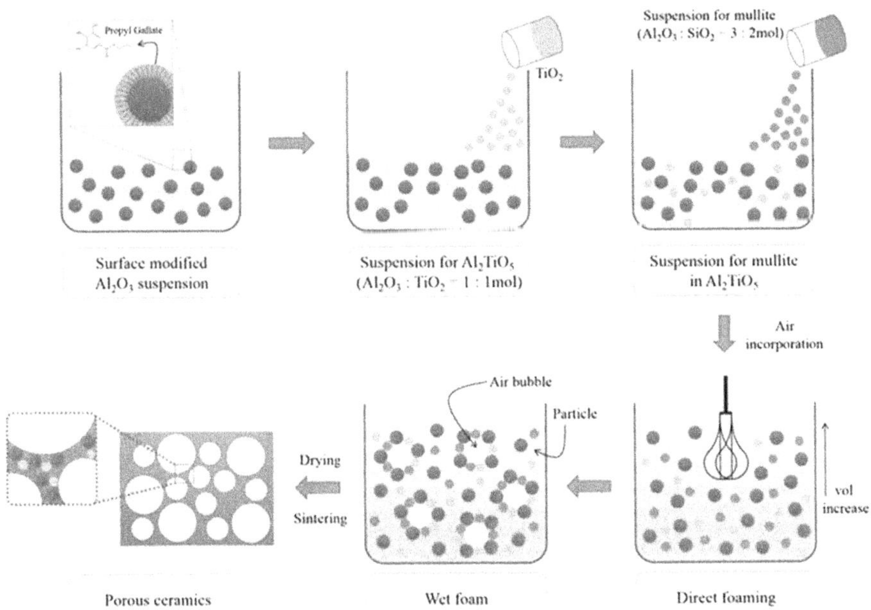

Fig. 6.22 Gas foaming process. (Sarkar et al. [12]. Adapted with permission from Elsevier)

(800 psi) to saturate the polymer. Under this condition, dissolved carbon dioxide becomes unstable and will phase separate from the polymer. Carbon dioxide molecules cluster to minimize free energy, and as a result, pore nucleation occurs. These pores cause the significant expansion of polymeric volume and decrease in polymeric density. A 3D porous structure (scaffold) is formed after completion of the foaming process. The porosity of the scaffolds is controlled by the use of porogens like sugar, salts, and wax. The polymer (e.g., PLGA) that expands in the foaming process fuses together around the porogen to create a continuous polymeric matrix and also entraps any other molecules which are present in the mixture. The polymer and porogen mixture is exposed to high pressure until it completes its saturation with carbon dioxide, followed by the foaming process. The porogen is then removed, and a highly interconnected pore structure is formed.

Freeze-Drying
This method (Fig. 6.23) freezes the material before reducing the surrounding pressure. Heat is then added to allow frozen water to sublime, leaving behind the formed ceramic. The freezing rate can be controlled to manage the porosity of the material. This can produce porosity up to 95%. This method allows for manufacturers to achieve a homogeneous powder formulation that performs better under certain processes such as hot pressing.

Fiber Bonding
Polyglycolide (PGA) fibers are incorporated into a polymer solution, and the polymer and fibers subsequently bind. This eventually produces a final product with a porosity as high as 80%. However, this technique has a number of limitations. It has poor structural and mechanical stability and can only be used to make small membranes. It is also difficult to vary or control the porosity of the material.

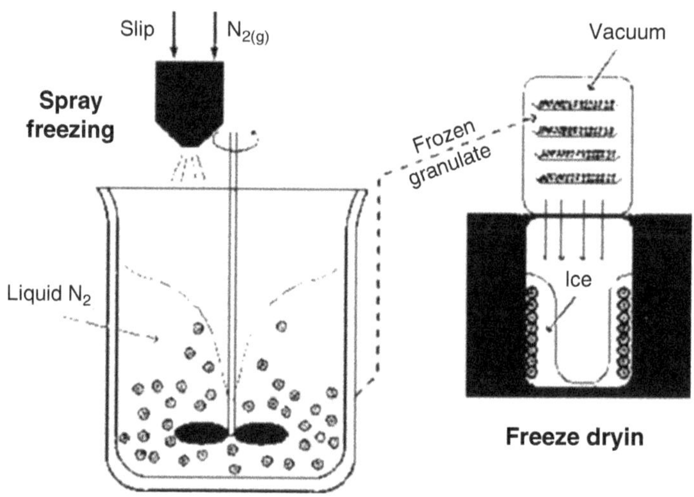

Fig. 6.23 Process of freeze-drying. (Lyckfeldt et al. [13]. Reproduced with permission from Elsevier)

Synthetic polymers, such as biodegradable poly (l-lactic acid) (PLLA) which consists of non-bonded PGA fibers embedded in PLLA matrix, can overcome these limitations for scaffold applications. Their scaffolds can be fabricated by bonding a collagen matrix to PGA polymers with threaded collagen fiber stitches during posttreatments at a temperature above the melting temperature of PGA. As a result, PLLA matrix made of the composite is removed by dissolving in methylene chloride agent due to the fact that PGA is insoluble in this solvent. This process yields the PGA fiber scaffolds that are bonded together by heat treatment. PGA mesh provides high-porosity, surface area to polymer mass ratio, and mechanical stability that allows for tissue ingrowths.

6.7 Process Compatibility

Different manufacturing processes are compatible for different materials depending on materials' properties, operating situation, tools involved, etc. Many factors needed to be considered when choosing the appropriate manufacturing process. Process compatibility accesses the variability of the output of a process using statistical methods. The process compatibility between different materials and basic processes is summarized in Fig. 6.24.

		Metals, ferrous	Metals, non-ferrous	Ceramics	Glasses	Elastomers	Thermoplastics	Thermosets	Polymer foams	Composites
Shaping	Die casting	●	●							
	Investment casting	●	●							
	Forging	●	●							
	Extrusion		●							
	Sheet forming	●	●							
	Conventional machining	●	●	●	●	●	●	●	●	
	Injection molding				●	●	●	●	●	
	Blow molding				●		●			
	Compression molding				●	●	●	●		
	Resin-transfer molding						●	●	●	●
Joining	Adhesives	●	●	●	●	●	●	●	●	●
	Welding, metals	●	●							
	Welding, polymers					●	●	●	●	
	Fasteners	●	●	●	●	●	●	●	●	●
Finishing	Grinding	●	●	●	●					●
	Polishing	●	●	●	●		●	●		●

Fig. 6.24 Material-process compatibility

Problems

Problem 6.1

Consider the injection molding operation of pre-cured silicone (viscosity, 100 Pa·s, and density, 965 kg/m^3) to continuously manufacture a container as shown in the Fig. 6.P1.

Here, we first focus on the cylindrical nozzle (shown below) in the machine with a diameter of 3 mm and 1 cm in length.

(a) Calculate the required flow rate of silicone such that the *Reynolds* number of the silicone flowing along the nozzle is ~10.
(b) Based on the flow rate calculated in (a), please estimate the pressure drop along the nozzle.
(c) Imagine an untrained worker accidentally turned off the machine, and the molded container was very hard to remove manually from the mold afterward. Can you guess possible reasons why? Do you have any suggestions on the product design in order to eliminate such problems?

Problem 6.2

The metal die casting operation of titanium was applied to manufacture the bone fixation plate at a temperature of ~1900 K (the viscosity of liquid titanium as a function of temperature is shown in the *left* of Fig. 6.P2). Here, we first consider that there was one cylindrical nozzle (*right* of Fig. 6.P2) in the die with a diameter of 5 mm and 1 cm in length. This nozzle has the highest fluidic resistance along the transmission path of molten titanium during the process. Calculate the required flow rate of liquid titanium such that the Reynolds number of the flow along the nozzle is ~10. What should the minimum gauge pressure of this process be?

Fig. 6.P1 A manufactured silicone container

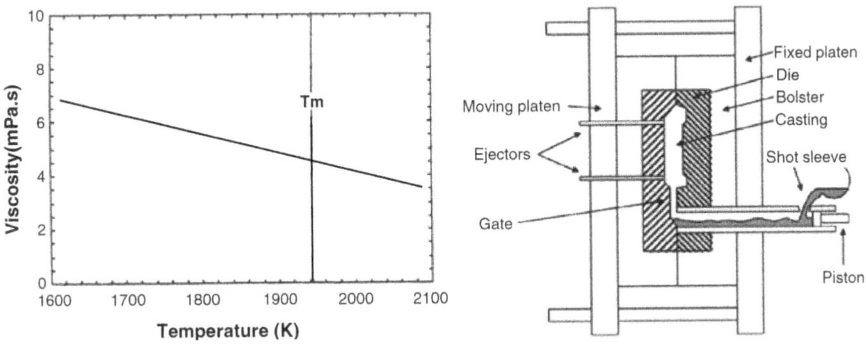

Fig. 6.P2 Viscosity of titanium against temperature (left) and die casting of titanium

Problem 6.3

(a) The patient is a 68 kg man who is momentarily subjected to a force which is five times his weight on one of his legs. Determine the average normal stress developed in the tibia *T* of his leg at the mid-section *a-a* (Fig. 6.P3). The cross section can be assumed to be circular with an outer diameter of 4.5 cm and an inner diameter of 2.5 cm (please note that a bone structure should be a hollow shape). Assume the fibula *F* does not support a load.

(b) Consider the tibia of the patient is being fixed by an external bone plate as shown in Fig. 6.P4 for which *L* is 35 cm and the radii of the four supporting rods (in *gray* color) are identical at 3 mm. Please estimate the shearing force shared by each rod under the external force *P*.

(c) Given a final shape of the fixation plate as shown in Fig. 6.P5, can you design the die? Please further mention the possible positions and modes of defects if you think there are any.

(d) The thickness of this bone fixation plate was 2 mm, and the multiple holes each had a diameter of 5 mm. The long hole located at the center was 5 cm in its total length (*L*). Based on the shear stress-shear strain curve (Fig. 6.P6), estimate the forces required to punch the different holes on the fixation plate.

Problem 6.4

Refer to Fig. 6.P7. Use average yield strength $Y = 400\ \text{N/mm}^2$, $h_0 = 2$ mm, $h_1 = 1.4$ mm, $v_0 = 10$ m/s, $\sigma_x = 180\ \text{N/mm}^2$, sheet width $w = 1000$ mm, and $R = 250$ mm.

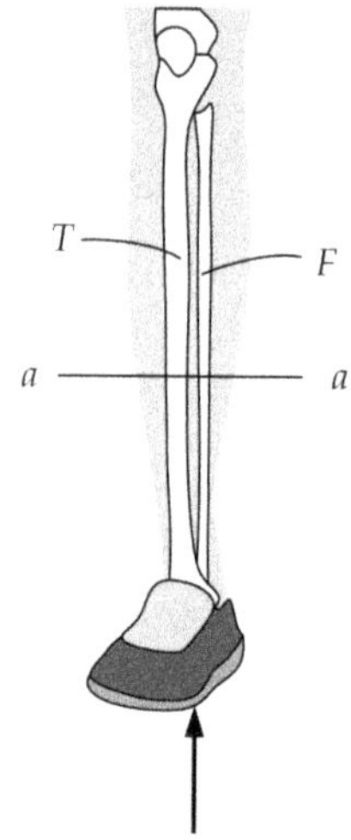

Fig. 6.P3 Momentary force acting on a leg

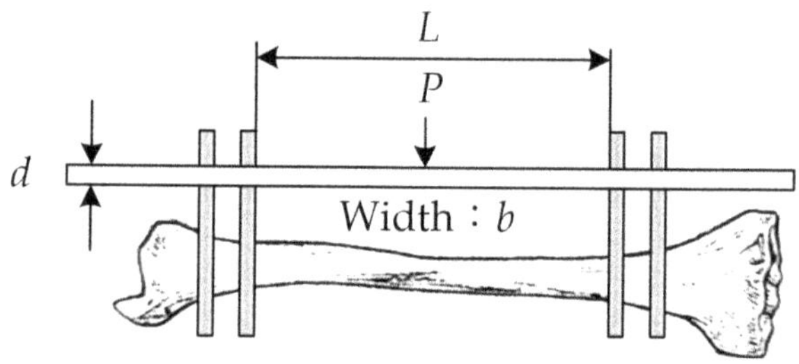

Fig. 6.P4 Configuration and key parameters of external bone fixation

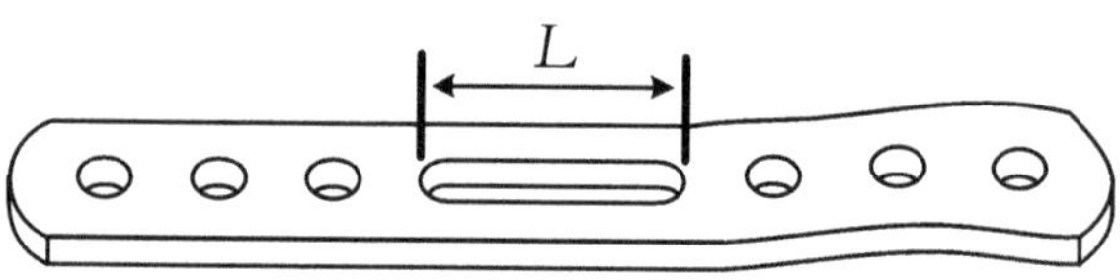

Fig. 6.P5 A bone fixation plate

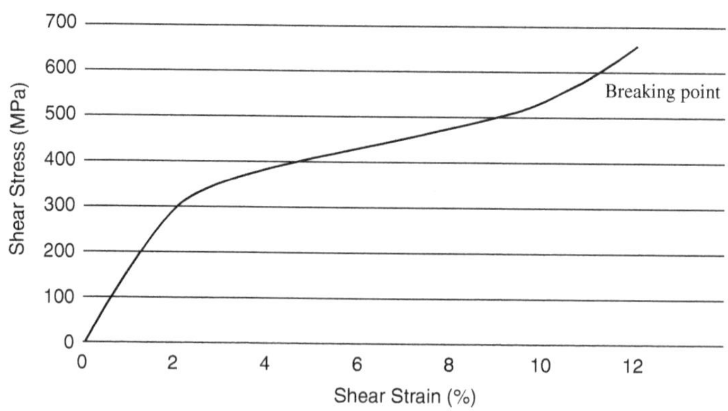

Fig. 6.P6 Shear stress versus shear stain for a material used in a bone fixation plate

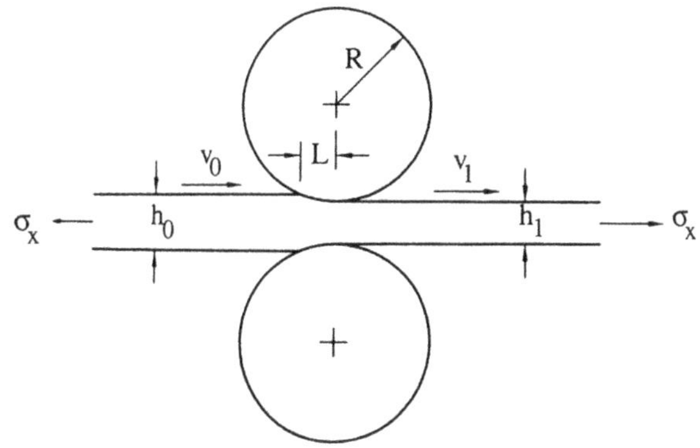

Fig. 6.P7 Key parameters during rolling

(a) In a stress-strain test of the material before the rolling process, the specimen, with an initial diameter $d_0 = 20$ mm, fractured with a diameter $d_f = 12$ mm. The engineering stress at fracture was determined as $S = 400$ N/mm^2. Determine the true strain ε_f and true stress σ_f (N/mm^2) at fracture.
(b) By assuming the Poisson's ratio of the material is 0.4 and the material deformation in the out-of-plane direction of Fig. 6.P7 is neglected, determine the exiting sheet speed v_1 (m/s), the length of compression zone L (mm), the compressive force F (N), and the torque on each roller T(Nm).
(c) Based on your results in (a), determine the total power P (kW), by assuming the average speed $v = (v_0 + v_1)/2$ and the temperature increase ΔT (°C) of the work following the material property: density × specific heat of capacity = 3.7 N/mm^2 °C.

Problem 6.5

Consider a drawing process as referring to Fig. 6.P8 and neglecting friction.

Let us consider the material yield strength Y is a function of strain ε, with the relation $Y = Y_0 + K\varepsilon = 280 + 320\varepsilon$ (N/mm^2). Also, $d_0 = 11$ mm, $d_1 = 6$ mm, $v_d = 4$ m/s, density × specific heat of capacity = 3.7 N/mm^2 °C. Determine ε_1, Y_0 (N/mm^2), Y_1 (N/mm^2), F_d (N), the drawing power P (kW), the increase of the work temperature, and the maximum possible ratio of d_0/d_1.

Problem 6.6

Consider a drawing process as referring to Fig. 6.P9 and neglecting friction.

It is given that the material yield strength Y is a function of strain ε, with the relation $Y = Y_0 + K\varepsilon = 250 + 300\varepsilon$ (N/mm^2), $d_0 = 6.5$ mm, and d1 = 4 mm.

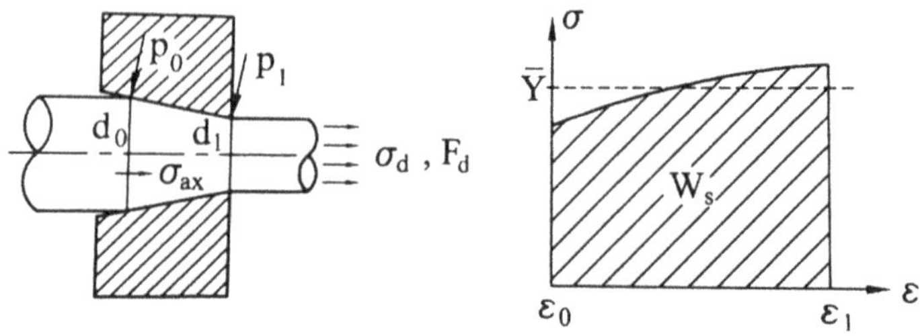

Fig. 6.P8 Key parameters during drawing

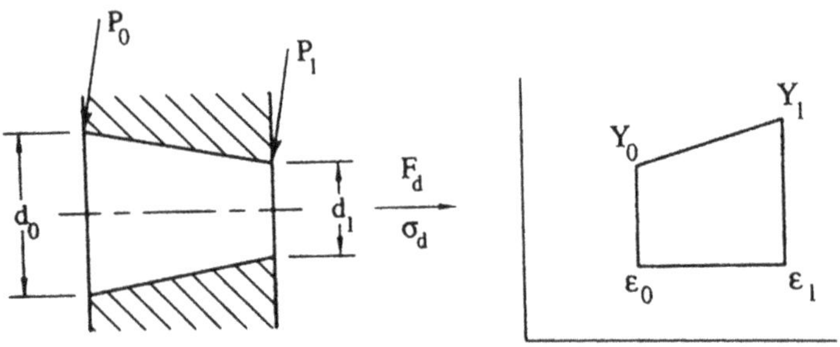

Fig. 6.P9 Key parameters during drawing

(a) Determine ε_1, Y_1 (N/mm^2), σ_d (N/mm^2), and f_d.
(b) Determine the pressure p_0 and p_1. Write out the yield criterion of the axisymmetric case as a relation between σ_x and p.

Problem 6.7

Please reconsider Figs. 6.P8 and 6.P9 as an extrusion process, with the parameters $d_0 = 18$ mm and $d_1 = 10$ mm. The yield strength of the material is $Y = 200 + 150\varepsilon$ (N/mm^2). Determine the extrusion stress σ_{ex} (N/mm^2) and pressure between work and die, i.e., p_0 and p_1 (N/mm^2).

Problem 6.8

A thin sheet of material is subjected to a biaxial stress field in the *xy*-plane, where *z* is the sheet thickness direction. The shear strains are zero and $\sigma_x = 9$ MPa, $\sigma_y = 6$ MPa, and $\sigma_z = 0$ MPa. If Young's modulus E of the material is 3 GPa and its Poisson's ratio $\nu = 0.3$, calculate the extensional strains ε_x, ε_y, and ε_z.

Problem 6.9

Consider a radial drawing process, a cylindrical cup of thickness $h = 1.0$ mm is being formed for an initial sheet diameter of D_o to a cylindrical cup diameter of D_{draw}. The material is non-strain-hardening, with a yield strength of $Y = 450$ MPa.

(a) Determine the radial stress σ_1 (MPa) in the flange at four radii: $r = D_o/2 = 10$ cm, $r = 8.5$ cm, $r = 6.5$ cm, and $r = D_{draw}/2 = 5$ cm for the initial position of the blank: $r_0 = D_o/2$.
(b) Determine the drawing stress σ_d (MPa) at the position $r = D_{draw}/2 = 5$ cm and the drawing force F_d (kN) required as the munch moves downward from its initial position to four subsequent positions of the blank: $r_0 = 9$, 8, 7, and 6 cm.
(c) (Optional) Repeat (a) and (b) by reconsidering the material with strain-hardening yield strength: $Y = 450 + 450\varepsilon^{0.3}$ (MPa).

Problem 6.10

Consider a shell of plastic material during the blow molding process. The material behaves as a neo-Hookean solid with an elastic modulus of 1.2 MPa. If the plastic after extrusion has a roughly spherical shell shape with a diameter of 2 cm and wall thickness 1 mm, before an additional pressure is applied, calculate the gauge pressure inside the plastic "balloon" during its expansion instantly up to a diameter of 2.2, 2.5, 3.5, and 10 cm.

References and Further Reading

1. Swift, K.G., Booker, J.D.: Manufacturing Process Selection Handbook. Butterworth-Heinemann, Oxford (2013)
2. Schey, J.A.: Introduction to Manufacturing Processes. McGraw-Hill, New York (1987)
3. Singh, R.: Applied Welding Engineering: Processes, Codes, and Standards. Butterworth-Heinemann (2015)
4. Bauser, M., Siegert, K.: Extrusion. ASM International, Materials Park (2006)
5. Altan, T., Erman Tekkaya, A.: Sheet Metal Forming: Fundamentals. ASM International, Materials Park (2012)
6. Francis, L.F.: Materials Processing: a Unified Approach to Processing of Metals, Ceramics and Polymers. Academic Press, Cambridge, MA, USA (2015)
7. Srivatsan, T.S., Sudarshan, T.S., Manigandan, K.: Manufacturing Techniques for Materials: Engineering and Engineered. CRC Press, Boca Raton (2018)
8. Rajput, R.K.: A Textbook of Manufacturing Technology: Manufacturing Processes. Firewall Media. Laxmi Publications, Boston, MA, USA(2007)

9. Groover, M.P.: Fundamentals of Modern Manufacturing: Materials Processes, and Systems. Wiley, Hoboken (2007)
10. Schrader, G.F., Elshennawy, A.K.: Manufacturing Processes and Materials. Society of Manufacturing Engineers, Dearborn (2000)
11. Lin, Q., Jho, J., Yee, A.F.: Effect of drawing on structure and properties of a liquid crystalline polymer and polycarbonate in-situ composite. Polym. Eng. Sci. **33**, 789–798 (1993)
12. Sarkar, N., et al.: Al_2TiO_5–mullite porous ceramics from particle stabilized wet foam. Ceramics Int. **41**(5), 6306–6311 Part A, (2015). Fig. 1. https://www.sciencedirect.com/science/article/pii/S0272884215000954
13. Lyckfeldt, O., et al.: Freeze granulation for the processing of Silicon Nitride Ceramics. Key Eng. Mater. **264–268**, 281–284 (2004). Figure 1. https://www.scientific.net/KEM.264-268.281

Chapter 7
Medical Imaging and Reverse Engineering

Abstract In the field of biomedical engineering, reverse engineering is used to reconstruct physical models in the form of computer data (virtual models) without the assistance of conceptual drawings, specifications, or technical drawings of the product design. This involves the phase of data collection which is achieved through the scanning of physical components by medical imaging techniques. Due to the demands of production time and complexity, rapid prototyping can then be applied to fabricate geometries matching those of the human body, which is often irregular, complex, and unique in shape. With rapid prototyping replacing some of the conventional manufacturing processes, the efficiency of the whole process of production can be boosted.

7.1 Introduction

A straightforward flow of product development can be started from a product idea fulfilling market demands, followed by product specification definition, prototype drafting, fabrication, assembling, and detail optimization. Such a product development flow is categorized as "forward engineering" which generally translates a high-level form of design (such as logical designs and data) into a low level form of structures (such as physical element). Sometimes, there are cases in which physical parts are required without any technical details of the product; in which cases, traditional forward engineering development approaches cannot fulfill the requirements. In some biomedical applications such as prosthesis manufacturing, custom implanted fixation devices need to be shaped and fabricated based on the bone damage pattern of a patient. However, the forward engineering approach is not applicable as the device shape and other details are determined by the patient's body status rather than market needs or any expected specifications. Furthermore, forward engineering normally requires a long duration of product development time which cannot generate the necessary implant device on time.

As an alternative approach, the idea of "reverse engineering" can fulfill these challenging requirements. In the field of biomedical engineering, reverse engineering is used to reconstruct physical models in the form of computer data (virtual

R. H. W. Lam, W. Chen, *Biomedical Devices*,
https://doi.org/10.1007/978-3-030-24237-4_7

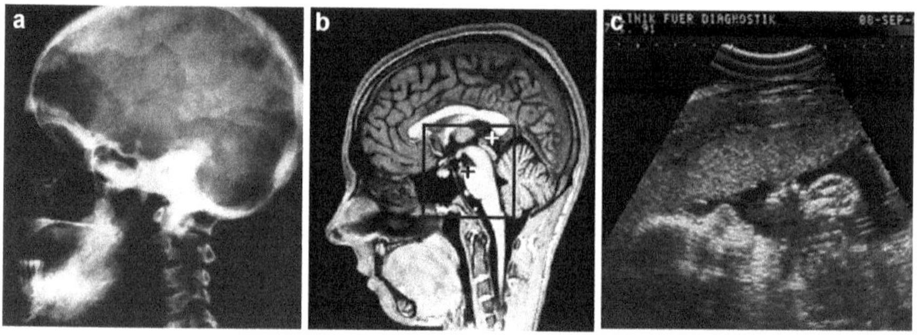

Fig. 7.1 Sample medical images of (**a**) X-ray (Du Boulay [7]. Adopted with permission from Butterworth-Heinemann), (**b**) magnetic resonance imaging (Ségonne [13], Neuroimage. Adopted with permission), and (**c**) ultrasound. Courtesy of Dr. Helena Lam

models) without the assistance of conceptual drawings, specifications, or technical drawings of the product design. Through capturing and analyzing the patient's body, the biomedical device designer can replicate the physical structures of a body part for implantation which is much harder to achieve using forward engineering. The designer can then identify the product shapes and their interrelationships within the patient based on geometry of the target body site of implantation in a digitalized form.

Typically, reverse engineering includes two phases: digitization and reconstruction. Digitization is the phase of data collection which is achieved through the scanning of physical components by medical imaging techniques, such as X-ray, magnetic resonance imaging, and ultrasound as shown in Fig. 7.1a–c, respectively. Normally, the firsthand acquired images of three-dimensional structures can be represented in three forms, which are point clouds, polygon models, and image series. Reconstruction is the phase which further refines the data and constructs the final three-dimensional model. During this process, noise and unnecessary data are removed in order to create a high-quality computer-aided design (CAD) model. Hence, reverse engineering also refers to the part-to-CAD process.

7.2 Ultrasound Imaging

Ultrasound is sound waves with frequencies from 20 kHz up to several gigahertz that are higher than the upper audible limit of human hearing. Ultrasound can be used as a nondestructive medical imaging technique to visualize muscles, tendons, and many internal organs in order to observe their size, structure, and pathological lesions with real-time imaging without the risk of tissue damage.

An ultrasound scan involves using an ultrasound transducer which applies an ultrasound sonic wave into the body and detects the resulting echoes as they bounce back from interfaces of internal structures. These sonic waves are generated through the vibration of a piezoelectric crystal in the ultrasound transducer which also detects the pressure of reflected ultrasound waves. The core working principle of ultrasound imaging is based on echoes. When an ultrasound wave travels from one medium to another medium, the wave is partially reflected at the interface. The amount of reflection depends on the difference between the acoustic impedances (Z_a) of the two media. Z_a of a medium also characterizes a medium that allows ultrasound to propagate. Z_a is defined as

$$Z_{\mathrm{a}} = \rho v_{\mathrm{a}}, \tag{7.1}$$

where ρ is the medium density and v_a is the speed of ultrasound in the medium. The unit of Z_a is kg m^{-2} s^{-2} or Rayl. Typically, the density and ultrasound speed in air are 340 m s^{-1} and 1.2 kg m^{-3}, respectively. The density and ultrasound speed in soft tissues/bioliquids are ~1600 m s^{-1} and ~1000 kg m^{-3}, respectively, and the density and ultrasound speed in bones are 3700 m s^{-1} and 1740 kg m^{-3}, respectively. In other words, the acoustic impedances of the air, skin, and bone are ~0.4 kRayl, ~1.6 MRayl, and 6.43 MRayl, respectively.

The intensity refection coefficient α_a is defined as the portion of ultrasound energy reflected at the interface relative to the incident ultrasound energy, with the relation

$$\alpha_{\mathrm{a}} = \frac{(Z_{\mathrm{a1}} - Z_{\mathrm{a2}})^2}{(Z_{\mathrm{a1}} + Z_{\mathrm{a2}})^2}, \tag{7.2}$$

where Z_{a1} is the acoustic impedance one of the media and Z_{a2} is that of the other medium. Notably, the proportion of transmitted ultrasound is $1 - \alpha$.

Echoes can be generated at the interfaces between air and soft tissues. Yet, ultrasound can also transmit across different soft tissues and bioliquids as these components contain mostly water. Most of an ultrasound wave is reflected at the tissue/bone interface because of the large acoustic impedance difference, and hence, it is hard to visualize the structure covered by bones using ultrasound imaging. Further, a gel with acoustic impedance similar to skin should be added on the skin as a coupling medium to eliminate any air between the ultrasound probe and the skin. Otherwise, this gas layer will reflect most of the ultrasound wave before it penetrates into the skin.

Ultrasound scans typically adopt the pulse-echo technique. A short pulse of ultrasound is sent from a transducer toward a material interface inside a human body as shown in Fig. 7.2. After a certain time t_a, the reflected pulse is detected by the probe. As the reflected sonic wave travels back and forth along the media body, the traveling distance is double the medium thickness, and the round trip time t_a can be estimated as

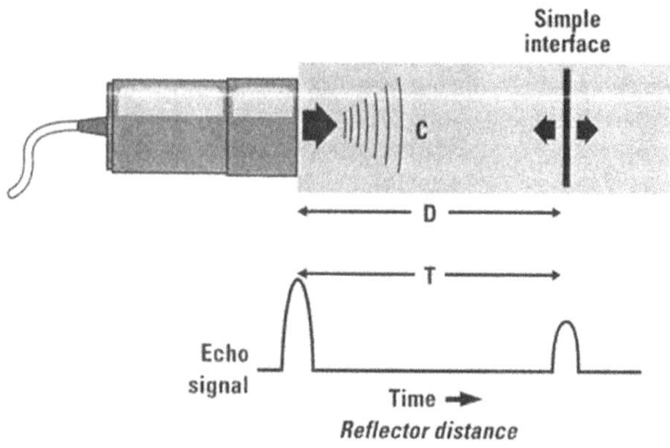

Fig. 7.2 Transmitted and reflected pulses detected by the pulse-echo technique

$$t_a = 2T_a/v_a, \tag{7.3}$$

where T_a is the medium thickness

In addition, during ultrasound imaging on the human body, the intensity of the sonic wave continuously dampens along the transfer distance in the same medium. Attenuation of ultrasound is caused by two processes: (1) absorption by the body medium such that ultrasound energy is converted to heat and (2) molecular scattering which occurs when the molecules in the medium absorb the energy and re-emit ultrasound in all directions. The working principle of scattering is different from reflection, yet part of the scattered ultrasound can be detected by the probe along with reflected signals, causing the measured echoes to become very noisy. Attenuation of ultrasound in a medium can be roughly represented as

$$I_a = I_{ao}e^{-\mu_a x_a}, \tag{7.4}$$

where I_a is the ultrasound intensity at the depth location of a media x_a; I_{ao} is the incident intensity where the ultrasound first enters the medium; and μ_a (unit: m^{-1}) is the linear attenuation coefficient depending of the material type. The values of μ_a are 0.00507 m^{-1} for water, 3.45 m^{-1} for blood, ~20 m^{-1} for soft tissues, 50 m^{-1} for muscle, and 327 m^{-1} for bone.

Ultrasound can generate either individual or time-lapsed images in real time at a relatively low cost. A widely adopted ultrasound imaging technique is "B-scan." B-scan produces a series of scan lines which are distances of interfacial points from the sensor. The scan lines combining together form a picture of objects inside the body. This scanning scheme can be further extended when using an array of transducers for an area scan such that a three-dimensional image can be captured.

Ultrasound induces neither pain nor ionizing radiation to the patients which is ideal for displaying and monitoring tissues like the utero, embryo, heart, liver, kidneys, breast, eye, large blood vessels, and gall bladder. It is a useful tool for detecting size, position, and abnormality of organ movements. However, ultrasonic images only offer relatively low image resolution because of acoustic signals from the environment which result in noisy measurements.

7.3 X-Ray Computerized Tomography

7.3.1 X-Ray

X-ray is one of the most well-known radiography methods and the earliest medical imaging method. Ever since X-rays were discovered in 1895 by the German scientist Conrad Roentgen, medical imaging has experienced generations of technological advancement. The discovery of X-rays enabled the first observation of bone and other structural features inside a human body. Conrad described the diagnostic capabilities of X-rays for imaging a human body and subsequently received the Nobel Prize in 1901. X-rays are a part of the electromagnetic spectrum with small wavelengths from 10 nm down to 0.01 nm. Due to such a small wavelength, they have great capability of straight penetration and transmission in the human body. X-rays can be used for diagnostic and therapeutic applications. Lower wavelength X-rays between 0.1 nm and 0.01 nm are used for diagnostic purposes.

In general, emission of X-rays relies on high-energy electrons with heavy target atoms such as tungsten or molybdenum. X-rays can be generated by an X-ray tube. In an X-ray generation tube, electrons are released by the source cathode and are accelerated toward the target anode in a vacuum under the potential difference ranging from 20 to 150 kV. When electrons collide with a metal target (anode), electromagnetic X-ray radiation can be generated due to the large drop in acceleration caused by the interaction between the anode and the electrons.

During X-ray imaging, patients are asked to lie or stand between the X-ray emitter and a film which captures the X-ray image (Fig. 7.3). An X-ray image can effectively visualize the bone shape features of the patient, including any cracks and abnormal growth of bones. When the radiation beams enter the human body, they are either absorbed by tissues or penetrated through the body with different organs/tissues having various capabilities of absorbing X-rays. This is because they are made of different elements. Heavier atom, e.g., calcium in the bone, can absorb more X-rays as they contain more electrons. The radiation beam is attenuated in the human body due to the mass attenuation coefficients of physiological structures. This relation can be represented by Eq. 7.4 as mentioned before. Here, we may consider that the captured X-ray intensity I_x has passed through N number of unit-thickness materials, with a linear attenuation coefficient $\mu_{x,i}$ (unit: m^{-1}) upon X-ray for the unit i. Then,

$$I_x = I_{xo} e^{-(\mu_{x,1}+\mu_{x,2}+\cdots+\mu_{x,N})\Delta x} \tag{7.5}$$

where I_{xo} is the incident X-ray intensity. For instance, the values of $\mu_{x,i}$ for X-rays are ~16 m^{-1} for water and ~48 m^{-1} for bone. The attenuation of radiation intensity is determined at each scan location by measuring the difference of intensity between the source and detector. A planar attenuation map, as obtained through scanning the object in the respective geometry, can be recorded on a radiographic film for an X-ray film radiograph.

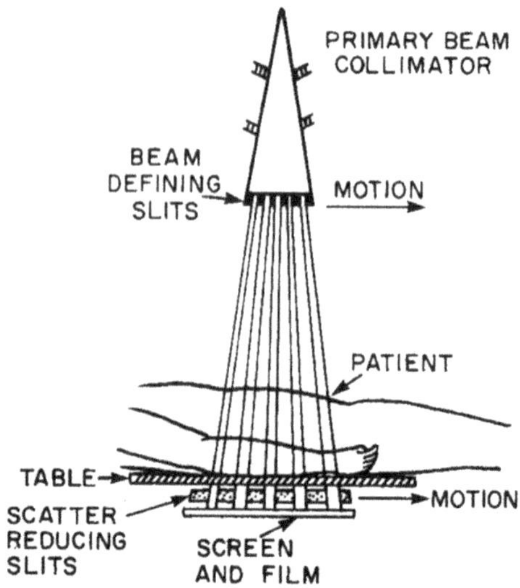

Fig. 7.3 Configuration by X-ray imaging

Although X-ray has the benefit of low cost and ease to use, X-ray radiation has the potential to damage organisms. Diagnostic X-ray exposure increases the risk of developmental problems and cancer. Normally, diagnostic X-rays only use a small amount of radiation to create images of the human body. The level of radiation exposure is considered safe for most adults. However, the risk of radiation exposure is greater to a fetus, so in pregnant patients, the benefits of an X-ray scan should be balanced with the potential hazards to the fetus. Moreover, it is hard to analyze if the organs/tissues have a similar absorption coefficient when the anatomical structure is unclear. It is also infeasible to deduce the three-dimensional structures of media from a single X-ray image because the intensity of each image point represents the integrated absorbance of all material points along a transmission path which cannot be decoupled back to the local attenuation coefficients for the individual material points.

7.3.2 Computerized Tomography

Computerized tomography (CT) scanning is a new radiography method transformed from X-ray technique for three-dimensional imaging. The CT scanning method captures a set of images of multiple cross section(s) at different angles in a helical scanner either in sequential or spiral format to obtain the three-dimensional structure information of the human body. CT is commonly used for the upper body (not including the head) to avoid potential damage to the brain. A CT scanner normally has three major components: (1) a scanner part consisting of an X-ray tube, detector, and rotary scanning frame (Fig. 7.4a); (2) a computer system which collects the X-ray images and computes them as the three-dimensional medium structures; and

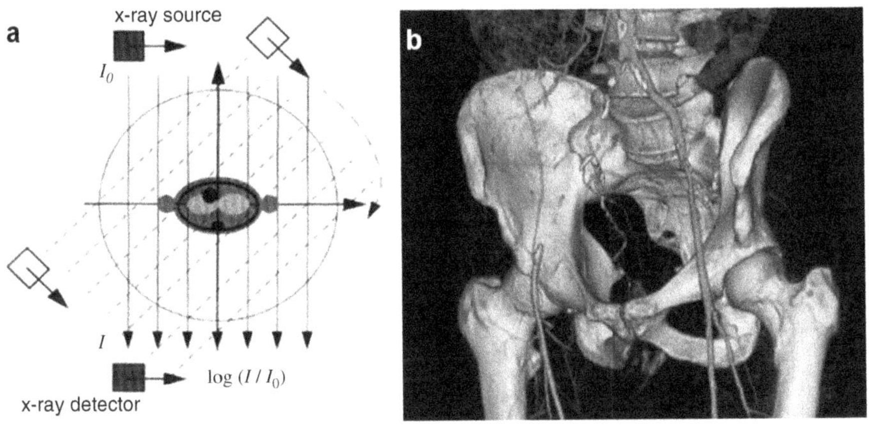

Fig. 7.4 (**a**) Configuration of the imaging in a CT chamber. (**b**) Clinical example of CT angiography. (Courtesy of D. Mears)

(3) a data storage and visualization system. After the computer analyzes those images/data, it can reconstruct the sequence of X-ray images in different orientations as a stack of high-resolution images. Each X-ray scan image, also called a CT slice, represents a cross-sectional view of the body. A stack of CT slices can further reconstruct the 3D geometry body features as shown in Fig. 7.4b.

The CT scanner is a ring-shaped machine with a tunnel through its center. The patient lies on a table that can move along the tunnel. Inside the ring is an X-ray unit which consists of an X-ray tube and an array of detectors located on the opposite side. The X-ray unit can rotate 360° around the patient. As the unit rotates, a series of parallel X-rays passes through the body and are attenuated and detected by a sensor mounted on the other side of the rotating component for capturing CT slices. After each complete rotation, the X-ray scanner records thousands of measurements which are fed into a computer program for image reconstruction. The computer then computes the linear attenuation coefficients μ_x of each volume pixel, called a voxel.

The total attenuation coefficient along an X-ray beam passing through the body is considered as the "single projection" P_x. Recalling Eq. 7.5, p_x has the expression as

$$p_x = \int \mu_x(x)dx = (\mu_{a,1} + \mu_{a,2} + \ldots + \mu_{a,N})\Delta x = \ln(I_{Xo}/I_X) \tag{7.6}$$

The single projection during CT scan for a CT slice should be the function of orientations and positions of the X-ray source and detector.

Mathematically, it is feasible to reconstruct the voxels of the μ_x values over a CT slice from the measured set of single projections for different positions and orientations. Such computation is called the back-projection. As an introduction, we consider an X-ray configuration with a simpler back-projection algorithm, which consists of the rotary module with both the X-ray source and detector moving together in the same direction in parallel as shown in Fig. 7.4a. In this case, we may let $f(x, y)$ be the two-dimensional map of local μ_x level, with $x = 0$ and $y = 0$ at the center of the rotary component; I_{xo} be the applied intensity from the X-ray source at a scan instant; and I_{out} be the detected attenuated intensity after the X-ray propagates through the object along the straight scan line $L = L(\theta, w)$; then

$$\int_{L(w,\theta)} f_\mu(x, y)ds = -\ln\left(I_x(w, \theta)/I_{xo}(w, \theta)\right) = p_x(w, \theta) \tag{7.7}$$

where θ is the projection angle and w is the source/detector position on the rotary component. First, we may consider a scan at the projection angle $\theta = 0$, the projection $p_x(w, 0)$:

$$p_x(w, 0) = \int_{-\infty}^{\infty} f_\mu(x, y)dy\Big|_{x=w} = \int_{-\infty}^{\infty}\int_{-\infty}^{\infty} f_\mu(x, y)dxdy \tag{7.8}$$

We may extend this expression to all orientations with the delta function $\delta(.)$, with an output value of 1 at an input value of 0 and an output value of 0 otherwise. Thus,

$$p_x(w, \theta) = \int_{-\infty}^{\infty}\int_{-\infty}^{\infty} f_\mu(x, y)\delta(x\cos\theta + y\sin\theta - w)dxdy \tag{7.9}$$

We further apply a mathematic transformation function called the two-dimensional Fourier transform on $f(x, y)$, resulting the function $F(u, v)$:

$$F_\mu(u, v) = \int_{-\infty}^{\infty}\int_{-\infty}^{\infty} f_\mu(x, y)e^{-j2\pi(ux+vy)}dxdy \tag{7.10}$$

For $v = 0$,

$$\begin{aligned} F_\mu(u, 0) &= \int_{-\infty}^{\infty}\int_{-\infty}^{\infty} f_\mu(x, y)dye^{-j2\pi ux}dx = \int_{-\infty}^{\infty}\int_{L(w,0)} f_\mu(x, y)dse^{-j2\pi ux}dx \\ &= \int_{-\infty}^{\infty} p_x(w, 0)e^{-j2\pi ux}dx \end{aligned} \tag{7.11}$$

which is the one-dimensional Fourier transform of $P_x(u)|_{\theta=0}$. (Here, we just use the format of Fourier Transform as a calculation trick, and hence, please try not to interpret its physical meaning.) Consider again for $u = 0$; then we have

$$F_\mu(0, v) = \int_{-\infty}^{\infty}\int_{L\left(w,\frac{\pi}{2}\right)} f_\mu(x, y)dse^{-j2\pi vy}dy = \int_{-\infty}^{\infty} p_x\left(w, \frac{\pi}{2}\right)e^{-j2\pi vy}dy, \tag{7.12}$$

which is the one-dimensional Fourier transform of $P_x(v)|_{\theta=\pi/2}$. Here, a serious mathematical derivation is skipped, and the result is that we can let $u = w\cos\theta$ and $v = w\sin\theta$ such that

$$P_x(w,\theta) = F_\mu(w\cos\theta, w\sin\theta) \tag{7.13}$$

where $P_x(w, \theta)$ is the one-dimensional Fourier transform of $p_x(w, \theta)$ with respect to w.

However, in the practical implementation, the function of the detected $p_x(w, \theta)$ lies on discretized w and θ. We may compute $P_x(w, \theta)$ with $N_w + 1$ (N_w is an even number) detection points along w with a total length W by

$$P_x\left(\frac{mW}{N_w},\theta\right) \approx \frac{1}{N_w+1}\sum_{k=-N_w/2}^{N_w/2} p_x\left(\frac{kW}{N_w},\theta\right)e^{-j2\pi(mk/N_w)}, \quad \text{for } m_w = -\frac{N_w}{2},\ldots,\frac{N_w}{2}. \tag{7.14}$$

The values of $P_x(m_wW/N_w, \theta)$ for all the discrete orientations should be computed. Clearly, based on Eq. 7.14,

$$P_x\left(\frac{mW}{N_w},\theta\right) = F_\mu\left(\frac{mW}{N_w}\cos\theta, \frac{mW}{N_w}\sin\theta\right), \tag{7.15}$$

which is on a radial grid. In the subsequent calculation, we would need to estimate the function $F(u, v)$ over a square grid by numerical interpolation. We can then estimate $f(x, y)$ by taking the discrete inverse Fourier transform of $F(u, v)$:

$$f_\mu\left(\frac{mW}{N_w},\frac{nW}{N_w}\right) = \sum_{m=-N_w/2}^{N_w/2}\sum_{n=-N_w/2}^{N_w/2} F_\mu\left(\frac{2m}{N_w},\frac{2n}{N_w}\right)e^{j4\pi(mx+ny)/N_w}, \tag{7.16}$$

A series of CT slices are captured and computed repeatedly until the range of the body scan is done. Then, by applying percentage darkness to each voxel according to its computed value of μ_x, the final stacks of CT slices can be constructed. These CT slices can show three-dimensional structures of the skeleton and some soft tissue organs because the difference in μ_x for different materials can be revealed by the minor different attenuations of multiple X-ray images from different orientations. The output of the program is a matrix of adjusted linear attenuation coefficients, known as CT numbers (unit: Hu), corresponding to the voxels of the CT slices. The "CT" number of a material is defined as

$$\text{CT} = \frac{\mu_x - \mu_{xw}}{\mu_{xw}} \times 1000 \tag{7.17}$$

where μ_{xw} is the linear attenuation coefficient of water. Typically the CT number is 3000 for bone, 0 for water, −1000 for air, and −100 to 60 for soft tissues and organs.

With increased hardware advancement, the voxel resolution of CT imaging has greatly improved in the past decades. Using ultrahigh-speed CT as an example, the scanning time can be as short as 40 m s or less, and multiple frames can be obtained per second. As this scanning time is very short, CT videos can now be acquired. CT also can avoid artifacts caused by body movements including heartbeats. Generally, the images computed from CT are high quality which can assist in visually separating tissues, organs, and bones. In addition, since the data is in digital form, it can be analyzed quantitatively. With these improvements, CT has become a multifunctional and reliable tool in medical imaging. However, significantly more data have to be collected, and it still poses a health risk, especially to soft tissues, due to its use of X-rays. Patients may absorb more radiation than a single X-ray imaging. Patients with acute trauma, cardiovascular, and pediatric diseases may also not be suitable for CT examination. Regardless, CT is still a powerful tool as it can collect massive amounts of information for a broad range of diagnosis.

7.4 Magnetic Resonance Imaging

Magnetic resonance imaging (MRI) is an imaging technique that reconstructs the electromagnetic signal generated by the resonance of the nucleus in the magnetic field. MRI as a medical imaging diagnostic technology has developed rapidly since the 1980s. MRI can produce the most detailed anatomical information in a harmless way. However, unlike X-ray or CT, MRI is used for capturing tissue images instead of the bone. MRI captures the signals from the nuclear magnetic resonance (NMR) of hydrogen atoms, found mostly in water molecules. NMR was discovered separately by Dr. Felix Bloch from Stanford University and Dr. Edward Purcell from Harvard University in the same year of 1946. Both were awarded the Nobel Prize in Physics together in 1952 for their work. Bioliquids and soft tissues contain mostly water and abundant hydrogen atoms, and therefore, MRI can capture the distribution of water molecules based on the different local densities of hydrogen atoms.

A further discussion on the mechanism and characteristics about NMR can help us better understand the working principle of MRI. Nuclei of hydrogen atoms contain single protons. The "spin" of these protons results in magnetic momentums. We might imagine such magnetic momentum as a small magnet. Without external disturbance, the orientation of such small magnet spin axis is arbitrary. Under a strong magnetic field in a static direction, the spin axis of such small magnet will be aligned with the magnetic field direction, meaning that the positively charged protons would then spin about an axis with the same direction of the magnetic field (i.e., $\varphi_n = 0$ in Fig. 7.5a). Such proton spin alignment happens in all hydrogen atoms in the human body under the condition of NMR. If there is a radio-frequency (RF) electromagnetic field added to the external magnetic field along a direction perpendicular to the magnetic field, the

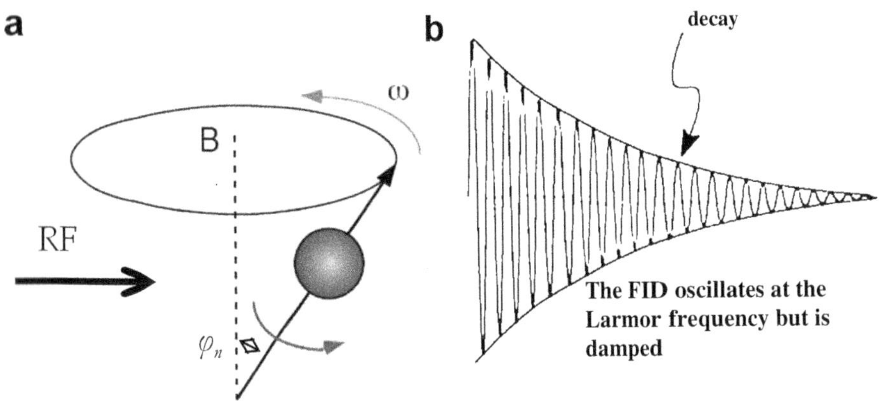

Fig. 7.5 (**a**) Rotation of nuclear momentum about its own axis (hidden line) and about the magnetic field axis (arrow passing through the sphere). (**b**) Expected electromagnetic signal detected along the initial axis of the nuclear momentum

nucleus might have a high energy absorption rate for a particular narrow range of frequency nearly ω_n according to the Larmor relation:

$$\omega_n \approx \gamma_n B_n \tag{7.18}$$

where γ_n is the gyromagnetic ratio with a value of 42.58 MHz/Tesla for hydrogen and B_n is the external magnetic field strength.

The absorbed energy in nucleus would then become a higher level of total nuclear momentum, reflected as a larger "deviation angle" (φ_n in Fig. 7.5a) from the magnetic field direction. However, once the additional RF electromagnetic wave is turned off, the nuclear spin would return back to natural energy level. If we measure the nuclear magnetic momentum at this moment onward, we should measure an RF oscillating magnitude M_{xy} along a direction in the x-y plane (the hidden line in Fig. 7.5a is normal to this plane) of the magnetic momentum decays over time t (Fig. 7.5b), that is

$$M_{xy}(t) = M_{xyo}e^{-t/T_{\text{fid}}}, \tag{7.19}$$

where M_{xyo} is the initial oscillating magnitude of the nuclear magnetic momentum and T_{fid} is called the horizontal relaxation time. Such detectable decaying signal is also called the free induced decay (FID), which is often referred to as the "MR signal." Indeed, the principle of MRI is based on the release of nuclear energy in the material with different attenuations during FID.

In MRI, it is necessary to distinguish the captured MR signals from different locations in the human body. However, under a consistent external magnetic field strength, all hydrogen atoms emit MR signals with the same frequency and in the same phase, and therefore, a three-dimensional image reconstruction is infeasible. In practice, a gradient magnetic field $\vec{B}_n$ in one of z-, x- and y-directions is applied in different stages during MRI:

$$\vec{B}_n(x, y, z) = z\frac{\partial \vec{B}_n}{\partial z} + \vec{B}_{nb}, \quad x\frac{\partial \vec{B}_n}{\partial x} + \vec{B}_{nb} \quad \text{or} \quad y\frac{\partial \vec{B}_n}{\partial y} + \vec{B}_{nb} \tag{7.20}$$

where $\vec{B}_{nb}$ the background magnetic field strength along the z-direction, which is along the human height or length direction of the patient table. The origin ($x = 0$, $y = 0$ and $z = 0$) is assumed to be at the center of the top surface of the patient table inside the MRI chamber. The x-direction is along the patient's shoulder direction or the width of the patient's table; and the y-direction is the normal distance from the surface of the patient table, which is a positive number. Through the addition of gradient magnetic field to detect the electromagnetic waves emitted, we can determine the spatial distribution of location and density of hydrogen atoms. First, a sufficiently large gradient of magnetic field strength is only applied along the z-direction. Such gradient is called the slice selective gradient because this will determine the level of cross section in the human body being imaged. Recalling Eq. 7.18, the NMR effect only occurs only over the cross-sectional layer at one z-level where the corresponding magnetic field strength level B_{nz} is close to a matching oscillating RF frequency, and hence, we can select the level of cross section of the image by adjusting the frequency of the additional RF electromagnetic wave to a value of ω_{nz}, causing only our interested cross-sectional layer to have the additional energy stored.

After both the magnetic field and electromagnet are turned off and a "short period of waiting time" (*we will come back and discuss about this in the next paragraph*), the FID process occurs and the MR signals can be detected. Meanwhile, another magnetic field with a gradient only along x is applied, with a background magnetic field strength adjusted to the level B_{nz}. The x-gradient in this stage is relatively much smaller such that the emission of MR signals from the material cross section is still present. The gradient magnetic field results in the MR signals with the frequency changing linearly along the x-direction according to the Larmor relation:

$$\omega_n(x) = \gamma_n\left(\frac{\partial B_n}{\partial x}x + B_{nz}\right) = \omega_{nz} + \left(\gamma_n\frac{\partial B_n}{\partial x}\right)x = \omega_{nz} + G_x x \tag{7.21}$$

As a result, material points with different x-positions would emit MR signals with different frequencies. In other words, we encode the x-position with the frequency of MR signals. This is why such application of x-gradient magnetic field is called "frequency encoding." In fact, the MR signals captured in this stage are for the computation of MR images.

Now, let's come back to discuss on the "short period of waiting time" as mentioned above. During this "waiting time," we can impose a magnetic field with a gradient only along the y-direction, given the position range of MR imaging along y is from 0 to Y_n, where Y_n is the upper limit of the measurable y-position. Apparently, according to Eq. 7.21, we can expect that MR signals would have different oscillating frequencies along the y-direction. If such a gradient magnetic field is only applied in an extremely short period of time T_{ny} (within a quarter period of the MR signal oscillation for the boundary voxel), then the phase change φ_{ny} of the MR signals for different y-positions should become

$$\begin{aligned}\varphi_{ny} &= \omega_n(y)T_{ny} = \gamma_n\left(\frac{\partial B_n}{\partial y}y + B_{nz}\right)T_{ny} = \varphi_{no} + \left(\gamma_n\frac{\partial B_n}{\partial y}\right)T_{ny}y \\ &= \varphi_{no} + G_y y,\end{aligned} \tag{7.22}$$

where $\varphi_{no} = B_{nz}T_{ny}$. This implies that we can induce MR signals from the cross-sectional layer with phase shifts linearly increasing with their y-positions. Therefore, the application of y-gradient magnetic field with a limited period is called "phase encoding." Afterward, the MR signals will then be collected during the subsequent frequency-encoding stage.

In the practical implementation, the MRI machine is typically configured as shown in Fig. 7.6a. The subject is lying on the patient table fitted inside the scanner chamber. It normally requires a primary superconducting magnet, a gradient coil, a main-field coil, and an RF coil. The primary magnet is used to generate a background magnetic field and force to align the spin direction of nuclei in hydrogen atoms. There is a horizontal tube that runs through the magnet called a bore. The magnet is extremely powerful, and its strength is measured in either "tesla" or "gauss" (1 tesla = 10,000 gauss). For comparison, most MRI magnets use a magnetic field of 0.5–2.0 tesla, whereas the Earth's magnetic field is only 0.5 gauss. There are three different pairs of gradient coils in the MRI machine which are located around the main magnet (Fig. 7.6b). Each pair of the gradient coils is responsible for generating a magnetic gradient along the Cartesian direction (x, y or z). They are weaker than the primary magnet as they are used to generate the magnetic gradient terms as described in Eq. 7.20. These gradient coils allow specific and different parts of the body to be scanned by altering and adjusting the magnetic gradients, coordinating simultaneously with the background magnetic field strength. A set of multiple RF coils is also equipped in the MRI scanner chamber. The basic function of the RF coils is to transmit additional RF electromagnetic waves to excite the aligned hydrogen nuclei. There are coils located inside the different places of MRI scanner to transmit waves into different body parts. These RF coils are also used for detection of the MR signals during the FID process. The MRI scan can be performed over the full body or a selected region. If a certain part of the body is specified, then only the corresponding RF coils are used for a small scan region and hence a higher voxel resolution.

The measured RF signal from the RF coil is actually a summation of MR signals from different material voxels (with each its density of water molecules) over a selected cross section. The map of water molecules density in voxels can be defined on the spatial x-y plane as $p_{\mathrm{M}}(x, y)$. The measured RF signal should be in the middle

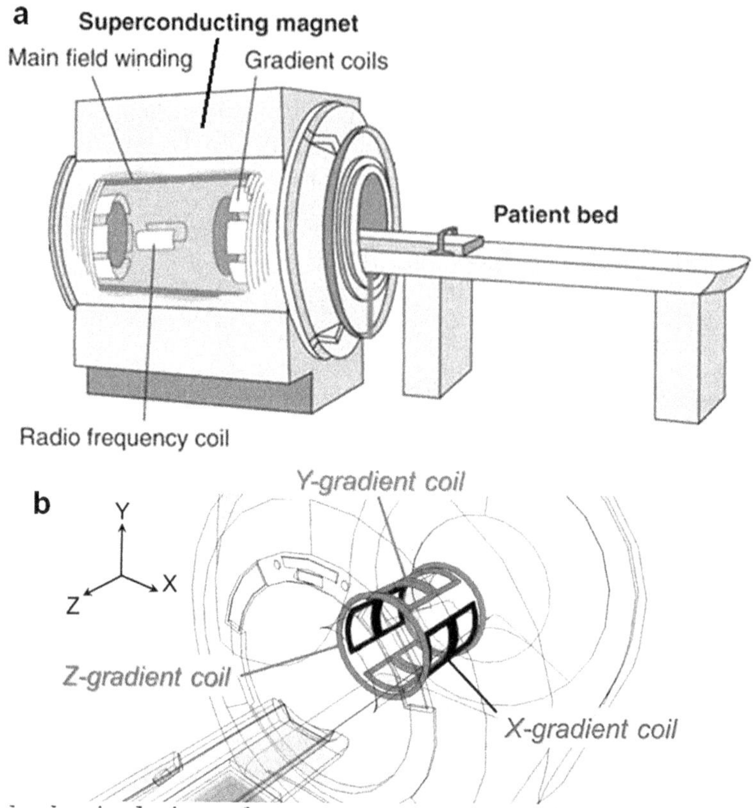

Fig. 7.6 (**a**) The basic design of an MRI scanner. Typically, it includes a 24-inch-wide chamber with a patient table transfer a patient into the chamber. Inside the MRI chamber, there are a magnet, a main-field coil, three pairs of gradient coils, and RF coils. The MRI scanner is connected to a computer system for signal processing. (**b**) A closer look of the gradient coils

of the FID process, and the signal amplitude reduces over time. Recalling Eq. 7.19, we can compensate such FID-induced signal reduction by magnifying the detected RF signal with a factor of $\exp(-t/T_{\text{fid}})$.

Additionally, in order to decouple the time-varying RF signal, multiple (N) times of measurements by the RF coils should be performed repeatedly for M times, except with each different value in the phase encoding stage, obtaining multiple continuous-time signals: $f_{\text{MR}1}(t)$, $f_{\text{MR}2}(t)$, …, $f_{\text{MR}N}(t)$. More specifically, we may consider a special case that for the ith measurement ($i_n = 0, 1, 2, \ldots, N$), $\partial \vec{B}_n / \partial y$ at the boundary voxel ($y = Y_n$) can be set to a level such that the corresponding phase difference is, for example, $i_n\pi/2$ ("$\pi/2$" can be changed to other values actually) from that of the voxel at $y = 0$ (φ_{no}). We may consider the measurement index i_n as a parameter of φ_{ny}, and thus we may redefine the phase difference as φ_n:

$$\varphi_n(y, i_n) = \frac{i_n \pi y}{2Y_n} + \varphi_{no} \tag{7.23}$$

We may consider the measurement index i_n as a parameter of the detected signal, which is the summation of MR signals with different frequencies (Eq. 7.21) and phase differences (Eq. 7.23), and obtain

$$f_{\mathrm{MR}}(t, i_n) = \sum\nolimits_x \sum\nolimits_y p_{\mathrm{M}}(x, y) \cos\left(\omega_n(x)t + \varphi_n(y, i_n)\right). \tag{7.24}$$

Further, to facilitate the following discussion, we define $f'_{\mathrm{MR}}(t, i_n)$ as a map of complex numbers such that

$$f'_{\mathrm{MR}}(t, i_n) = \sum\nolimits_x \sum\nolimits_y p_{\mathrm{M}}(x, y) e^{j\omega_n(x)t} e^{j\varphi_n(y, i_n)} \text{ and} \tag{7.25}$$

$$\mathrm{Re}\{f'_{\mathrm{MR}}(t, i_n)\} = f_{\mathrm{MR}}(t, i_n), \tag{7.26}$$

where j is $(-1)^{1/2}$. Each of these RF signals are then further processed by Fourier transform in the x-dimension with the computer, inducing the map of $F_{\mathrm{MR}}(\omega, i_n)$. If we assume the captured RF signals are continuous over time t and substitute Eq. 7.21, the Fourier transform can be described as

$$F'_{\mathrm{MR}}(\omega_n, i_n) = \int_{-\infty}^{\infty} f'_{\mathrm{MR}}(t, i_n) e^{-j\omega_n t} dt = \sum\nolimits_y p_{\mathrm{M}}\left(\frac{\omega_n - \omega_{nz}}{G_x}, y\right) e^{\varphi_n(y, i_n)}. \tag{7.27}$$

In practice, as the measured RF signal is sampled as discrete time points along t, we should apply the discrete Fourier transform. We further substitute Eq. 7.23 in the above equation:

$$F'_{\mathrm{MR}}(\omega_n, i_n) = \sum\nolimits_y \left[p_{\mathrm{M}}\left(\frac{\omega_n - \omega_{nz}}{G_x}, y\right) e^{j\varphi_{no}}\right] e^{j\frac{\pi}{2Y_n} y i_n}. \tag{7.28}$$

Next, we can then apply the discrete Fourier transform of i_n to the above relation in order to transfer the signal information from the i_n-domain to the y- or φ_{ny}-domain:

$$F_{\mathrm{MR}}(\omega_n, \varphi_{ny}) = \sum_{i_n=0}^{N} F'_{\mathrm{MR}}(\omega_n, i_n) e^{-j\frac{\pi}{2Y_n} y i_n} = p_{\mathrm{M}}\left(\frac{\omega_n - \omega_{nz}}{G_x}, \frac{\varphi_{ny} - \varphi_{no}}{G_y}\right) e^{j\varphi_{no}}. \tag{7.29}$$

Considering also that $p_M(x, y)$ must be real numbers and $\left|e^{j\varphi_{no}}\right| = 1$, we can compute the map of water molecules over a body cross section as

$$p_{\mathrm{M}}(x, y) = \left|F_{\mathrm{MR}}(\omega_n(x), \varphi_{ny}(y))\right|. \tag{7.30}$$

7.5 Rapid Prototyping

7.5.1 Introduction

In some medical devices, the geometries of the products must match those of the human body, which is often irregular, complex, and unique in shape. Conventional manufacturing methods may not be able to support such highly irregular requirements. In the recent past, the rapid development of functionally graded materials has been noticed by medical device developers, especially in the areas of custom medical device applications. Due to the demands of production time and complexity, in the mid-1980s, rapid prototyping was introduced and has begun to replace traditional manufacturing methods as it can effectively increase the product development. With rapid prototyping, the efficiency of the whole process of production can be boosted. This is because:

- It can be implemented at the modular or even system level which may become a significant part of the product without a complicated assembly process.
- It can be applied when the object geometry is digital form and use "reverse engineering" with support for a wide range of materials (e.g., alloys and composites).
- It helps with the work put into development, whether it be for design evaluation (a physical three-dimensional geometry analysis), function verification (mimics and studies of the geometric functions such as the kinematics and aerodynamics), or further manufacturing processes (decision of processes and sequences such as adhesive joining and assembling).

Briefly, rapid prototyping has two production styles: material removal and material addition. For material removal, one typical example is computer numerical control (CNC) machining which removes materials along a predefined pathway and sequence. On the other hand, the material addition techniques often involve adding materials layer by layer. This process normally goes through four major steps: (1) geometric modeling, (2) slicing shapes, (3) fabricating object layer by layer, and (4) post-processing. To model the geometry of the final product, designers can use computer-aided design (CAD) or computer-aided manufacturing (CAM) programs to generate and arrange the model geometry with precise data/parameters. Alternatively, designers can get the CAD model through geometric scanning or medical imagining techniques which may reduce the time duration of product design. Then, the geometry file can be transformed and stored in the "stereo lithography" (STL) format which can allow designers using CAD software to position, scale, and even optimize the STL model by checking and fixing shape features such as holes, overlaps, and offsets. CAD software can then transform the geometry into multiple slices of two-dimensional cross sections (Fig. 7.7). During this step, designers can configure the slicing procedures with the machine parameters such as the layer thickness, printing position, object position, and supporting mode. Designers should also select the printing materials. If necessary, support materials

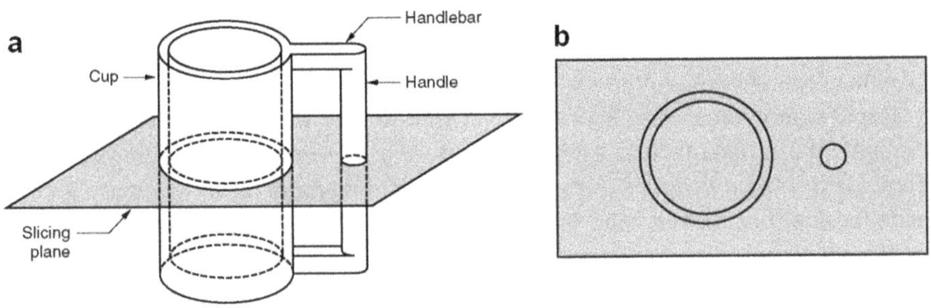

Fig. 7.7 (**a**) A sample three-dimensional geometry. (**b**) An extracted slice as a two-dimensional shape

Fig. 7.8 Printed object (black) and supporting material (white) during fabrication. (Reproduced with permission from Ultimaker)

can be added underneath the product shape as temporary supports during the fabrication process (Fig. 7.8). After printing the material as defined by the STL model (and the support material), post-processing should then be conducted, including taking the object out of the printing board, removing the support material, cleaning the printed object, further curing (e.g., by additional ultraviolet exposure for photo-curing materials), surface coating, assembling, and final finishing such as polishing over the object surface.

7.5.2 Role in Product Development Cycle

Rapid prototyping fabricates some components with their geometry defined by computer readable formats. Hence, designers can easily modify the shape of parts and increase the complexity and geometries of parts without influencing manufacturing time. Rapid prototyping can shorten the product development cycle and enhance productivity since the resultant prototype is largely similar to the desired part or assembly in terms of its geometry. The resultant prototype can mimic the functions of the final product which can avoid misinterpretations in the later production stages.

It can then be applied for preliminary product testing to detect any design errors. Accordingly, the design can be optimized before being transferred from the manufacturing process to the real production line.

Rapid prototyping can also assist and improve the manufacturing process. For example, it can manufacture a cast for molds (e.g., for investment casting) directly. Some parts in a product can even be manufactured by rapid prototyping as the final parts. In this case, the demand of tooling and handling material can be minimized. Parts with complex geometry can be manufactured directly instead of being assembled from multiple sub-parts, reducing the number of parts and tool wear compared to fabrication processes which use more conventional techniques. Furthermore, replacing machining processes with rapid prototyping can eliminate machining sequence planning. This is due to the fact that a whole prototype can be made through a one-time, layer-by-layer material "printing" process instead of using different equipment and processes to machine. This helps to reduce time and cost since there is no need to account for specimen transportation, blank geometry and feature design, tooling machines, and large material loss.

There are still some limitations that rapid prototyping cannot overcome. The basic principle of rapid prototyping technology has shown that it is still difficult to achieve as good surface quality and accuracy as the more traditional processes. Since most rapid prototyping use "layering manufacturing technology," some produced surfaces can be rough, e.g., a staircase-alike surface profile generated for an inclined plan. Post-processing would then be required to finely tune the surface conditions, yet it is still challenging to post-process the interior structures. Precision of manufacturing is another limitation because materials are deposited, solidified, or welded with a defined volume unit bit by bit during the process. Such volume unit would determine the precision (or the feature size) of the final prototype, and it is typically much lower than the precision induced by many conventional processes. On the other hand, the mechanical performance is highly constrained because only limited material candidates are compatible with the rapid prototyping process. The physical function of the fabricated parts can be very different from the desired parts fabricated with the chosen materials.

7.5.3 Stereolithography

Stereolithography (SL) is a rapid prototyping technology based on the photopolymerization effect. In general, an SL machine (Fig. 7.9) contains a motorized platform which can move along the vertical (z) axis, an ultraviolet (UV) laser emitter which can move/rotate along x- and y-axes, and a tank full of liquid-based material which contains some prepolymers and photopolymers (resin) such as epoxies and acrylates.

During the printing operation, the platform is located lower than the liquid level which covers and exposes around 0.05–0.15 mm over the platform surface. Then, the UV laser emitter generates a laser whose energy is absorbed by the resin.

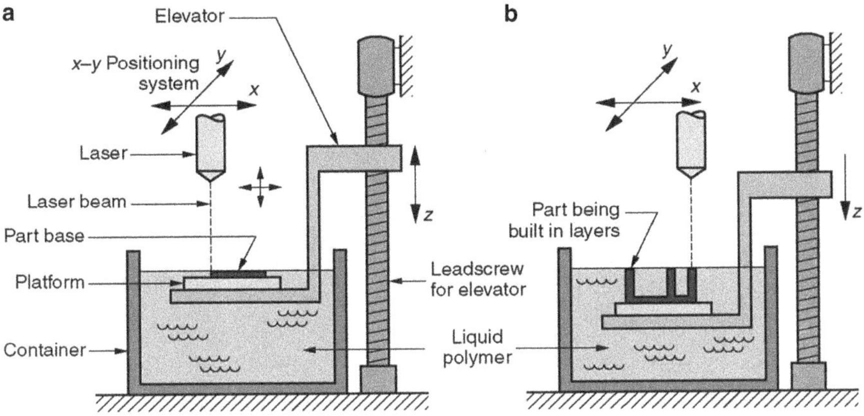

Fig. 7.9 Configuration of a stereolithographic machine

Because of the photopolymerization effect, the polymer (liquid) hardens upon laser exposure. By "drawing" a one-layer pattern of the product (a cross section with an auto-generated support) with the laser source moving along a defined path on the *x*-*y* plane on the platform surface, the resin solidifies into the desired pattern. After completing a layer, the platform is moved to a lower level with another layer of liquid covering the formed prototype layer, and the production step repeats again. After the hardening process, the "fresher" overhead layer solidifies and bonds to the lower existing layer. Then, the layer fabrication repeats until the entire product is generated. Afterward, the product is taken out of the tank and rinsed with resin solvent in order to remove excess resin (after some time, the resining solution can also remove the support materials). Finally, the product is baked in an ultraviolet oven to ensure thorough curing of the photopolymer.

7.5.4 Fused Deposition Modeling

Fused deposition molding (FDM) is a kind of 3D printing technology developed after the SL process, as shown in Fig. 7.10. The FDM technology was invented by Scott Crump in 1988 who founded his own company, Stratasys, based on the technology. In 1992, Stratasys introduced the world's first FDM-based 3D printer which also marked the transition of FDM technology into the commercial phase. Unlike other rapid prototyping technologies, FDM machines can be installed in the office environment which can significantly increase design and production efficiency in many industries.

Fused deposition heats filamentous hot melt material and squeezes the material through an extruder with a fine nozzle. The nozzle moves in the direction of the *z*-axis, while the table moves along the *x*-axis and *y*-axis (of course, the design of the mechanical structure of different equipment may not be the same). When the printing

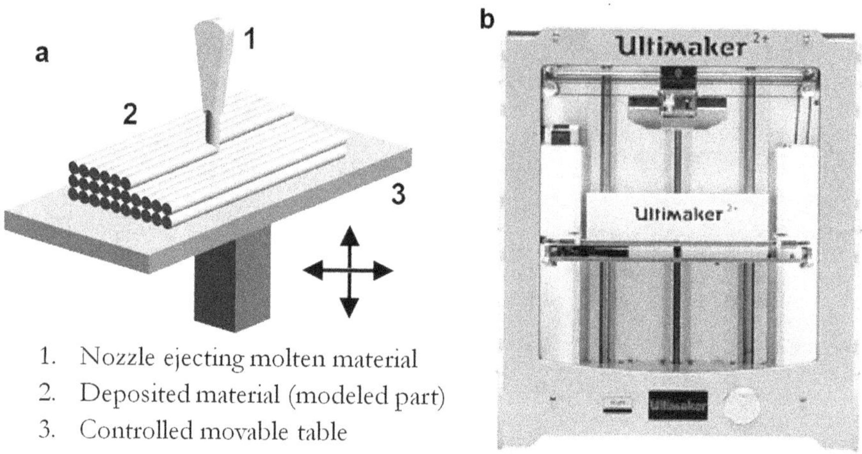

Fig. 7.10 (**a**) Procedures of fused deposition modeling and (**b**) a tabletop FDM machine. (Ultimaker 2+, reproduced with permission)

job starts, the printing material is unwound from a coil and sent into the extrusion nozzle. Before the extrusion, the material is melted in the nozzle. Then, the material "wire" will be extruded from the nozzle. Meanwhile, the stepper motor controls the nozzle and moves across the pathway (both product and support) which is generated by computer-aided manufacturing software in order to paint/outline the platform. Printed material "wires" commonly have a thickness of 0.178–0.356 mm based on the required quality. Once printed, the material is cooled down and hardens on the platform. Then, the nozzle continuously extrudes the material over another layer, repeating the above steps until the workpiece is fully formed. Finally, the product will be cleaned through solvent and machine processes or by hand.

FDM mainly uses thermoplastics for its solid-based printing, but myriad materials can also be the printing materials for FDM. ABS, PLA, polycarbonate, and rubber are some widely adopted material candidates. Depending on the design geometry, FDM can offer two kinds of printing materials: the molding material and the support material. Newer FDM equipment adopts the design of double nozzles. The first nozzle is responsible for extruding the molding material, and the second nozzle is responsible for extruding the support material. The support material can be chosen to be dissolved in a solvent which cannot dissolve the molding material. Hence, the support material can be removed by dipping the printed solid into solvent in order to specifically remove the support material.

In the extruder, the hot-melt wire (usually ABS or PLA material) is wound on the feed roller and rotated by the stepper motor drive roller. Then, the wire is fed to the extruder nozzle by the friction of the rollers. Between the feed roller and the nozzle, there is a guide sleeve. The guide sleeve is made of low-friction material so that the wire can be sent smoothly and accurately from the feed roller to the inner cavity of the nozzle. The top of the nozzle has a resistance wire heater which heats the material to the molten state. Then, the now molten material is squeezed through the extruder onto the platform.

Advantages of FDM:

- The operating environment is clean and safe and can be carried out in an office environment (no danger of gas or chemicals and no use of laser).
- Process is clean, simple, and easy for production and does not produce garbage.
- High-dimensional accuracy, good surface quality, easy assembly. Quickly build open or hollow parts, e.g., bottle.
- Raw materials are provided in the form of reel filaments, easy handling, and quick replacement.
- Low cost of raw materials.
- High material utilization.
- Can use a variety of materials, such as dyed ABS and medical ABS, PC, PPSF, casting wax, and synthetic rubber.

Shortcomings of FDM:

- Low precision; it is difficult to build complex parts or structures. Highest FDM precision is not as good as SL.
- The direction of the direction perpendicular to the cross section is small.
- The forming speed is relatively slow, not suitable for building large parts.

7.5.5 Solid Ground Curing

Solid ground curing (SGC) is another liquid-based rapid prototyping technology. Its working principle is based on the photopolymerization effect which is similar to SL. However, the fabrication procedures of SGC are notably different. It is an additive manufacturing technology which is widely used to manufacture models and patterns. SGC equipment does not need a laser beam to curve each layer, and the entire layer is exposed to the light though a mask over the liquid polymer. Since the main principle of SGC is based on the exposure of each layer of the model by means of a lamb though a mask, the time of process is independent on the complexity of each layer. For example, an ultraviolet (UV) emitter will hit the whole printing area. Also, SGC may involve the milling/grinding process, which is definitely special in rapid prototyping field.

For the operation, the SGC machine uses the ionic graphic process to generate an electrostatic image with recyclable toner on a reusable glass plate which forms a mask at the end (Fig. 7.11). The operation procedures are illustrated in Fig. 7.12. This mask is placed over the liquid photopolymer. A UV lamp then shines. UV light cannot penetrate the toner pattern, meaning that UV only shines through the unmasked part (the outline) and causes the polymer to harden along the outline. Next, the liquid polymer is removed, while the interspace is covered with molten

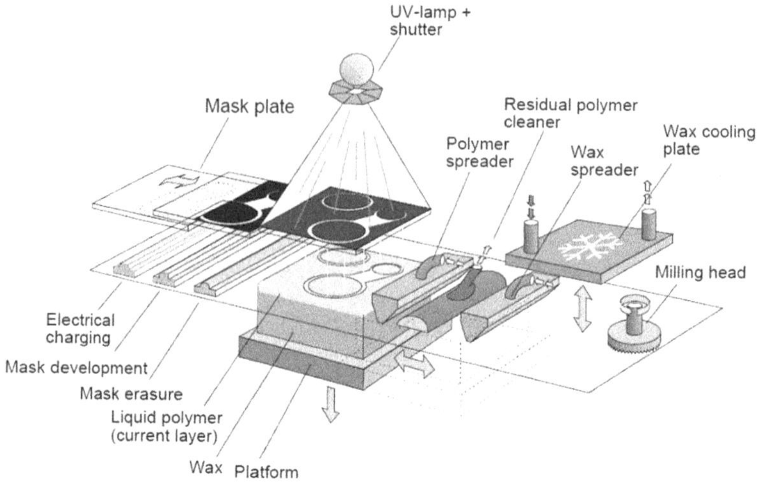

Fig. 7.11 The diagrammatic figure: processing of solid ground curing

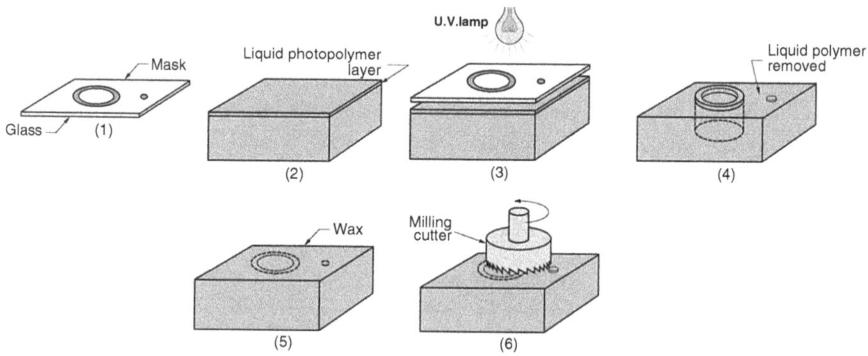

Fig. 7.12 Process flow of solid ground curing

wax again. A cold plate is then placed over the wax, cooling down the wax such that it solidifies. A cutter mills the wax at a specific flatness and thickness. Afterward, the platform is moved lower such that it is covered again by the liquid photopolymer. The glass mask can be replaced with a new one for other cross-sectional features. At the end, a polymer "block" is generated, and the surrounding uncured photopolymers can be cleaned up to finish the manufacturing process.

The advantages of this process lie in the facts that the wax is only used to fill the voids and the system does not need another type of polymer as a support material. Each layer is milled after each successive light-exposure process which allows for higher accuracy in z-direction compared to FDM and SL. The xy outline is defined by the photomask, and thus, the precision is also very high (in the scale of sub-microns). The main drawback of SGC is its relatively much larger waste generation, despite its convenience and high throughput.

7.6 Application Examples

7.6.1 Surgery Planning

Reverse engineering can be applied to construct patient-specific models and functional parts. With developments in computer-aided design and manufacturing technologies, a patient's health information obtained from medical imaging can be applied in mimicking surgery conditions. Surgeons can first acquire data from medical imaging techniques like MRI or CT, followed by 3D printing parts by rapid prototyping. These printed prototypes can then help in diagnosis and surgery planning.

For instance, rapid prototyping has been adopted in surgery planning for cerebral arteriovenous malformation (CAVM). CAVM is a tangle of abnormal blood vessels connecting arteries and veins in the brain. It occurs most often in the brain or spine. CAVM patients may suffer from headaches, seizures, backaches, neckaches, and eventual nausea. In serious cases, the blood vessels rupture, causing bleeding within the brain (intracranial hemorrhage). As shown in Fig. 7.13a, the human brain and spinal nervous system structures are very complex yet very small. Therefore, direct surgery is very difficult. If the operation fails, it can cause serious physical trauma to patients.

CT angiography is a dynamic, real-time diagnosis method that not only demonstrates the presence of a CAVM site but also shows vascular in-transient time. Diagnostic angiography is uniquely able to delineate the size and number of feeding arteries (Fig. 7.13b), and it can define the pial, dural, or mixed origin of the CAVM. CT angiogram can be used to identify areas of acute hemorrhage, and these results can suggest a vascular malformation, particularly with the judicious use of color contrast. It can also reveal the certain features that may correlate with an increased risk of bleeding, many of which are related to CAVM. Rapid prototyping technology can quickly create a neural model of a patient's operation site prior to surgery (Fig. 7.13c), greatly improving the success rate of neurosurgery. Surgeons can apply this reconstructed model to plan and simulate their surgical procedures in order to maximize surgical success rates.

7.6.2 Custom Implant Devices

Apart from the fabrication of surgical models, rapid prototyping can be used to manufacture customized implantable devices. Rapid prototyping provides a feasible and novel way to fabricate a human organ/tissue model which either cannot be produced or is hard to be manufactured by conventional approaches. Rapid prototyping was first used for biomedical applications as a means for producing biomaterials for implants as its process allows for patient-specific models based on medical scans. Rapid prototyping techniques in conjugation with CT imaging are an

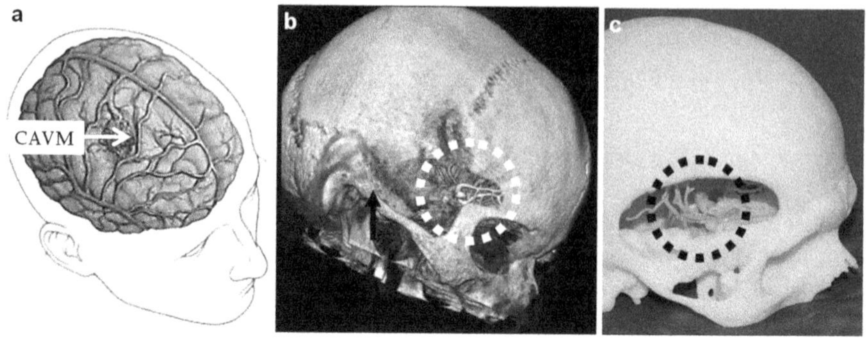

Fig. 7.13 (**a**) Abnormal blood vessels. (**b**) CT Angiogram of CAVM. (**c**) Reconstructed model of a CAVM site. (Wurm [14], Surgical invention. Reproduced with permission)

effective combination for the manufacturing of prostheses and implants. For example, a custom implant for facial reconstruction can be obtained by imaging the skull and reproducing the damaged bones based on the undamaged side as reference.

Taking a real case of a patient who had trauma and deformity to the right side of his face after a tumor removal as an example, scientists have applied a custom 3D printed implant made of biodegradable polymer for fixing an incorrect eye position. After the surgical removal of a tumor on a patient's face, the right side of the patient's face was traumatized, causing his eyes to be misaligned. The surgical team in charge of this challenging procedure was composed of doctors all from South Korea: Professor Dong-Woo Cho of the Department of Mechanical Engineering at POSTECH; Professor Jong Won Rhie of the Department of Plastic Surgery, College of Medicine, at the Catholic University of Korea; and Won-Soo Yun of T&R Biofab Company.

An X-ray CT was first implemented to obtain the geometry of the patient's skull. By subtracting the left half by the right half of the skull, doctors determined the missing amount of bone. The geometry of this missing part was further modified by adding a replacement eye socket in order to ensure the correction of his eye ball positioning after implantation (Fig. 7.14a). The team used a US Food and Drug Administration (FDA)-approved biodegradable polymer, polycaprolactone, to make the implant by the rapid prototyping technology. Afterward, this 3D printed object (Fig. 7.14b) was then placed into the patient's bone-missing region and corrected his eye position. It is expected that similar medical applications can be implemented to other bone/tissue replacements.

7.6.3 *Scaffolds*

Rapid prototyping can be applied in scaffold fabrication for tissue engineering applications (please see further information in Chap. 11). For instance, the scaffold

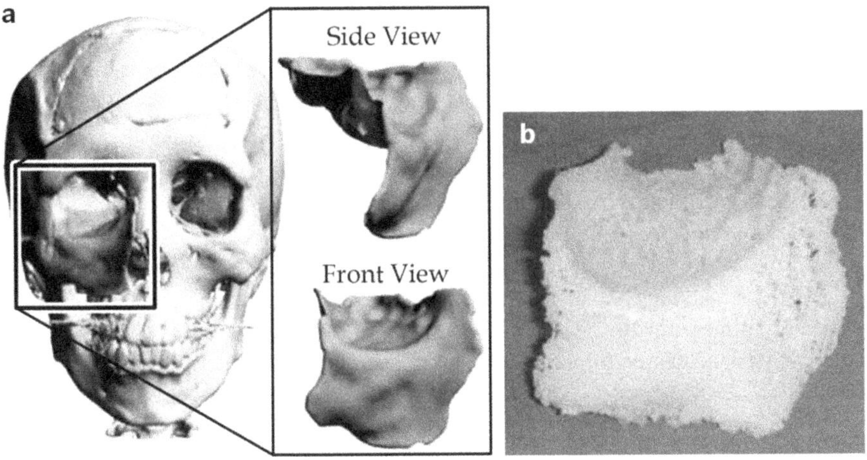

Fig. 7.14 (**a**) Design of the implant fixing eye misalignment of a patient. (**b**) The 3D printed implant. (Reproduced with permission from 3Dprint.com: https://3dprint.com/20851/3d-printed-scaffolding-implant/)

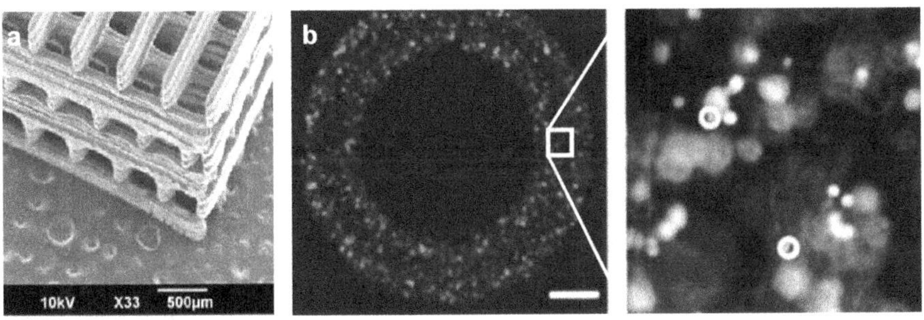

Fig. 7.15 (**a**) Scanning electron micrographs of hydroxyapatite-embedded poly(propylene fumarate) PPF scaffolds with square pores. (Lee [15], Microelectronic engineering. Adopted with permission.) (**b**) Poly(ethylene glycol)-dimethacrylate hydrogels with encapsulated cells prepared by stereolithography. Live and dead human dermal fibroblasts encapsulated in the gel are shown as bright dots and the dead cells are highlighted by circles in the right inset. Scale bar: 1 mm. (Arcaute [16]. Adapted with permission from Springer)

of artificial bone with bioactivity can be fabricated based on stereolithography techniques of hydroxyapatite-embedded UV-curable polymers as shown in Fig. 7.15a.

The flexible microstructural architecture fabricated by rapid prototyping can offer excellent mechanical properties and biocompatibility as well as for adhesion and growth of osteoblasts after implantation. Furthermore, FDM can deposit polymers, including live cells as well as biochemical molecules such as related growth factors, for use in creating structures. Various studies have reported on the use of water-soluble poly(ethylene glycol)-dimethacrylate (PEG-DMA) to create structured, cell-containing hydrogels for the regeneration of human dermal tissues (Fig. 7.15b).

Problems

Problem 7.1

Please answer the following questions according to Table 7.P1.

(a) Complete Table 7.P1 by filling in the missing values.
(b) Find the intensity reflection coefficient at the interface between A and B.
(c) An ultrasound wave of intensity 0.08 W m^{-2} in medium A comes across the interface between A and B. Calculate the intensity of the ultrasound wave entering medium B.
(d) An interface can be formed by any two media above. Which of these interfaces will cause the greatest proportion of the incident ultrasound to reflect?

Problem 7.2

Please answer the following questions according to Table 7.P2.

(a) Calculate the acoustic impedance of the kidney tissue.
(b) Ultrasound waves traveling through kidney tissue in the body encounter fat tissue. Show that the proportion of the incident pulse being reflected at the boundary expressed in decibel is −22 dB. Also, find the proportion of the incident pulse being transmitted at the boundary. Express your answer in decibel.
(c) Explain why ultrasound of low frequency should be used in examining the kidney.
(d) Explain why the resolution of the ultrasound images of kidney is usually low.
(e) A research study is being carried out to examine the density of bone in a body using ultrasound. Explain why this is theoretically possible.

Table 7.P1 Ultrasound-related material properties for media A, B and C

	Density [kg m^{-3}]	Speed of sound [m s^{-1}]	Acoustic impedance [MRayl]
Medium A	952		1.38
Medium B		1560	1.63
Medium C	1740	3700	

Table 7.P2 Density and speed of sound for different common human tissues

Medium	Density [kg m^{-3}]	Speed of sound [m s^{-1}]
Fat tissue	952	1450
Kidney tissue	1038	1560
Muscle tissue	1076	1580
Bone	1740	3700

Problem 7.3

A block of tissue consists of two layers, A and B, with thicknesses of 1 cm and 2 cm, respectively. The linear attenuation coefficients of the layers are $\mu_A = 18\ \text{m}^{-1}$ and $\mu_B = 12\ \text{m}^{-1}$, respectively. The intensity reflection coefficient at the interface of the layers is 0.058. An ultrasound wave travels through A, crosses the interface, passes through B, and eventually leaves the tissue. The intensity of the ultrasound wave in A just after entering A is $0.2\ \text{W m}^{-2}$.

(a) Calculate the intensities of the ultrasound:

- In A just before hitting the interface between A and B
- In A just after reflected by the interface
- In B just after passing through the interface
- In B just before leaving the tissue

(b) Find the ratio of ultrasound intensity just before leaving the tissue to that just after entering the tissue. Express your answer in decibel.

Problem 7.4

The figure (Fig. 7.P1) below shows a sample screen display of an A-scan using an ultrasound machine which is connected to a computer. The eye examination procedures are carried out. The speed of ultrasound in cornea will be taken to be $1532\ \text{m s}^{-1}$. After the ultrasound scan, data received form the transducer will be analyzed. The computed positions of the components of the eye will be shown using a scale marked in mm. On the display, spikes C1 and C2 are caused from the cornea, spikes L1 and L2 are caused from the lens, and spike R is caused from the retina.

(a) Estimate the length of the eyeball from the display.
(b) Explain why two spikes are obtained from each of the cornea and the lens.
(c) In the study, the computed thickness of the cornea is found to be 576 μm.

- Estimate the time interval between receiving spikes C1 and C2 by the computer.
- What is the maximum pulse length required to produce the above result?
- Calculate the minimum frequency used by the ultrasound machine.
- Sketch the resulting display if the frequency of ultrasound used is too low.

(d) A student said that due to attenuation, the pulses received later should have lower peaks. Suggest a reason to explain why the peaks of the pulses do not decrease gradually.
(e) A series of pulses appears on the right side of R. Suggest a reason for this.

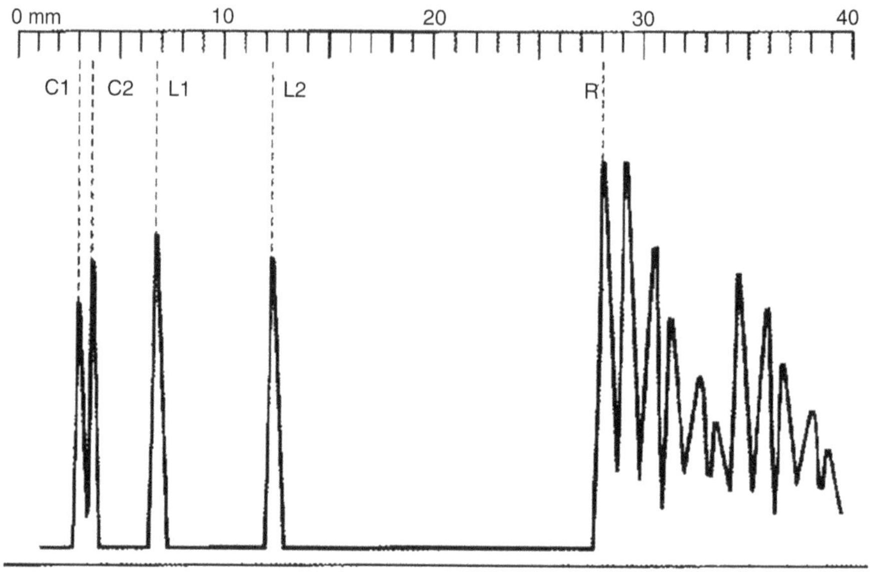

Fig. 7.P1 Sample screen display of an A-scan using an ultrasound machine

Problem 7.5

(a) State two causes of the attenuation of ultrasound in a medium.
(b) Find the half-value thickness and the linear attenuation coefficient of the medium for the ultrasound.
(c) Write down an equation relating I and x.
(d) Determine the intensity of the ultrasound at a depth of 100 mm.
(e) Using the given axes, sketch the I-x graph if, separately,

- The initial intensity of the ultrasound is 0.5 W m^{-2}.
- The soft tissue is replaced by water.
- Ultrasound of a higher frequency is used.

Problem 7.6

Please reconsider Fig. 7.P2 again as the variation of intensity I of an X-ray against depth x in a medium.

(a) Prove that the half-value thickness $x_{1/2}$ is related to the linear attenuation coefficient μ by $x_{1/2} = \ln 2/\mu$.
(b) Find the linear attenuation coefficient (in m^{-1}) of the medium.
(c) Write down an equation relating I and x.
(d) Calculate the depth required when the intensity falls to 0.01 W m^{-2}.

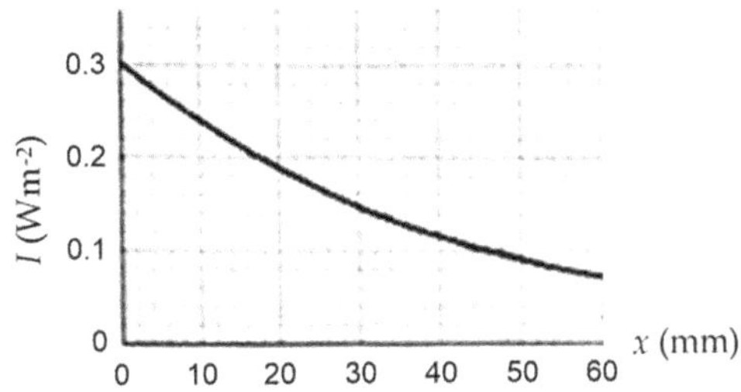

Fig. 7.P2 The variation of intensity I of a 120 kHz ultrasound against depth x in a soft tissue

Problem 7.7

A beam of diagnostic X-ray of intensity 20 W m^{-2} and photon energy 80 keV is produced by an X-ray tube. The beam is incident on a thick lead block. The linear attenuation coefficient of lead for the X-ray photons is 58 m^{-1}.

(a) All X-ray machines are remotely controlled by operators behind a shielded wall. By referring to the property of X-ray, explain why this is necessary.
(b) Write down an equation for the intensity I of the X-ray at depth x in the lead block.
(c) Calculate the half-value thickness of lead for the X-ray photons.
(d) Calculate the thickness of the lead block needed to reduce the intensity of the emerging X-ray to 1%.
(e) State what happens to the linear attenuation coefficient:

- If aluminum is used instead of lead.
- If the X-ray photon energy is reduced to 5 keV.

(f) In practice, an X-ray tube used in radiographic imaging will produce both high energy and low energy X-ray photons. Explain why the low energy photons have to be filtered out; and suggest how to filter out these low-energy photons.

Problem 7.8

An X-ray machine produces a parallel beam of X-ray of intensity 500 W m^{-2}. The cross-sectional area of the beam is 0.1 mm^2. The energy of each X-ray photon is 90 keV.

(a) If the efficiency of the X-ray machine is 0.8%, calculate the rate of electrical energy consumed.
(b) What is the minimum working voltage across the X-ray machine?
(c) Express the energy of an X-ray photon leaving the X-ray machine in Joule.

(d) The working voltage of the machine is now increased. Given that the magnitude of charge on an electron to be 1.6×10^{-19} C, state and explain what happens to the ionizing power of the X-ray, the intensity of the X-ray leaving the X-ray machine, and the half-value thickness of lead for the X-ray.

Problem 7.9

Two rectangular blocks, A and B, with the arrangement and dimensions shown in Fig. 7.P3, are placed in the path of an X-ray. μ_A and μ_B are the linear attenuation coefficients of the blocks. If there is no image contrast between A and B, roughly sketch the result on the film. Write down mathematically the condition for such situation to occur.

Problem 7.10

Figure 7.P4a shows an X-ray of intensity I_o entering into a structure which is formed by six cubic blocks (A–F), each with a side length x. The linear attenuation coefficients of the blocks are μ_A, μ_B, …, μ_F, respectively.

(a) Find the intensity I of the emerging X-ray in terms of I_o, μ_B, …, μ_F and x.
(b) Figure 7.P4b shows the variation of intensity of X-ray with the distance along the path. You may assume that after passing through each block, the intensity falls by 0.6 unit in the plot.

- Which block has the largest linear attenuation coefficient? And which has the lowest linear attenuation coefficient? Justify your answer.

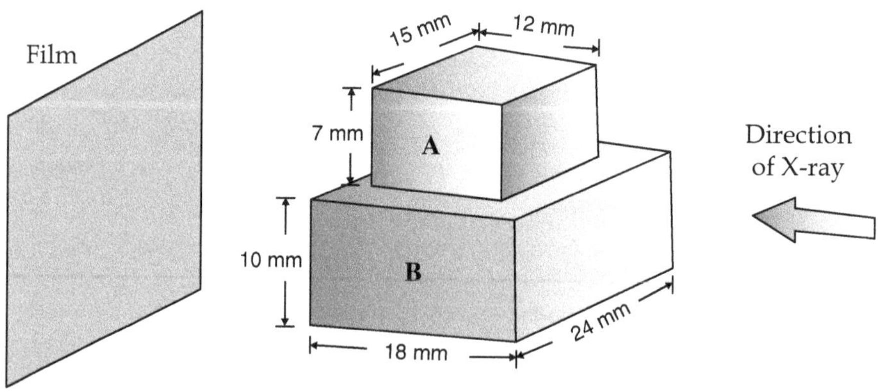

Fig. 7.P3 An X-ray path blocked by two rectangular blocks A and B

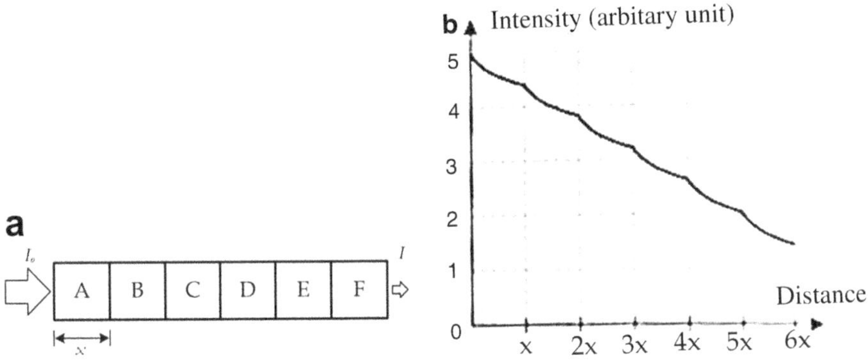

Fig. 7.P4 An X-ray passing along six identical cubic blocks (left) with its intensity dropping throughout the path (*right*)

- Rough sketch the variation of intensity with distance if the same X-ray enters from F and emerges from A.
- If we consider all the six blocks as a single block with a uniform linear attenuation coefficient μ such that the resultant effect (i.e., I/I_o) is the same, what would this value of μ comparing to μ_A?

Problem 7.11

Given that the gyromagnetic ratios with values of 42.58 MHz/Tesla for ^{1}H, 10.71 MHz/Tesla for ^{13}C, and 40.05 MHz/Tesla for ^{19}F, please suggest a range of the resonance frequencies for an external magnetic field strength ranging from 0.1 to 2.0 Tesla. Under the same magnetic field strength, please suggest also the corresponding resonance frequencies for ^{13}C and ^{19}F.

Problem 7.12

If it takes 300 m s to obtain a single voxel in an MRI operation, how much time is required to obtain five slices with a matrix size of 256 $\times$ 256?

Problem 7.13

(a) If a material has a horizontal relaxation time constant $T_{\text{fid}} = 200$ m s, what will the MR signal be 200 m s after an RF pulse has applied to trigger the bulk magnetization into the transverse plane?

(b) For a material having a T_{fid} of 80 m s, what is the relative magnitude of the MR signal at 320 m s after an RF pulse is applied, compared to its initial maximum signal?

Problem 7.14

Consider the printing of polycarbonate using a nozzle with a diameter of 0.3 mm and filament with a diameter of 1.8 mm. The slice height or gap is set at 0.18 mm. (**a**) Find the minimum feeding rate needed to print at average liquid flow velocity of 127 mm/s (**b**). Estimate the maximum shear rate on exit from the nozzle. Assume that polycarbonate has a viscosity of 500 Pa·s and is approximately Newtonian.

References and Further Reading

1. Rooppakhun, S., Nattapon, C., Kriskrai, S.: Advanced Medical Imaging and Reverse Engineering Technologies in Craniometric Study. IntechOpen (2011)
2. Mikla, V.I., Mikla, V.V.: Medical Imaging Technology. Elsevier, Waltham, MA, USA (2013)
3. Hendee, W.R., Russell Ritenour, E.: Medical Imaging Physics. Wiley, New York (2003)
4. Lisle, D.A.: Imaging for Students. Hodder Arnold, London (2012)
5. Dance, D.R., Christofides, S., Maidment, A.D.A., McLean, I.D., Ng, K.H.: Diagnostic Radiology Physics: A Handbook for Teachers and Students. International Atomic Energy Agency, Vienna (2014)
6. Lantada, A.D.: Handbook on Advanced Design and Manufacturing Technologies for Biomedical Devices. Springer, Boston (2013)
7. Du Boulay, G.H.: Principles of X-Ray Diagnosis of the Skull. Butterworth-Heinemann. Oxford, UK (2016)
8. Wong, K.K.L.: Methods in Research and Development of Biomedical Devices. World Scientific, Singapore (2013)
9. Chua, C.K., Leong, K.F., Lim, C.S.: Rapid Prototyping: Principles and Applications. World Scientific, Singapore/Hackensack (2010)
10. Gibson, I.: Advanced Manufacturing Technology for Medical Applications: Reverse Engineering, Software Conversion and Rapid Prototyping. Wiley, Chichester (2006)
11. Bushberg, J.T., Boone, J.M.: The Essential Physics of Medical Imaging. Lippincott Williams & Wilkins, Philadelphia (2011)
12. Gibson, I., Rosen, D.W., Stucker, B.: Additive Manufacturing Technologies. Springer, New York (2014)
13. Ségonne, F., Dale, A.M., Busa, E., Glessner, M., Salat, D., Hahn, H.K., Fischl, B.: A hybrid approach to the skull stripping problem in MRI. NeuroImage. **22**, 1060–1075 (2004)
14. Wurm, G., Lehner, M., Tomancok, B., Kleiser, R., Nussbaumer, K.: Cerebrovascular biomodeling for aneurysm surgery: simulation-based training by means of rapid prototyping technologies. Surg. Innov. **18**, 294–306 (2011)
15. Lee, J.W., Ahn, G., Kim, D.S., Cho, D.W.: Development of nano-and microscale composite 3D scaffolds using PPF/DEF-HA and micro-stereolithography. Microelectron. Eng. **86**, 1465–1467 (2009)
16. Arcaute, K., Mann, B.K., Wicker, R.B.: Stereolithography of three-dimensional bioactive poly (ethylene glycol) constructs with encapsulated cells. Ann. Biomed. Eng. **34**, 1429–1441 (2006)

Chapter 8
Laser Metal Processing

Abstract Laser material processing includes the application of lasers for achieving different manufacturing goals. The advantages include an accessible spacing between the sample surface and the laser source, enabling real-time inspection of the process, simplifying the quality control process, and relatively lower costs. In particular, selective laser sintering is a manufacturing process that uses laser and powder to generate a three-dimensional model based on the concept of rapid prototyping. This chapter covers basic working principles and instrumentation settings of these laser processes.

8.1 Introduction

Lasers play a major part in the material processing used in engineering and manufacturing, including those used in biomedical device manufacturing and other biological applications. Laser material processing includes the application of lasers for achieving different manufacturing goals such as material shaping, welding, and surface optimizing. Different lasers as a focused light beam with a defined wavelength can be created for material processing by different methods and materials such as solids, liquids, gases, or semiconductors. For example, an Nd-YAG solid-state laser can be generated by a hot lattice crystal of yttrium aluminum garnet (YAG) with the chemical composition $Y_3Al_5O_{12}$, where the Nd^{3+} ions (0.1–2%) substitute for some of the Y_3 ions. Briefly, this laser is generated in an optically resonant cavity including an activating crystal in two mutually parallel mirrors, one of which reflects light-reflected part of the light from the other mirrors such that photons can accumulate between the mirrors with an increased light intensity. In this optical cavity, the axially propagating monochromatic spectrum travels back and forth within the cavity, inducing a self-excited oscillation with the accumulation of the light energy, i.e., a monochromatic spectrum with a maintained wavelength but a higher laser light intensity. Ideally, a half energy level of the generated laser light is able to penetrate via the transmission mirror and is emitted as a continuous laser. Applications of Nd-YAG lasers include laser surgery and material processing such as welding, cutting, drilling, and surface modification. A gas laser such as a carbon

R. H. W. Lam, W. Chen, *Biomedical Devices*,
https://doi.org/10.1007/978-3-030-24237-4_8

dioxide (CO_2) laser typically uses a gas mixture primarily consisting of CO_2, N_2, and He. The CO_2 laser is a kind of molecular laser which can manifest a variety of energy states depending on how it vibrates and rotates. Plasma is applied to the mixed gas at a low pressure (usually 30–50 torr) for the electron emission. When colliding with a hollow wall or while naturally exuding, CO_2 molecules at a high energy state can accidentally lose energy and spontaneously emit photons. These emitted photons propagate down the cavity along the optical axis and swing in the resonant cavity to generate the laser. CO_2 lasers are relatively inexpensive and are available in a wide range of output power, ranging from 10 W to 10 kW. The output laser is a series of emission lines from 9.3 to 11.0 μm in the infrared region with a band center around 10 μm. These emission lines can be filtered for a specific wavelength depending on their application. For instance, 9.3 μm is more efficient for polymer processing, and 10.6 μm is better for ceramic processing.

Laser material processing has different standard parameters for users to follow based on a variety of factors. In general, two lasers commonly used for operations are Nd-YAG and CO_2 which have wavelengths of 1064 nm and 10.6 μm, respectively. For different operations and final shape features, laser material processing requires a specific range of laser intensity with an appropriate heat factor for specific processing purposes. Typically, drilling requires 10^5–10^7 W mm^{-2}, cutting requires 10^5–10^6 W mm^{-2}, welding requires 10^3–10^5 W mm^{-2}, and heat treatment requires 10^3–10^4 W mm^{-2}. The power distribution of a laser beam as a circular spot with a beam radius exposed on a surface can be modeled as a Gaussian distribution function (Fig. 8.1a). Also, there are two modes of applying a laser: continuous and pulsed modes. The continuous mode maintains a constant power supply with a typical range of 100 W to 20 kW. On the other hand, the pulsed mode applies the laser with a defined peak laser power (P_{peak}) over short activation duration (t_{p}) and repeats after a longer period of time (T_{period}), as shown in Fig. 8.1b. The typical pulse length t_{p} is around 1 ms down to 1 ns. The repetition rate (f_{p}), which is the reciprocal of T_{period}, is typically around 0.1 Hz to 1 MHz. The energy per pulse (E_{p}) in the pulsed mode

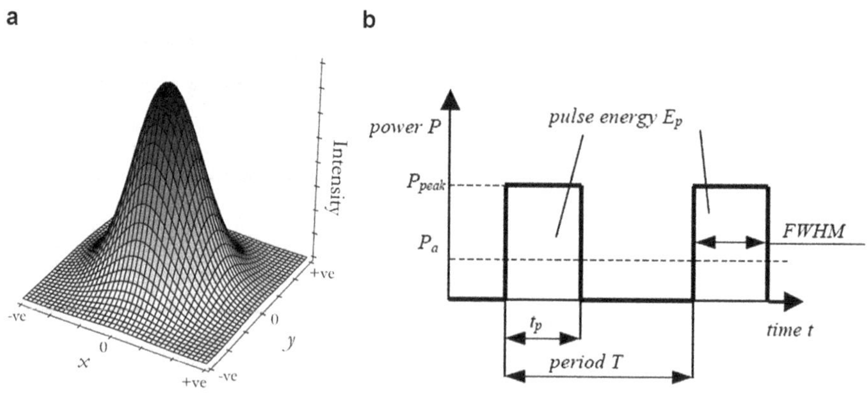

Fig. 8.1 (**a**) Spatial profile of a laser beam. (**b**) Power variation over time of a laser applied with the pulsed mode

can be calculated by $E_p = P_{peak} \times t_p$, with a typical range of 1 mJ to 1 kJ. Therefore, the equivalent laser power (P_a) for the pulsed mode is

$$P_a = \frac{P_{peak} t_p}{T_{period}} \tag{8.1}$$

For Nd-YAG lasers, the power output in continuous mode varies from 150 W to 6 kW, while the pulsed mode is of the order of 50 MW with pulse duration of about 20 ps at a repetition rate of 1–100 Hz.

Moreover, when a laser beam exposes on a material surface, the laser power will split into multiple portions: reflection from the material surface, absorption by the material body, and transmission through the material (Fig. 8.2a). The absorbed laser power will create a temperature increment over the exposure area on the material surface. Here, we define the ratio of the absorbed laser power as the absorption coefficient (A). As shown in Fig 8.2b, A is a function of the laser wavelength and the material type. For example, steel has an A of ~0.35 for Nd-YAG laser (wavelength: 1064 nm).

Lasers can be applied in a wide range of manufacturing processes. For example, laser cutting works by directing the output of a high-power laser at a material, most commonly through the use of optical focusing. The focused laser beam is directed at the material which then either melts, burns, vaporizes away, or is blown away by a jet of gas, leaving an edge with a high-quality surface finish. Industrial laser cutters are used to cut flat-sheet material as well as structural and piping materials. A CNC (computer numerical control) machine can manipulate the laser along a defined process path. A typical commercial laser for cutting materials involves a motion control system to follow a CNC or G-code of the pattern to be cut onto the material. Laser welding employs the high power density (on the order of 1 MW/cm^2) of lasers,

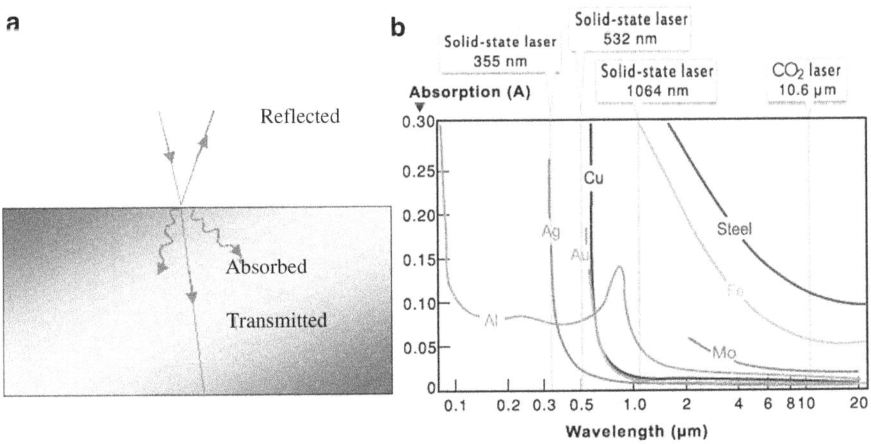

Fig. 8.2 (**a**) Energy absorption of a local material region under laser exposure. (**b**) Absorption of wavelength for different materials

resulting in small heat-affected zones and high heating and cooling rates. Laser welding works for many materials, e.g., carbon steels, stainless steel, aluminum, and titanium. Due to high cooling rates, cracking is a concern. The speed of welding is proportional to the amount of power supplied. However, it also depends on the type and thickness of workpieces. The high-power capability of gas laser makes them especially suitable for high-volume applications. The spot size of the laser can vary between 0.2 mm and 13 mm, though only smaller sizes are used for welding. The depth of penetration is proportional to the amount of power supplied. Such depth is also dependent on the location of the focal point; penetration is maximized when the focal point is slightly below the surface of the workpiece. Millisecond-long pulses are used to weld thin materials, while continuous laser systems are employed for deep welds.

In addition, a beam of finite diameter is focused by a lens onto a plate with a transverse electromagnetic mode pattern "000" as shown in Fig. 8.3. The individual parts of the beam striking the lens can be imagined as point radiators of a new wavefront. The lens will draw the rays together at the focal plane, and constructive or destructive interference will take place there. When two rays arrive at the screen and are half a wavelength (denoted as λ) out of phase, then they will destructively interfere with a minimal light intensity. Thus, if the ray AB (Fig. 8.3) is $\lambda/2$ longer than ray CB, the point B will represent the first dark ring. The central maximum will contain approximately 86% of the total beam power. The radius of this central maximum will be the focused beam radius (R_{beam}).

Considering the initial beam radius before focused is R_{l}, the distance of AB longer than CB would be $\lambda/2$, which can be related to the lens radius as $\lambda = 2R_{\text{l}}\sin\varphi$, where φ is the angle of ABC. Besides, the beam radius R_{beam} can be related to the focal length of lens f_{l} as $R_{\text{beam}} = f_{\text{l}}\tan\varphi$. For very small φ, sin $\varphi \approx$ tan φ; and hence,

$$R_{\text{beam}} \approx f_{\text{l}}\lambda/(2R_{\text{l}}) \tag{8.2}$$

For example, the expected spot size of a CO_2 laser beam is 22 mm in diameter before focused. It is focused by a 125 mm focal length lens. R_{beam} is expected to be $0.125 \times (10.5 \times 10^{-6})/2/0.022 = 30$ μm.

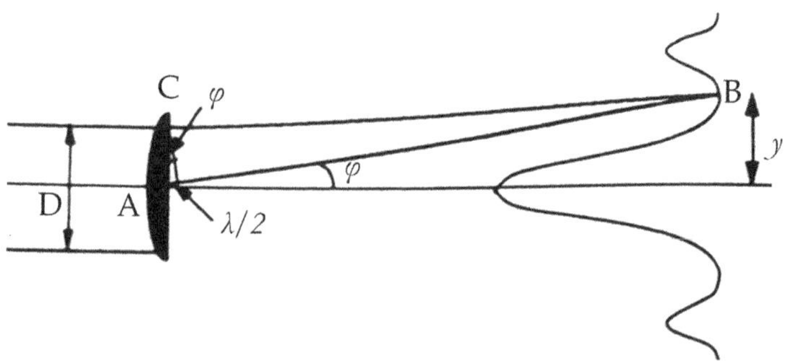

Fig. 8.3 Diagram illustrating the diffraction-limited spot size

8.2 Basic Instrumentation

The general arrangement for laser processes is shown in Fig. 8.4. The principle components are the laser itself with some shutter controls, a beam guidance train, focusing optics, and a mechanism of moving the laser source or workpiece relative to each other. The shutter is usually a retractable mirror which blocks the laser beam path and diverts the beam. When the beam is required, the shutter mirror is rapidly switched to release the laser beam. The beam then passes through the beam guidance train which directs the beam to center on a focusing optic. The included guidance mirrors are used to reflect or redirect laser beams in laser applications such as beam steering or beam folding. They are designed with a high degree of reflectivity for a specific wavelength or range of wavelengths using various substrates, coatings, or a combination of the two. They are also ideal for laser applications where space is limited, as a beam can be precisely directed multiple times to fit within a particular area. Additionally, high damage mirrors are special durable mirrors that possess high damage threshold. A wide selection of laser line mirrors is available for use with a range of popular laser wavelengths.

A polarizer is an optical filter that lets light waves of a specific polarization pass and blocks light waves of other polarizations. It can convert a beam of light of undefined or mixed polarization into a beam of well-defined polarization. The two most common types of polarizers are linear polarizers and circular polarizers. Polarizers are used in many optical techniques and instruments, and polarizing filters have found applications in photography and liquid crystal display technology. Polarizers can also be made for other types of electromagnetic waves besides light, such as radio waves, microwaves, and X-rays.

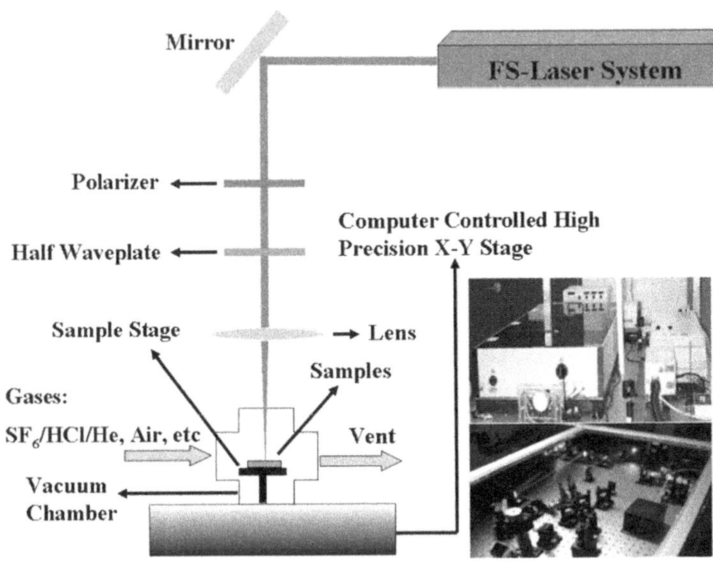

Fig. 8.4 Configuration of a laser processing machine

A wave plate or retarder is an optical device that alters the polarization state of a light wave traveling through it. Two common types of wave plates are the half-wave plate, which shifts the polarization direction of linearly polarized light, and the quarter-wave plate, which converts linearly polarized light into circularly polarized light and vice versa. A quarter-wave plate can be used to produce elliptical polarization as well. Wave plates are constructed out of materials such as quartz or mica for which the index of refraction is different for different orientations of light passing through it. The behavior of a wave plate (i.e., whether it is a half-wave plate, a quarter-wave plate, etc.) depends on the thickness of the crystal, the wavelength of light, and the variation of the index of refraction.

The focusing optic can be either transmissive or reflective. Transmissive optics is made of ZnSe, GaAs, or CdTe lenses for CO_2 lasers or quartz lenses for YAG lasers. Reflective optics consists of parabolic off-axis mirrors. Laser lenses are used to focus collimated light from laser beams in a variety of laser applications. Many different lens types are available in a wide range of wavelengths. Laser lenses can also include a range of lens types including PCX lenses, aspheric lenses, cylinder lenses, or laser generator lenses. Depending on different laser tasks, laser lenses are designed to focus light in several different ways, such as focusing down to a point, a line, or a ring. The focused beam then passes through a nozzle for laser processing.

A motorized *x-y* table helps provide horizontal planar motion for automated machinery such as assembly robots in manufacturing facilities. Robotic arms and other automated machinery only have a limited range of motion, while their bases remain stationary; *x-y* tables allow these bases to move horizontally along *x-* and *y*-axes. Also known as *x-y* stages, *x-y* tables are motorized linear slides with linear motion based in bearings which are driven by a drive mechanism, typically a linear motor. *x-y* tables are built and configured to provide high-performance positioning along multiple axes.

8.3 Laser Surface Treatment

Laser processes can perform surface treatment for improving a material's surface characteristics. In general, laser surface treatment can improve material hardness, toughness, and wear/corrosion resistance, via either/both thermal and thermochemical processes. Overall, the purpose of laser surface treatments is to obtain better performance of the workpiece whenever it is submitted to conditions such as wear, high temperature, and corrosive media. Surface treatments enhance material surface properties to address any particular requirements while keeping the bulk of the workpiece in its original structural geometry. Thermal processes such as laser cutting, welding, and melting do not affect surface composition, and the material's chemistry remains the same. On the other hand, thermochemical processes like laser-induced addition polymerization, cladding, and alloying can change surface composition.

In recent years, industrial lasers have become available for metalworking uses, including surface hardening. In various industrial fields, laser surface hardening is gradually replacing other traditional crafts such as induction hardening and chemical heat treatment. A laser can generate very intense energy fluxes at the workpiece surface, and the resulting temperature profiles in the workpiece usually can be made steep enough to negate the need for external quench media. The material surface can be accurately hardened with a fast processing speed and a small heating zone. Laser hardening is a surface-hardening process commonly used for complicated shapes or large objects because it allows for absolute control of the surface's hardness and texture. Laser hardening consists of the rapid heating of a material surface by laser beam, a short hold of the target temperature, and intensive cooling due to the high thermal conductivity of the metallic material. "Self-quenching" occurs when the cold interior of the workpiece constitutes a sufficiently large heat sink to quench the hot surface by heat conduction to the interior at a rate high enough such that carbon atoms do not have time to diffuse out of the material surface and form "Martensites." This process causes higher hardness on the surface than the rest of the material, leaving the interior of the workpiece essentially unaffected. The control of hardened layer depth and uniformity is an important part of the study of the laser surface-hardening process.

Laser beams can remove surface materials by techniques such as vaporization and melt-and-blow (Fig. 8.5). Such material removal can define holes, roughness, and surface profile. For vaporization, the focused laser beam heats the surface of the material above its boiling points and induces local vaporization and cavitation of the exposed spot. Vaporization cutting can perform drilling micro holes. This method can also apply to non-melting material such as carbon composites and thermoset plastics. Furthermore, melt-and-blow (or fusion cutting) uses high-pressure inert gas to blow molten material from the cutting area, greatly decreasing the power requirement compared to the vaporization approach. Typically, laser melting offers surface

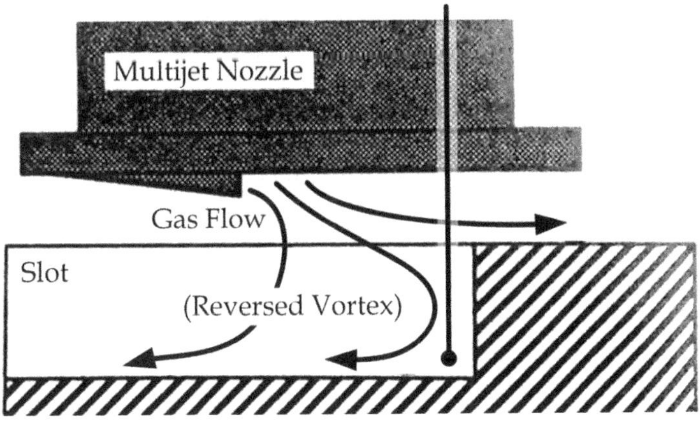

Fig. 8.5 Melt-and-blow surface material removal process

finishes within 25 microns, and such high precision reduces the finishing work required. Using this procedure, one can maintain homogeneous properties on the bulk body due to little thermal penetration, resulting in little distortion, smooth surfaces, and reduced work after processing. As other laser processes, this melt-and-blow process can be automated, and hence, laser surface melting is of industrial interest for materials such as cast irons, stainless steels, and titanium.

Laser cladding is a surface processing technique used for adding one material to the surface of another in a controlled manner that will produce a coating with 0.3–1 mm thick by fusion bonding. It generates a clad track by inputting powder particles to a molten pool made by moving laser (Fig. 8.6). The adding powder material can be the same or different from the bulk sample material. This enables the applied material to be deposited selectively wherever it is required. Nowadays, there are still demands for improvements on surface wear resistance. Conventional manufacturing processes such as press molds, stamping tools, casting forms, shafts, and other machinery elements usually come with the process of surface wear which significantly affects the lifetime of cutting tools and machines. Laser cladding is one of the most effective ways to improve surface characteristics, including corrosion resistance, wear resistance, and heat resistance. Furthermore, the material can be cladded on worn out surfaces for repair or protective coating purposes.

In particular, surface modification as a means of enhancing the surface energy level and the related tissue integration on bone-implanted devices can have significant advantages including less marginal bone resorption, predictable aesthetic outcome, improved soft tissue stability, and barrier formation against bacterial leakage. For instance, laser-roughened nickel-titanium alloy surfaces can modulate cell viability, proliferation, and adhesion on implanted devices made of titanium. Previous research has exhibited that modified titanium surfaces with different surface profiles can act as an environmental stimulation to human mesenchymal

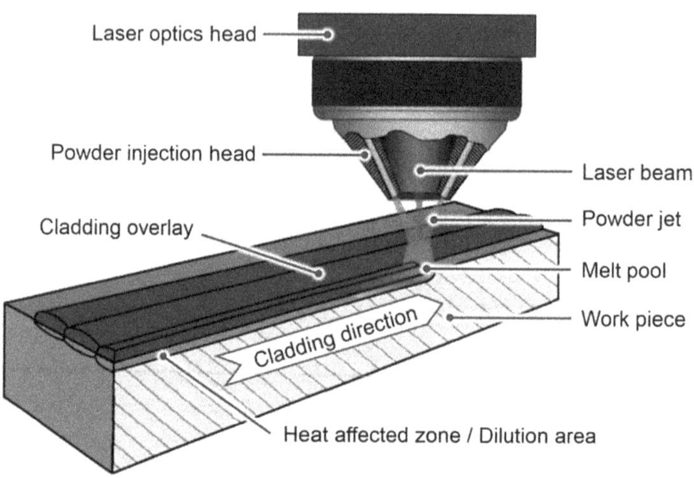

Fig. 8.6 Working principle of laser surface cladding

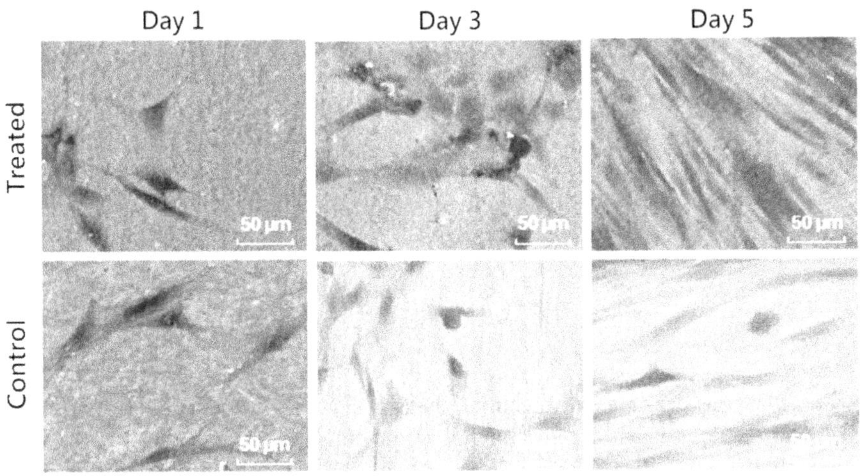

Fig. 8.7 Inference of laser treatment on cellular growth and morphology on nickel-titanium alloys. (Zhou [8], Journal of laser micro-/nanoengineering)

stem cells. Laser beam (Yb-YAG) treatment is a controllable and flexible approach to modifying surfaces. Laser beam irradiation at various fluencies (~200 J/cm^2) could be adopted to treat commercially pure titanium discs to create complex surface topographies. This process creates a complex surface topography with micro- and nanoscale patterns as shown in the results for a study in Fig. 8.7. The surface roughness was quantified by the arithmetical mean roughness R_a for each sample. The R_a values for sandblasted surfaces were between 0.55 and 0.67 μm, and those for laser-treated surfaces were between 2.09 and 2.51 μm. Laser-treated surfaces showed significantly different roughness values. The analysis of cell proliferation over a 5-day period revealed a difference between the behavior of osteoblasts on raw and laser-treated titanium materials. Generally, the mean proliferation and spreading area were larger for laser-treated titanium materials.

8.4 Spatial Temperature Profiles

8.4.1 Basic Heat Transfer Relation

The laser thermal treatment involves a basic heat transfer process which strongly affects the surface treatment performance. Consider the differential element shown in Fig. 8.8. The heat balance on the element can be considered as the difference between the "heat in" and "heat out" to be equal to the total of generated heat and accumulated heat in the element. In this chapter, we will only consider conduction as the dominant heat transfer scheme, and we will assume there is no heat generated from the element.

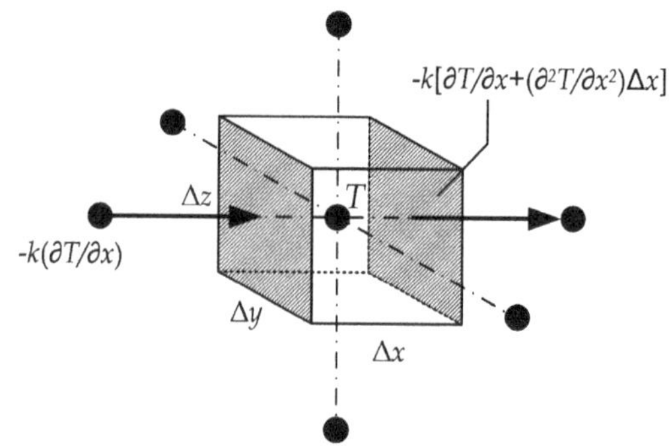

Fig. 8.8 Heat flow through a differential element in Cartesian coordinates

We first consider only one direction (x) of the Cartesian coordinates (x, y, z) on the heat portion along x-axis q_x, contributing to the total heat flow q_h ($=q_x + q_y + q_z$), by conduction:

$$q_x = \left\{-k\frac{\partial T}{\partial x} + k\left[\frac{\partial T}{\partial x} + \left(\frac{\partial(\partial T/\partial x)}{\partial x}\delta x\right)\right]\right\}\delta y\delta z = k\frac{\partial^2 T}{\partial x^2}\delta x\delta y\delta z \tag{8.3}$$

where T is temperature of the element and k is an intrinsic material property called thermal conductivity. Clearly, the total conduction in all the three dimensions is

$$q_h = k\left(\frac{\partial^2 T}{\partial x^2} + \frac{\partial^2 T}{\partial y^2} + \frac{\partial^2 T}{\partial z^2}\right)\delta x\delta y\delta z \tag{8.4}$$

Also, the accumulation of heat in the element is

$$q_h = \rho C_p\frac{\partial T}{\partial t}\delta x\delta y\delta z, \tag{8.5}$$

where ρ is the material density and C_p is the specific heat of capacity. Thus

$$\frac{\rho C_p}{k}\frac{\partial T}{\partial t} = \frac{1}{\alpha}\frac{\partial T}{\partial t} = \frac{\partial^2 T}{\partial x^2} + \frac{\partial^2 T}{\partial y^2} + \frac{\partial^2 T}{\partial z^2}, \tag{8.6}$$

where α ($=k/(\rho C_p)$) is called the thermal diffusivity.

8.4.2 Instantaneous Point Source on a Plane

Consider the case where an instantaneous point source is applied on a material surface at a position "i" in the Cartesian coordinates (x_S, y_S, z_S) with a total energy of

Q and time $t = 0$. The differential equation for conduction of heat in a stationary medium, assuming neither convection nor radiation, is

$$\frac{1}{\alpha}\frac{\partial T_i}{\partial t} = \frac{\partial^2 T_i}{\partial x^2} + \frac{\partial^2 T_i}{\partial y^2} + \frac{\partial^2 T_i}{\partial z^2} \tag{8.7}$$

where T_i is the temperature profile in the material body. If we let the environmental temperature to be 0, the solution of T_i as a function of time t and Cartesian position (x, y, z) should be in the form

$$T_i(x, y, z, t) = \Psi \exp\left\{ -\frac{(x - x_S)^2 + (y - y_S)^2 + (z - z_S)^2}{4\alpha t} \right\} \tag{8.8}$$

where Ψ is a constant. The total quantity of heat absorbed in the material should be equal to Q and it can be expressed as

$$\begin{aligned}
&\int_0^{\infty}\int_{-\infty}^{\infty}\int_{-\infty}^{\infty} \rho C_p T(x,y,z,t)\,dxdydz \\
&= \rho C_p \Psi \int_{-\infty}^{\infty} \exp\left\{-\frac{(x-x_S)^2}{4\alpha t}\right\} dx \int_{-\infty}^{\infty} \exp\left\{-\frac{(y-y_S)^2}{4\alpha t}\right\} dy \int_0^{\infty} \exp\left\{-\frac{(z-z_S)^2}{4\alpha t}\right\} dz \\
&= \frac{\rho C_p \Psi}{2} \int_{-\infty}^{\infty} \exp\left\{-\frac{(x-x_S)^2}{4\alpha t}\right\} dx \int_{-\infty}^{\infty} \exp\left\{-\frac{(y-y_S)^2}{4\alpha t}\right\} dy \int_{-\infty}^{\infty} \exp\left\{-\frac{(z-z_S)^2}{4\alpha t}\right\} dz \\
&= 4\rho C \Psi (\pi \alpha t)^{3/2} = Q
\end{aligned} \tag{8.9}$$

where ρ is the material density and C_p is the specific heat of capacity. Hence, the temperature profile is

$$T_i(x, y, z, t) = \frac{Q}{4\rho C_p (\pi \alpha t)^{3/2}} \exp\left\{ -\frac{(x - x_S)^2 + (y - y_S)^2 + (z - z_S)^2}{4\alpha t} \right\} \tag{8.10}$$

If we convert the spatial temperature profile in the spherical coordinates with a point source considered as the origin in the coordinates on the material surface such that T is a function in time and the radial position r [$= ((x - x_S)^2 + (y - y_S)^2 + (z - z_S)^2)^{1/2}$]:

$$T_i(r, t) = \frac{Q}{4\rho C_p (\pi \alpha t)^{3/2}} \exp\left\{ -\frac{r^2}{4\alpha t} \right\} \tag{8.11}$$

Although an instantaneous laser point source is hard to achieve in practice, Eq. 8.11 describes a fundamental relation for developing more useful temperature profiles for (1) a point source applying at a static position over time and (2) a moving point source, as discussed in the following sections.

8.4.3 Stationary Point Source on a Plane

In practice, the laser should be applied on a material surface with a steady heating power of q continuously starting from $t = 0$. If we substitute the total heating energy from an instantaneous point source Q in Eq. 8.11 as the heating power of a point q, the temperature profile $T_c(r, t)$ can be calculated by the time integral:

$$T_c(r,t) = \frac{q}{4\rho C_p(\pi\alpha)^{3/2}} \int_0^t \exp\left\{-\frac{r^2}{4\alpha t}\right\} \frac{dt}{t^{3/2}} \tag{8.12}$$

We may let $\tau = 1/t^{½}$,

$$T_c(r,t) = \frac{q}{2\rho C_p(\pi\alpha)^{3/2}} \int_{1/\sqrt{t}}^{\infty} \exp\left\{-\frac{r^2\tau^2}{4\alpha}\right\} d\tau = \frac{q}{2\pi\rho C_p \alpha r} \operatorname{erfc}\left(\frac{r}{\sqrt{4\alpha t}}\right) \tag{8.13}$$

where erfc(.) is called the error function whose value can be found from a pre-calculated table or computed using a computer. For a rough estimation, the part erfc(Θ) can be approximated as $\exp(-1.10\,\Theta - 0.76\,\Theta^2)$, where $\Theta = r/(4\alpha t)^{½}$ is a positive number. Accordingly,

$$T_c(r,t) \approx \frac{q}{2\pi\rho C_p \alpha r}\left[1\text{-}\tanh\left(\frac{\pi r}{\sqrt{24\alpha t}}\right)\right]. \tag{8.14}$$

For pulsed lasers, $t \sim 10^{-10}$ s and $\alpha \sim 10^{-5}$ m^2 s^{-1} and therefore, for $r > 100$ nm,

$$T_c(r,t) \approx \frac{q}{\pi\rho C_p \alpha r} \exp\left(-\frac{\pi r}{\sqrt{6\alpha t}}\right). \tag{8.15}$$

8.4.4 Continuous Moving Point Source on a Plane

Recall Eq. 8.10. Substituting the instantaneous heating energy Q with the heating power q, the temperature profile $T_m(x, y, z, t)$ can be expressed by the time integral:

$$T_{\mathrm{m}}(x,y,z,t)=\int_{0}^{t}\frac{q}{4\rho C_{\mathrm{p}}(\pi\alpha t)^{3/2}}\exp\left\{-\frac{(x-x_S)^2+y^2+z^2}{4\alpha t}\right\}dt \tag{8.16}$$

where t is process time; (x_S, 0, 0) is position of the laser source.

Now, we may consider a point laser source applied on a material surface along a straight line along the x-direction. We may consider that the laser source is at a moving origin along x with a speed v. We consider another time scale of τ such that the laser process starts at $\tau = t$ and is being performed currently at $\tau = 0$. In other words, τ is the time scale of heat transfer from an instantaneous point source at a previous time t. We may further assume that the source location at a past moment is a function of τ, i.e., $x_S = -v\tau$, and if the current laser source is always located at the new moving origin, then

$$\begin{aligned}T_{\mathrm{m}}(x,y,z,t)&=-\int_{0}^{t}\frac{q}{4\rho C_{\mathrm{p}}(\pi\alpha\tau)^{3/2}}\exp\left\{-\frac{(x+v\tau)^2+y^2+z^2}{4\alpha\tau}\right\}d\tau\\&=\frac{q}{2\rho C_{\mathrm{p}}(\pi\alpha)^{3/2}}\exp\left(-\frac{xv}{2\alpha}\right)\times\int_{0}^{t}\frac{-1}{2\tau^{3/2}}\exp\left\{-\frac{\tau v^2}{4\alpha}-\frac{r^2}{4\alpha\tau}\right\}d\tau\end{aligned} \tag{8.17}$$

where $r = ((x - x_S)^2 + (y - y_S)^2 + (z - z_S)^2)^{1/2}$. We may let $\xi = 1/\tau^{1/2}$, $d\xi = -1/(2\tau^{3/2})$, then

$$T_{\mathrm{m}}(x,r,t)=-\frac{q}{2\rho C_{\mathrm{p}}(\pi\alpha)^{3/2}}\exp\left(-\frac{xv}{2\alpha}\right)\int_{1/\sqrt{t}}^{\infty}\exp\left\{-\frac{r^2}{4\alpha}\xi^2-\frac{v^2}{4\alpha}\xi^{-2}\right\}d\xi \tag{8.18}$$

If we consider the temperature profile of a stabilized laser process around the current laser source, we may assume a quasi-steady condition that the process duration is $t \to \infty$[1]:

$$\begin{aligned}T_{\mathrm{m}}(x,r)&=-\frac{q}{2\rho C_{\mathrm{p}}(\pi\alpha)^{3/2}}\exp\left(-\frac{xv}{2\alpha}\right)\times\left(-\frac{\sqrt{\pi\alpha}}{r}\right)\exp\left(-\frac{vr}{2\alpha}\right)\\T_{\mathrm{m}}(x,r)&=\frac{q}{2\pi\rho C_{\mathrm{p}}\alpha r}\exp\left(-\frac{v(r+x)}{2\alpha}\right)\end{aligned} \tag{8.19}$$

Notably, the variable r is the distance from the moving point of exposure.

[1] $e^{2ab}erf\left(ax+\frac{b}{x}\right)+e^{-2ab}erf\left(ax-\frac{b}{x}\right)=\frac{4a}{\sqrt{\pi}}\int\exp\left(-a^2x^2-\frac{b^2}{x^2}\right)dx$, where $erf(.)$ is the error function.

8.5 Laser Processes for Bulk Materials

8.5.1 *Laser Drilling and Cutting*

Laser drilling and cutting rely on vaporization. A focused laser beam first heats up the material surface to its boiling point and generates an initial vaporization at the exposed spot. The localized material removal induces the formation of a cavity which is called a "keyhole" on the surface spot. The keyhole causes a sudden increase in the absorptivity of heat due to the multiple reflections back and forth over inner surfaces of the keyhole, resulting in the quick deepening of the hole subsequently. As it deepens, material vapor is generated and escapes from the surface. Pulsed lasers are often used in drilling or cutting of materials as the laser heating duration is relatively short. In this case, heat dissipates to farther positions with a more evenly distributed profile over a bulk material body, rather than accumulating locally at the cutting/drilling site and causing excessive melting/ vaporization along the material interfaces. Though we have obtained some exact solutions for a few laser heating cases in the previous section, the rate of penetration of a laser beam into the workpiece can be estimated from a lumped heat capacity manner. In particular, if a pulsed laser is applied with a very high power density within a very short period of time, we may ignore the heat conduction effect because the rate of heat absorption is faster than the rate of conduction. Hence, the volume removal per second per unit surface area equals to the penetration velocity V_p (ms^{-1}) during the laser exposure with the relation

$$V_p \approx \frac{p_v}{\rho\left[L_v + C_p(T_v - T_o)\right]} \tag{8.20}$$

where p_v is the absorbed power density (unit: Wm^{-2}), L_v is the latent heat of fusion and vaporization ($J\ kg^{-1}$), T_v is the vaporization temperature, and T_o is the initial temperature of material.

Notably, for the onset of laser drilling/cutting, we need to take the absorption coefficient (A) into account and ensure that the absorbed energy can induce a local temperature rise beyond the vaporization temperature. We must also consider L_v in order to set a higher target value of the temperature rise (i.e., set T_c as a virtual level of $T_c^* = L_v/C_p + T_v - T_o$ in Eq. 8.15). We need also to consider P_{peak} (rather than P_a) in Eq. 8.1 to estimate the value of p_v. In practice, a laser beam has a radius R_{beam}, instead of acting as a point source in the ideal case as considered in the previous section and $p_v \sim P_{peak}/(\pi R_{beam}^2)$, accordingly. The estimation of p_v for different materials can be calculated using the values listed in Tables 9.2 and 8.1. Applying the vaporization temperature in such estimation would represent the necessary level of the required power density p_v^*.

For the following drilling/cutting process, we would assume the absorption coefficient (A) to be ~1. We may then adopt a defined p_v (it must be higher than the necessary level $A\ p_v^*$) to approximate the keyhole radius R_v. Using Eq. 8.15

Table 8.1 Thermal properties of common metals used in biomedical applications

	Titanium	Stainless steel (304)
L_f (kJ kg^{-1})	437	~300
L_v (kJ kg^{-1})	9000	6500
C_p (J kg^{-1} °C^{-1})	519	500
T_M (°C)	1668	1450
T_v (°C)	3260	3000
α ($\times 10^{-5}$m^2 s^{-1})	0.81	0.50

again, we can find the circular region radius R_v with a sufficiently high temperature for the material vaporization. Furthermore, it is suggested that the relative laser activation duration t_p/T_{period} to be small enough (e.g., < 0.5) such that the material can cool down to be close to the original temperature before the next application of a pulse laser. The average speed of keyhole deepening can then be considered as $V_p P_a t_p/(P_{peak} T_{period})$. For a special case where $R_v \approx R_{beam}$, according to Eq. 8.11 and the "diffusive length" to be $(4\alpha t)^{1/2}$, the temperature for conduction in all direction inside a material body (T') would be

$$T'(r \approx 0, t) \sim \frac{p_v(\pi \times 4\alpha t) \times t}{4\rho C_p(\pi \alpha t)^{3/2}} = \frac{p_v}{\rho C_p}\left(\frac{t}{\pi \alpha}\right)^{1/2}. \tag{8.21}$$

Yet, in the configuration of a laser shooting on a material surface, the conduction only appears as a hemispherical shape, and the temperature should be about

$$T(r \approx 0, t) = 2T'(r \approx 0, t) = \frac{2p_v}{\rho C_p}\left(\frac{t}{\pi \alpha}\right)^{1/2}. \tag{8.22}$$

Therefore, the essential cutting time t_v can be estimated as

$$t_v \approx \pi \alpha \left(\frac{\rho C_p T_c^*}{2p_v}\right)^2. \tag{8.23}$$

From this simplified relation, it can be seen that the peak power P_{peak} and radius R_{beam} of the beam are very important since $p_v = P_{peak}/R_{beam}$. In drilling with a YAG laser, considerable attention is paid to the design of the pulse shape with time. The aim is usually to have a short sharp pulse for cutting as opposed to a longer pulse with a reduced initial peak for welding. The shape of the pulse can be critical as in the drilling of glass, for example. In general, it is not particularly important except for very short pulses (<100 fs). The resulting laser ablation for micromachining operations (drilling and cutting) shows negligible head affected zones but little debris sometimes.

It should be rather obvious that drilling and cutting by laser use the same working principles. The only difference between the two micromachining operations is the path of laser tracing on the material surface.

8.5.2 Laser Welding

As with laser cutting, welding relies on a finely focused beam to achieve penetration and localized heating at a spot on the material surface. A laser can be used as a high-speed, high-quality welding tool. The advantages in laser welding, for example, include an accessible spacing between the sample surface and the laser source, enabling real-time inspection of the process, simplifying the quality control process, and relatively lower costs. Compared to the traditional approaches such as tailored blank welding for the car industry, laser welding has great demand in areas requiring heat-sensitive components such as heart pacemakers.

The basic operation of laser welding is illustrated in Fig. 8.9. Conduction-limited welding occurs when the power density is insufficient to cause boiling – and therefore generate a keyhole – at the given welding speed. The weld pool has strong stirring forces driven by Marangoni-type forces resulting from the variation in surface tension with temperature. There are two principle areas of interest in the mechanism of keyhole welding. The first is the flow structure since this directly affects the wave formation on the weld pool and hence the final frozen weld bead geometry. This geometry is a measure of weld quality. The second is the mechanism for absorption within the keyhole which may affect both this flow stability and entrapped porosity. The absorption of the beam occurs through the absorption of the laser during reflection from a surface which depends on the slope of the face, mode structure of original incident beam, polarization effects, and focal position. Generally, the molten metal has a surface tension which reduces with temperature, and hence, the surface of the tailing keyhole has a resultant tension pulling the molten material away from the exposure site.

In this welding mode, we can consider applying a continuously moving laser source with a moving velocity v_w. We also take the latent heat of capacity L_m into consideration in order to set a higher target value of the temperature rise T_w during welding (i.e., set T_w as a virtual level of $T_w^* > L_m/C_p + T_M - T_o$ in Eq. 8.19, where T_M

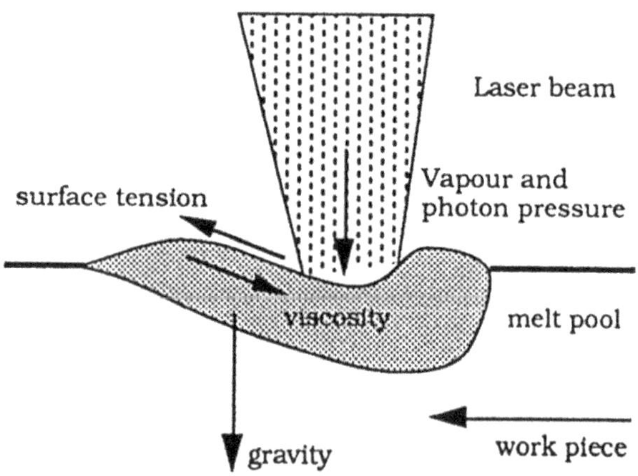

Fig. 8.9 Side-view illustration of the melt pool shape and liquid motion

is the melting point). Finding the relationship between the laser beam power and the width of welding is relatively straightforward. We may adopt a continuous laser (i.e., $t_p = T_{period}$) with power P_{peak}. We may recall Eq. 8.19 at the x-position equals to zero and substitute the laser power as AP_{peak} in order to obtain the cross-sectional radius of the welded material part (R_w):

$$\frac{AP_{peak}}{2\pi\rho C_p \alpha R_w} \exp\left(-\frac{v_w R_w}{2\alpha}\right) = T_w{}^* \tag{8.24}$$

For the case where the moving velocity of the laser source is slow enough such that $v_w R_w/(2\alpha) \ll 1$:

$$R_w \approx \frac{2\alpha AP_{peak}}{T_w{}^* 4\pi\rho C_p \alpha^2 + v_w AP_{peak}} \tag{8.25}$$

In laser welding, a term has been used to define the joining efficiency η_{join} as

$$\eta_{join} = v_w R_w / P_{peak} \tag{8.26}$$

The higher the value of the joining efficiency, the less energy is spent in unnecessary heat generation in heat-affected zone or distortion. Besides, a higher power would provide a larger operating window for regulating other process parameters. There are technical boundaries for a good weld for a given laser power such as lack of laser power penetration to the inner material interface and the "falloff," which describes the material melts from a welding site flowing to the non-welding sites along the material interfaces.

The alternative mode is "keyhole" welding in which there is sufficient energy per unit length to cause evaporation and hence a hole in the melt pool. The "keyhole" behaves like an optical black body in that the radiation enters the hole and is subject to multiple reflections before being able to escape with an absorption coefficient $(A) \approx 1$ during the cutting process discussed before. We may apply Eq. 8.19 again to calculate for the welded cross-sectional radius by setting the target temperature rise T_w as a virtual level of $T_w{}^* > L_m/C_p + T_M - T_o$. Under a sufficient slow-moving velocity of the laser source,

$$R_w \approx \frac{2\alpha P_{peak}}{T_w{}^* 4\pi\rho C_p \alpha^2 + v_w P_{peak}} \tag{8.27}$$

On top of the continuous wave approach, the use of pulsed power allows two more variables to be considered: (1) pulse repetition frequency and (2) percentage of spot overlapping length between two consecutive exposures. The applied laser pulse is usually longer for welding than for drilling/cutting purposes and is shaped to have a lower peak power. The welding speed is decided by the spot size ($=2R_{beam}$) × pulse repetition frequency ($=1/T_{period}$) × (1 – overlapping %). The pulse mode allows for better control over the flow in the weld pool. It can also reduce the porosity in thick section welding by pulsing at the oscillation frequency of the weld pool, leading to a resonance effect. In fact, the welding speed is independent of power. Penetration is a function of power, and hence, if the peak power is raised by pulsing or modulating the beam, there can be greater penetration for a given average power. This effect can be more marked than expected. For example, pulsing with a square pulse form at 100–500 Hz and a peak power of twice the average power, an improvement of 30% penetration is reported when welding 304 stainless steel. The increased peak power also means better welding of reflective material since the keyhole is initiated quicker. Using a 1 kW average power of Nd-YAG laser at 500 Hz and 2 kW peak power, 6181 aluminum was welded at nearly three times the speed of continuous wave welding at 1 kW or with an improvement of 60% in penetration. Higher peak power also means greater tolerance to focal position, and pulsing means less energy deposited in the workpiece leading to reduced distortion.

Similar to the special case of the cutting process, when $R_v \approx R_{beam}$, the essential welding time t_w for each application of the laser pulse can be approximated as

$$t_w \approx \pi\alpha\left(\frac{\rho C_p T_w{}^*}{2AP_{peak}}\right)^2, \tag{8.28}$$

assuming that there is negligible vaporization induced.

8.6 Selective Laser Sintering

Selective laser sintering has been a successful product development tool since the early 1990s. Europe and the United States, Japan, and other regions have gradually recognized laser sintering as the standard for next-generation rapid manufacturing technology. Developed based on the concept of rapid prototyping, laser sintering is a method that uses laser and powder to generate a three-dimensional model which has the benefit of not requiring a supporting material.

Laser sintering utilizes laser power as a heat source on the traditional powder compact sintering technology. This technique has unique advantages that are not easily achieved by conventional sintering furnaces. Due to a laser beam's concentration of heat and penetration, suitable for small areas, thin products can be sintered. It is easy to sinter powder or flake compacts different from the matrix composition. The typical process is illustrated in Fig. 8.6. Briefly, this process is based on assembling materials by heating microbeads of known materials to their melting temperature in order to have the microbeads attach together into the defined product geometry. Laser sintering can use nearly all kinds of material as a "base," no matter metal, polymer, alloy, or ceramics as products with very flexible geometry. This technology generates 3D parts by selectively fusing thermoplastic, ceramic, or metallic powders with the localized heat from an infrared laser, usually a CO_2

laser. The continuous wave laser model is often applied in laser sintering, and therefore, the process modeling is similar to the laser welding process discussed previously in Sect. 8.5.2.

Similar to the other rapid prototyping methods, designers are first required to design the product construction and data preparation through CAD and convert the design into an STL format. The STL is then sliced into multiple 2D layers and the laser pathway tracing generated as the G-code over the entire volume of the product. For a new material slice during laser sintering, the powder bed moves a level lower, and microbeads refill the power bed. The thickness is defined by removing excessive microbeads with a roller. Next, a movable high-power laser pulse of the laser machining system fires a layer of the microbeads following the G-code. Because the material will absorb the laser and change its state via processes such as melting and deforming, the material can bond together and become a solid layer. The fabrication process is similar to SL in that a thin layer is placed on a plate (Fig. 8.10a), and a pattern is solidified in it using a laser tracing along the required region of a product cross section after which another layer is laid on top. The process is repeated until the object is built slice by slice. In this case, each thin layer is a layer of fine powder which is spread over the surface to a precise depth by a plough. As shown in Fig. 8.10b, the laser beam from a 50 W CO_2 laser is then scanned or rastered by rotating the driving mirrors in the required *xy* pattern taken from the STL file. Typically, it is common to melt 20–60-μm-thick layers of material with the maximum laser power of 200 W. Then, the laser moving speed is optimized to meet the range of 225–400 mm/s based on the design production time and operation. In

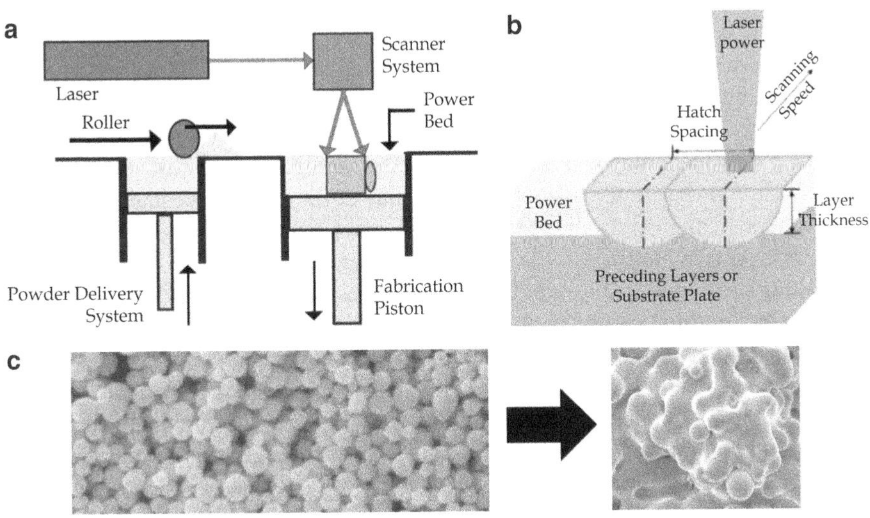

Fig. 8.10 (**a**) Deposition mechanism of a new microbead layer and (**b**) laser welding during the 3D laser sintering. (**c**) Aggregation of microbeads after laser sintering. (Wang [10], Applied sciences)

addition, the laser radius and intensity should not exceed 40 μm and 2×10^8 W m^{-2}. Furthermore, surface roughness is roughly the bead diameter (Fig. 8.10c). The accuracy is in the range of ±0.05–0.25 mm. To minimize the required laser energy and speed up the process, the chamber is heated and may be filled with an inert gas, such as nitrogen, to avoid burning or oxidation of the powders. The completed product is withdrawn from the unsintered powder bed and bead blasted to remove unsintered adhering particles. The finished parts are in some cases functional and testable prototypes as well as conceptual models. The unsintered particles can be recycled in the next process.

The use of laser can achieve a high temperature beyond the melting points for polymeric, metallic, and ceramic bonding. Lasers can weld polymers such as nylon or polystyrene; metals including steel, titanium, and alloy mixtures; and composites. Laser sintering works with single-component powders where the laser melts only the outer surface of the particles (surface melting), fusing the solid non-melted cores to each other and to the previously assembled layer. This process does not usually require support structures. It also works with two-component powders, typically either coated powder or a powder mixture. The physical process can be full melting, partial melting, or liquid-phase sintering. Depending on the material, up to >80% density can be achieved with material properties comparable to those from conventional manufacturing methods. These usually suggest 26% porosity. In many cases, large numbers of parts can be packed within the powder bed, allowing for very high productivity. Laser sintering of components allows for good dimensional precision, low cost, high production speed, and material diversification. Compared with machining and other conventional manufacturing processes, it is possible to get fully functional prototypes, molds, or models in just a few days based on 3D CAD data, helping to significantly reduce time to market. Laser sintering supports a wide range of production with a complex geometry of products. By avoiding the time and cost associated with tooling, the use of laser-targeted products can lead some industries to gain higher competitiveness. To date, laser sintering has already become a key technology in electronic manufacturing, making fast, flexible, and cost-effective production possible directly from CAD files. In particular, selective laser sintering can manufacture products with small quantities of high-quality parts. Therefore, custom biomedical devices are ideal applications.

Making prosthetics using highly flexible processes such as rapid prototyping is ideal. Laser sintering can be applied in the manufacturing of biomedical implants. For example, a patient suffering from chest wall sarcoma, a tumor which grows around the ribs, has successfully been implanted with a titanium 3D printed sternum and rib cage designed by an Australian company in 2012. The implant design was based on the unique rib geometry of the patient obtained from an X-ray CT-scan. The 3D digital CAD file was then downloaded to a selective laser sintering machine, and the product was fused from titanium powder layer by layer.

In 2014, Peking University Hospital doctors successfully implanted an artificial, 3D printed vertebra replacement in a 12-year-old young boy with bone cancer. Currently, the standard procedure for this kind of operations involves removing the bone and inserting a titanium tube held in place by screws and surgical cement.

However, the tube can become detached over time. In this example, doctors used a combination of CT scans and engineering software to create a perfect replica of the piece of the patient's spine perfectly matching the anatomical structure that needed to be replaced. The surgeons then removed a tumor from patient's spine before implanting the 3D printed vertebra. Applying selective laser sintering, the implant was made from titanium powder and included a series of tiny pores which allowed the bone to grow and bond to the vertebral structure as it heals. With the optimized shape matching with the vertebra, the resultant implantation was much stronger and more convenient than traditional methods. It resulted in a faster recovery for the patient and increased mobility after he healed. One month after surgery, the patient was on the road to recovery, demonstrating such reverse engineering approaches using laser sintering manufacturing as a new treatment option for spinal replacement surgery patients. In the future, it is expected that problematic bone sites in patients can be scanned immediately in the hospital, and a scaffold/prosthetic device can be printed next to the operation table for the implantation surgery. Notable target examples include metallic components in an artificial knee, an acetabular cup in the artificial hip joint, and skull replacement.

Problems

Problem 8.1

You are drilling titanium, 1.5 mm thick, hole diameter $d = 0.5$ mm, absorbance 50%, pulse length 2×10^{-5} s. Determine the instantaneous power P (W).

Problem 8.2

Considering line welding of stainless steel using a laser source with a power (P_a) of 1 kW and moving velocity (v) of 2 mm/s, (i) please determine the width of welded steel w over the steel surface, given that the melting temperature of steel (T_M) is 1425 °C and the environmental temperature (T_o) is 20 °C. The thermal properties of steel include thermal diffusivity $\alpha = 4.4$ mm^2/s, thermal conductivity $k = 15$ °CN/s, and density × specific heat of capacity $\rho c = 3.6$ N/(mm^2 C). Please be noted that the cross-hatched area enclosed by the 1425 °C isotherm over the steel material indicates the cross section of the weld pool. (ii) Further, please express w with a symbolic expression, as a function of P_a, v, T_M, T_o, and material properties of the welded material.

Problem 8.3

Recalling Eq. 8.19 with the origin ($x = 0$ and $r = 0$) of coordinates moving with the laser source, i.e.,

$$T(x, r) = \frac{q}{2\pi\rho C_p \alpha r} \exp\left(-\frac{v(r + x)}{2\alpha}\right)$$

Please find the steepest cooling gradient and the corresponding location behind the moving laser source. The cooling gradient can be considered as $-dT/d\tau = v \times dT/dx_S$.

Problem 8.4

Considering that laser welding of alloy steel with thermal diffusivity $\alpha = 10.2\ \mathrm{mm^2/s}$ and thermal conductivity $k = 38$ °CN/s, melting temperature $T_M = 1510$ °C, and environmental temperature $T_o = 20$ °C, the laser source is moving with a traversing velocity $v = 5$ mm/s for generating a weld width (w) on the material surface of 10 mm.

(a) Determine the power input q.
(b) Determine the distance behind the arc (x_c), at which the temperature cools down to 550 °C.
(c) Determine the cooling rate, i.e., $-dT/d\tau$, at this (moving) point.

Problem 8.5

Consider the laser welding along a line on carbon steel, stainless steel, and aluminum, as three individual processes. The traversing velocity for carbon steel is $v = 3$ mm/s, for stainless steel is 2 mm/s, and for aluminum is 5 mm/s. The weld width on material surfaces is 6 mm in all these three cases. The thermal properties are given in Table 8.P1.

(a) Determine the power input q (mW) for welding.
(b) Calculate the heat input H (kJ/mm) for three cases as well as the thermal efficiency η, defined as the ratio between the effective energy inducing material melting and the heat input.
(c) Plot the heat input H (kJ/mm) versus speed v (mm/s) in the range 0.1–10 mm/s for a weld width of $w = 6$ mm, for the three materials. Plot all three cases in one plot.

Table 8.P1 Thermal properties of carbon steel, stainless steel and aluminum

	α (mm^2/s)	k (°CN/s)	ρC (N/(mm^2 °C))	T_M (°C)
Carbon steel	12	43	3.7	1480
Stainless steel	4.4	15	3.6	1425
Aluminum	60	140	2.3	546

Problem 8.6

Let's consider that case of the "moving-point" laser welding of two very thin stainless steel plates. Accordingly, if we consider a large enough time scale, Eq. 8.7 would be simplified as a "planar" relation:

$$\frac{1}{\alpha}\frac{\partial T_i}{\partial t} = \frac{\partial^2 T_i}{\partial x^2} + \frac{\partial^2 T_i}{\partial y^2}.$$

This is because the temperature variation along the z-direction can settle much faster than along x- and y-directions. We may neglect the derivative of temperature with respect to z as a reasonably good approximation.

(a) Please derive the temperature profile as a function of x and r for the case of a continuous moving point source on a plane, similar to Eq. 8.19. (You answer might be expressed with the Bessel function.)

In the following parts, please use any missing parameters as defined in Problem 8.4.

(b) Repeat Problem 8.4 as the planar laser welding of two alloy plate, by assuming each the plate has a thickness of 100 μm, to obtain the power input q, the distance behind the arc (x_c), and the cooling rate at x_c.
(c) Please adopt the power input q you obtain from (b). Then, find out and plot the corresponding weld width w against different plate thicknesses from 50 to 200 μm.
(d) Please fix the weld width $w = 10$ mm, and calculate the power output q for different plate thicknesses. If we further consider a plate to be "thin" when the welding process for the same weld width w requiring only $\leq$10% of the power output q for the thick materials as what you calculated in Problem 8.4(a), please suggest the condition for the "thin plate" approximation (and also the planar welding simplification).

Problem 8.7

The governing equation for conduction heat flow in a solid with a coordinate system fixed at a stationary origin relative to the solid may be expressed as

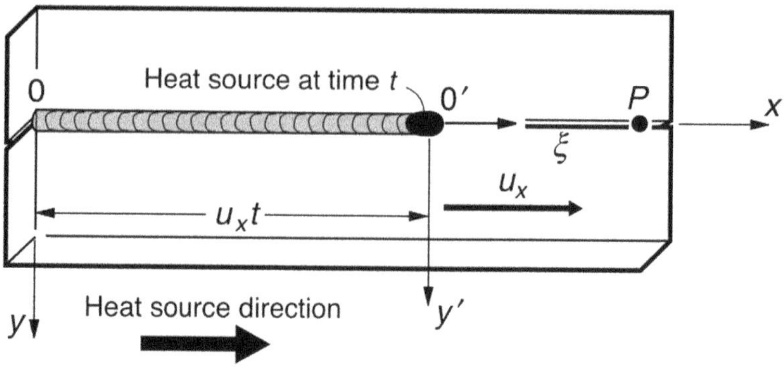

Fig. 8.P1 The coordinate system moving together with a translating heat source long the x-axis

$$\frac{\partial T}{\partial t} = \alpha \left(\frac{\partial^2 T}{\partial x^2} + \frac{\partial^2 T}{\partial y^2} + \frac{\partial^2 T}{\partial z^2} \right)$$

Show that for the moving heat source case with a coordinate system moving with the heat source along the x-axis at a velocity u_x, as shown in Fig. 8.P1, the corresponding governing equation obtained by a coordinate transformation from the plate to the heat source, with x replaced by ξ, that is, $\xi = x - u_x t$, is given by

$$\frac{\partial T}{\partial t} = \alpha \left(\frac{\partial^2 T}{\partial \xi^2} + \frac{\partial^2 T}{\partial y^2} + \frac{\partial^2 T}{\partial z^2} \right) + u_x \frac{\partial T}{\partial \xi}$$

Problem 8.8

The temperature distribution in a semi-infinite plate for a moving point heat source is given as in Eq. 8.19. Derive an expression for the cooling rate at points of the plate where the cooling rates are highest. Where do these occur?

Problem 8.9

Considering the laser processing on thin materials such as welding of two thin plates, we may assume that $\partial T / \partial z \approx 0$ for all material positions. Derive the temperature profile of a continuous moving laser source on a plate with a speed v along x. If we let $\xi = x - vt$, the temperature profile should be in the form

$$T(\xi, r) = \Upsilon_1 \exp(\Upsilon_2 v) K_0(\Upsilon_3 r)$$

where Υ_1, Υ_2, and Υ_3 can be expressed by other variables, r is the radial distance from the current position with laser exposure, and $K_0(\cdot)$ is the modified Bessel function of the second kind of order zero.

References and Further Reading

1. Joshi, S.N.: Lasers Based Manufacturing. Springer, New Delhi, India (2016)
2. Crafer, R., Oakley, P.J.: Laser Processing in Manufacturing. Springer, Dordrecht (2013)
3. Sugioka, K., Meunier, M., Piqué, A.: Laser Precision Microfabrication. Springer, Berlin (2010)
4. Ion, J.: Laser Processing of Engineering Materials: Principles, Procedure and Industrial Application. Elsevier, Amsterdam (2005)
5. Kruth, J.-P., Levy, G., Klocke, F., Childs, T.H.C.: Consolidation phenomena in laser and powder-bed based layered manufacturing. CIRP Ann. **56**, 730–759 (2007)
6. Santos, E.C., Shiomi, M., Osakada, K., Laoui, T.: Rapid manufacturing of metal components by laser forming. Int. J. Mach. Tools Manuf. **46**, 1459–1468 (2006)
7. Bass, M.: Laser Materials Processing. Elsevier, New York, NY, USA (2012)
8. Zhou, R., Lu, X., Lin, S., Huang, T.: Laser texturing of NiTi alloy with enhanced bioactivity for stem cell growth and alignment. J. Laser Micro/Nanoeng. **12**(1), 22–27 (2017)
9. Burakowski, T., Wierzchon, T.: Surface Engineering of Metals: Principles, Equipment, Technologies. CRC Press, Boca Raton (1998)
10. Wang, D., Yu, C., Zhou, X., Ma, J., Liu, W., Shen, Z.: Dense pure tungsten fabricated by selective laser melting. Appl. Sci. **7**, 430 (2017)

Part III
Design Techniques

Chapter 9
Biocompatible Material Selection

Abstract Material selection under the guidance of the general principles must be considered from the material's own properties and the available manufacturing processes. Manufacturing designers should consider a broad range of factors affecting the requirements and feasible options for the optimal functions, biocompatibility, and manufacturing outcomes of the target biomedical or healthcare products. Sometimes, relationships of the design factors can be transformed as a combined expression called the material performance index.

9.1 Product Design and Material Selection

9.1.1 General Product Design Flow

Material selection is a critical step in the process of designing any product, including biomedical devices. In the context of product design, the main goal of material selection is to minimize cost and production time while meeting product performance goals. Systematic selection of the best material for a given application begins with the properties and costs of candidate materials. For example, a thermal blanket must have low thermal conductivity to minimize heat transfer for a given temperature difference. An incorrect or inappropriate selection may result in poor results. An incorrectly chosen material can lead not only to failure of the part but also to unnecessary life cycle cost. A poorly chosen material can add to manufacturing cost and unnecessarily increase the cost of the part. Besides, the properties of the material can be enhanced or reduced by different material processes, and these may affect the resultant product performance.

Material selection under the guidance of the general principles must be considered from the material's own properties and the available manufacturing processes. Material selection not only acts as a downstream step after product design but also plays an important role in the feedback loop of product refinement and innovation. Formulation of technical specifications for a biomedical device can be achieved through a breakdown of the complete device into assemblies and components, followed by considering the requirements for each of them. Material and process

R. H. W. Lam, W. Chen, *Biomedical Devices*,
https://doi.org/10.1007/978-3-030-24237-4_9

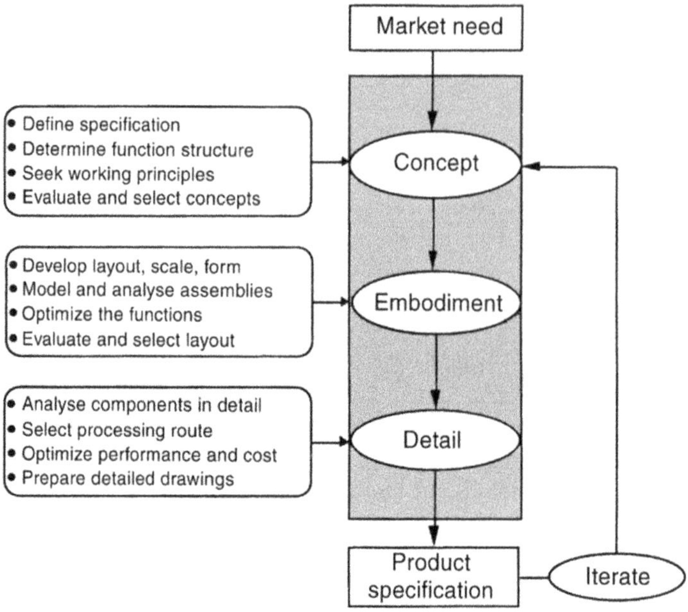

Fig. 9.1 The general product design flow. The design proceeds from an identification and clarification of task through concept, embodiment, and detailed analysis to a product specification

selections can first be decided at the component level and then be integrated with some further refinements as more comprehensive requirements for the device are decided. The general product design flow contains four main stages (Fig. 9.1):

- Market need: Market research should be first carried out with the purpose of investigation, especially targeted market research. The contents of a market investigation should be closely related to product design. There are many methods of market research, such as on-the-spot investigation, online consultation, and consulting materials. For instance, home-based care and health monitoring devices need to pay closer attention to their product appearance and user-friendliness. These devices, used by patients or family members without direct clinical supervision, may require extra concerns, such as improved ergonomics. The use of color, special notice effects, and textures can help patients intuitively understand and follow proper usage parameters. Lightweight materials without compromised strength or durability can increase the device's portability.
- Concept formulation: This is defining the target specifications of a product. Besides, it is also required to discuss the functional structure of a product. Seeking working principles is also needed during this stage. The final step is to evaluate and select concepts of the product.
- Embodiment: It is suggested to draft multiple possible layouts, geometric features, and other brief descriptions of the product which all fulfill the basic product specifications. Based on these brief descriptions, we may further model the

assemblies and define the sub-parts and components. It is expected that there are multiple options and designs for the targeted product. Hence, we then need to evaluate and select the "best" brief design of the product.

- Detailed design: After the above three steps, the components can be analyzed in detail in order to finalize the component shapes, the materials to be used, and the chosen manufacturing processes. As multiple material and manufacturing processes can be involved in producing one component, we need to arrange and configure the processing route (including assembly of all the components) and operation parameters in order to optimize performance and minimize the production cost and time. After all the product and process details are confirmed, detailed drawings of each component of the product can also be prepared.

The process from the concept formulation to the detailed design can be integrative. Once the detailed design is ready, it can be further examined on whether or not it can meet the target product specifications. If there are any necessary modifications realized at this stage, the product design process may need to be revised from the concept formulation stage.

9.1.2 *Design Flow of Biomedical Products*

For designing biomedical products, designers need to consider both the material property and biocompatibility requirements; a simplified decision flow is shown in Fig. 9.2. First, we need to ensure the chosen material can fulfill the physical requirements of a product component following the product design process. Furthermore, we need also to check whether the material is biocompatible. Biocompatibility is the capability of a material to function in an in vivo environment for an acceptable period of time with no detrimental effect or only an acceptable level of effects on the host. If there is no related record available, we will need to test whether or not the material would initiate any effects at the expected implant site. In particular, since the immune response and regenerative functions of the body are very complicated, it is not adequate to describe the biocompatibility of a single material in relation to a single cell type or tissue. We would need to go through a series of animal experiments and subsequent clinical trials to determine the biocompatibility of the material.

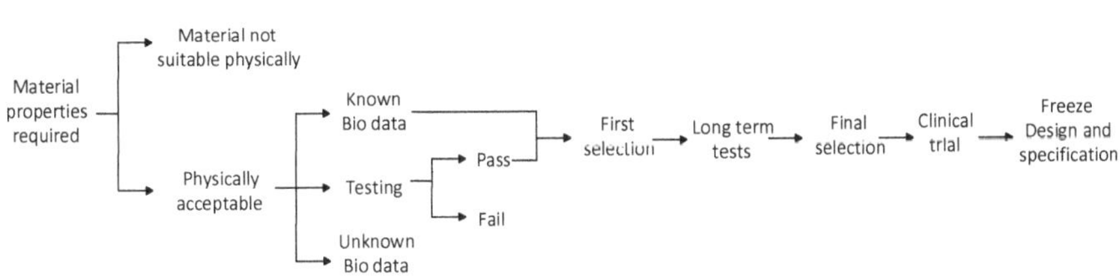

Fig. 9.2 Decision tree for biomedical device development

Establishing the biocompatibility of medical devices and their materials is of key importance in assuring product safety. Central to the issue of biocompatibility is the role of chemistry and material characterization. Presently, the biological evaluation of medical devices is governed by the set of standards developed by ISO and known as the ISO 10933 series of standards. This document establishes the need for toxicological risk assessment based on chemical and material characterization prior to initiating any biological testing.

Although regulatory bodies certify the safety and efficacy of finished medical devices rather than their component materials, a detailed master access file for a certain material can often help smooth the road to device approval. A manufacturing partner with a dedicated regulatory practice can help device companies identify medical-grade plastics, elastomers, and composites that meet US Food and Drug Administration (FDA) and United States Pharmacopeia (USP) Class VI requirements as well as those for international regulations.

Materials characterization forms the basis for understanding the composition of a medical device material and its potential to have an adverse biological effect when the device is put into clinical use. Materials selection and risk analysis are integral components of the design process for medical devices and play critical roles in evaluating biological safety. Just as important as the testing, the biological safety evaluation plan should be designed and performed to demonstrate the achievement of special criteria for safety. Clearly, the process presented in ISO 10933-1, 10933-18, and 10933-17 may aid in the selection of optimal materials and in the control of the uniformity of those materials throughout the lifetime of the device. If appropriate care is taken to minimize critical controlling factors, in vitro materials testing can frequently accurately predict clinical outcomes.

The evaluation of risk management includes identification of all hazards and the estimation of associated risk. A major component in hazard identification is materials characterization. These following steps can be identified:

1. Define and characterize each material, including suitable alternative materials.
2. Identify hazards in materials, additives, processing, and cleaning aids.
3. Estimate exposure (total or clinically available amounts).
4. Review toxicology and other biological safety data (published and available).

The risk management approach emphasizes that conducting animal testing for risk evolution should only be considered after all alternative courses of action (review of prior knowledge, chemical characterization, in vitro evolution) have been established. Several clauses and subclauses in ISO 10933-1 ask the user to conduct chemical characterization of the device undergoing biological evaluation.

A variety of techniques are available for performing chemical and materials characterization. The tests may be carried out directly on material samples or on material extracts prepared under specified conditions. These tests, which have evolved over many years, are relevant, sensitive, rapid, and inexpensive, yet they provide extremely valuable information to establish material safety and biocompatibility. The extent to which a material needs to be characterized depends upon the type of material, the end use of the device, and the function of the material within the

Table 9.1 Categorization of medical devices

Nature of contact and tissue type	Duration of contact	Types of devices	Degree of characterization
Surface Skin Mucous membranes	Limited	Exam gloves, tape, blood pressure cuff dental dams, endoscopes	Minimal
External communication Blood, indirect	Prolonged	Dialysis, cardiopulmonary bypass	Intermediate
Implant Blood, direct tissue contact	Permanent	Shunts or grafts orthopedic implants	High

device. The more critical the role of the device and the more important the properties of its materials are to device performance, the more detailed the characterization program should be. Table 9.1 lists the various categories of devices defined in ISO 10993-18 and the proposed degree of characterization necessary.

9.2 Considerations of Material Characteristics

9.2.1 Basic Factors

Manufacturing designers should consider a broad range of factors affecting the requirements and feasible options for the optimal functions, biocompatibility, and manufacturing outcomes of the target biomedical or healthcare products. Biocompatibility is ultimately a function of a completed device, so it encompasses all of its components, assembly processes, and overall design. However, thoughtful material selection can help ensure the overall biocompatibility of a finished product. Even when materials incorporated into a medical device have been assessed for their biocompatibility, new risks can be introduced through manufacturing and postproduction processes that can have an adverse effect on the device. Properly clean assembly and hygiene processes help to lower these risks. Here, we briefly exemplify other common factors in most product design processes.

Generally speaking, material cost is a critical aspect of any medical device. In the broader sense, the manufacturing designers should look beyond the simple expenses of raw materials and adopted processes in order to assess a device's true life cycle costs. The labor wage or the tool life during assembly should also be taken into consideration. Excluded from the material cost are all indirect materials, such as cleaning supplies used in the production process. The cost of labor to produce the product is separate from this cost. The total cost of a product and its eventual sale price are arrived at by combining the material cost with the cost of labor. Additionally, the costs have time factors. Different materials can cause costs to increase or decrease over time. It is important to consider how each candidate material can add or reduce the unit cost of a device throughout the full life cycle. If necessary, manufacturing processes should also be tweaked to reduce the required energy and cost. However, sometimes a more expensive, higher-performing material can save money in the long run.

The material selection process is to match performance capabilities fulfilling the device specifications. Performance capabilities can range from mechanical properties, such as strength, flexural modulus, and high-temperature resistance, to chemical resistance, particularly exposure to harsh antibacterial cleaners. Furthermore, the sustainability of medical devices comes into play primarily during design and manufacturing as they cannot be easily disposed of or replaced once implanted into a human body. In the design phase, sustainability considerations include strategies such as the use of lower-density and more biocompatible materials. They also include the demands of common usage scenarios, including resistance to breaking/cracking from impact. Some materials are inherently strong, flexible, or chemically resistant, while others can be enhanced with additives or reinforced with glass fiber.

Apparently, the material properties discussed in Chaps. 2, 3, 4, and 5 are major concerns as they directly determine whether the product is functional and biocompatible. Taking an artificial joint product as an example, a good artificial joint material is necessary. For the weight-bearing joints, the related material ought to bear stress of 700 MPa with a safety factor of 10 because the ultimate tensile strength of bone is about 70 MPa. Elastic modulus is clinically important because it indicates that the selected biomaterial has similar deformable properties with the material it is going to replace. (A number of methods are available for determining the tensile strength of materials, such as the bending flexural test, the biaxial flexural strength test, and the Weibull analysis.) These force-bearing materials require high elastic moduli with low deflection. As the elastic modulus of a material increases, fracture resistance decreases. It is desirable for the biomaterial elastic modulus to be similar to bone. This is because if it is greater than bone's elastic modulus, then load will be borne by the material only, while the load is borne by bone only if it is less than the material.

The fracture strength, which is the stress at which a specimen begins to crack, should also be analyzed. Strength of a biomaterial is an important mechanical property. In particular, bioceramics are often brittle. Once cracks are formed on a material body, the cracks can then easily propagate when the material is subject to a much lower loading stress than the material strength. This importance is clear for bioceramics which have higher ductility on top of their high strength. For example, there are a number of other ways that cracks can be produced in bioceramics, such as thermal sintering and heating.

In addition, fatigue is also an important parameter for biomaterials because cyclic load can be applied during their life cycle in a human body with repeating and continuous body movements. Fatigue is defined as failure of a material due to repeated/cyclic loading or unloading (tensile or compressive stresses). In this cyclic loading condition, microcracks/flaws may be generated. These microscaled cracks can initiate permanent plastic deformation which results in large crack propagation or failure. During cyclic loading, several factors also contribute to microcrack generation such as frictional sliding of the mating surface, progressive wear, residual stresses at grain boundaries, and stress due to shear.

On the other hand, electrical properties of materials are key factors. For example, an artificial hand with a myoelectrical control is a new type of dynamic artificial hand and a model "man-machine system" which is controlled by biological electricity.

9.2.2 *Material Degradation*

The life span of biomedical products which are capable of offering required operational performance is extremely important. The degradation of materials has a significant impact on the performance and reliability of an implanted device. There are reasons that cause the degradation of materials. For metals and alloys, degradation is mainly caused by electrochemical reactions with bioliquids in a human body, such as blood or gastric acid. Degradation of biopolymers may include (1) alteration of the covalent interatomic bonds in polymer chains by chain scission or further crosslinking, (2) the alteration of the intermolecular interactions between chains by incorporation or loss of low-molecular-weight compounds, and (3) chemical reactions in the aqueous environment such as oxidation and hydrolysis which can change the properties of implanted polymers. Sometimes, methods of sterilizing biopolymers (e.g., autoclaving and radiation) can significantly alter their properties, resulting in an accelerated degradation rate after implantation. The degradation of ceramics includes a preferential dissolution of impurities that results in crack propagation. Bioactive ceramics and glasses degrade via their reactions with adjacent tissues. For example, mineral components such as calcium phosphates can be resorbed by osteoblasts.

9.2.3 *Ion Release of Metals and Alloys*

The surface biochemical properties of biomaterials are very important since the surface interactions with surrounding tissues and fluids may affect that the biocompatibility of the materials and mechanical survival of the implanted device. When metallic implants are placed in the electrolytic environment of the human body (pH about 7.35 to 7.45, temperature about 38 °C, etc.), electrochemical reactions may result in the release of metal ions coupled with corresponding reduction reactions of constituents in the aqueous environment to maintain charge neutrality. This usually has a negative effect on mechanical factors such as pre-existing cracks, surface abrasion, film adhesion, etc. The quantity of these metal ions released is a function of the corrosion resistance of the metals. The compositions and ion release rate of some metals and alloys widely used in biomedical devices are shown in Table 9.2.

Metal ions generated by implants and their wear debris may remain in local tissues, while some may bind to protein moieties that are then transported through the blood stream and lymphatics to remote organs. Toxic effects of iron, titanium,

Table 9.2 The compositions and ion release rate of metals and alloy

	Compositions (weight ratio)	Ion release rate (μg/cm^2/week)				
		Ti	Ni	Fe	Cr	Co
Stainless steel	Mainly Fe, 17% Cr, 10% Ni	–	1.15	8.4	0.01	–
Cobalt-chromium	Mainly Co, 30% Cr, 0.8% Fe	–	–	0.6	0.03	0.1
Titanium	–	23	–	–	–	–
Titanium alloys	Mainly Ti	3	–	–	–	–

cobalt, chromium, and nickel in implant alloys and their tolerance in human body have been reported in the past decades of clinical research. Such tolerance is quantified as the maximum allowable daily intake of metal ions per liter of blood. (On average, an adult human male who weighs 70 kg has a blood volume of about 5 liters.) The allowable daily intake of iron ions is 200 μg/l/day. Iron is essential for the functioning of hemoglobin in red blood cells, helping to prevent anemia. A man needs an average daily intake of 7 mg of iron and a woman 11 mg, which can be fulfilled by a normal diet with an allowable daily intake of 35 μg/l/day. However, excessive iron ions in a human body may cause conjunctivitis, choroiditis, and retinitis. Cobalt is part of vitamin B12 which is essential for making red blood cells and properly maintaining the nervous system. Yet, excessive exposure of cobalt can cause an excessive level of red blood cells (polycythemia). Excessive cobalt is toxic to the heart muscle and can cause heart muscle disease (toxic cardiomyopathy) and even congestive heart failure. It can also cause enlargement and reduced activity of the thyroid gland. The maximum allowable daily intake of chromium is 28 μg/l/day. The most common health problems of excessive chromium intake involve the respiratory tract, and chromium compounds are carcinogenic. The daily intake of nickel ions must not exceed 400 μg/l/day. Nickel may lead to renal effects due to carcinogenesis and hypersensitivity. Other serious harmful health effects from exposure to nickel include chronic bronchitis and reduced lung function. On the other hand, a person can tolerate titanium in relatively large doses. There is not yet any biological role in human body known for titanium. As much as 0.8 mg of titanium can be in the human body without causing any known adverse health reactions. In fact, most titanium passes through the body without being adsorbed. However, titanium dioxide may induce a toxic effect on glial cells in the brain, suggesting that exposure to titanium dioxide may cause brain injury and pose a health hazard. Regardless, titanium has relatively very good biocompatibility.

9.2.4 *Wear of Materials*

Material wear often occurs in artificial joints and at any moving parts with direct physical contacts in biomedical devices. Wear phenomena are intimately linked to frictional processes. Roughly speaking, when the shearing stress is beyond the shear strength of either material of the interface, the material bodies can slide on each other. Such stress level is sufficiently high to cause the interfacial materials to gradually become damaged as the shear stress is sufficiently high. Furthermore, considering that most engineering surfaces are rough, the real contacting surfaces between single asperities and the material particles come off from the damaged asperities. The loss of material volume V_W from one surface due to grinding between two material surfaces along a grinding distance L_W under a normal load F (Fig. 9.3) has been characterized by the wear effect with the steady-state wear equation:

$$V_W = K_W \frac{F L_W}{3 H_B} \tag{9.1}$$

where K_W (which is a dimensionless quantity) is the standard wear coefficient of the surface with the volume loss V_W; and H_B is the Brinell hardness for the surface of the softer material surface. Notably, the volume loss can occur on both sides of the grinding interface.

To a certain extent, the ratio F/H_B reflects the real contact area between the materials as a softer material would deform further on the asperities of the hard material surface with a larger contact area. Hence, the frictional force can be viewed as the product of the shear strength and the real contact area of the softer material, and therefore, the frictional force/stress is basically proportional to the normal loading force/stress. In practice, while all asperity contacts contribute to friction, even a very small fraction of the contacts can result in wear and the loss of volume. The physical meaning of K_W is the ratio of the volume removed V_W caused by wear per unit sliding distance L_W to the real interfacial area of contact. K_W can also be interpreted as the proportion of the plastically deformed volume that is removed by the wear process.

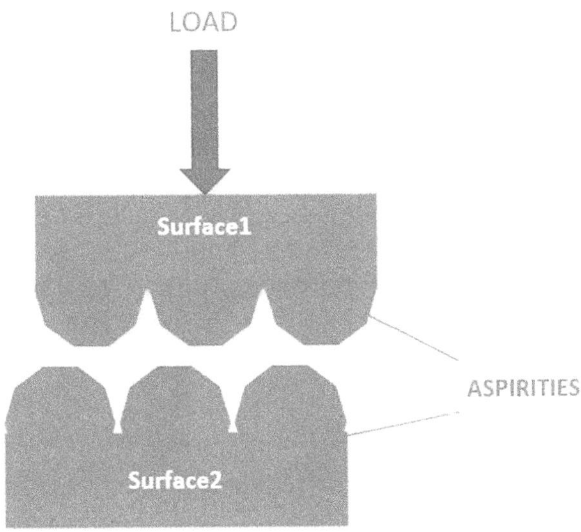

Fig. 9.3 Material wear and volumetric loss

Taking the artificial total knee joint (Fig. 9.4a) as an example, this device is composed of a femoral prosthesis, a tibial prosthesis, and a patellar prosthesis. It is made of two classes of materials: metal and plastic. The metal part includes the femur, tibia, and patella joint of titanium alloy or cobalt-chromium alloy. The plastic part is made of polyethylene. This ultrahigh-molecular-weight polyethylene can be manufactured to attach to the metal part of the tibial and patellar joint in order to reduce the friction with the femoral articular surface. After many years of wear due to join movement, extensive wear, delamination, and cracks can be found on the plastic spacer component of the total knee replacement device (Fig. 9.4b). In a very bad case in which the bone dimensions and geometry change over years for some patients, the lower limb force line changes, and the prosthesis wear will accelerate, causing the product life to be shortened. More critically, the resulting pieces of plastic from the wear could be released into the human body and may cause severe health problems for the patients, such as osteolysis and other tissue necrosis.

9.2.5 *Surface Roughness*

Surface structure plays a key role in governing cell-material interactions. For example, it has been reported that human mesenchymal stem cells (hMSCs) on polycaprolactone with a surface roughness of ~2–3 μm can facilitate osteogenic (bone cell) differentiation. Similarly, osteogenic differentiation of hMSCs should occur on tricalcium phosphate ceramics with a surface roughness of 2.78 μm. Over

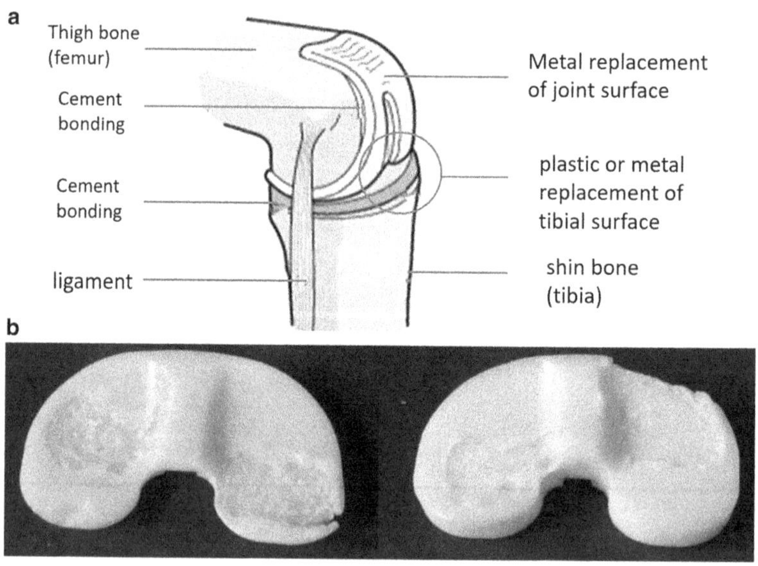

Fig. 9.4 (**a**) A total knee replacement. (**b**) Two implanted plastic spacers after years of wear. (Musib [5], American Journal of Biomedical Engineering)

this range of surface roughness, the cells demonstrate higher survival rate (live cells-to-dead cells ratio) and closer bone cell-related behaviors (e.g., osteocalcin production). Higher contents of calcium, phosphate, and silicon ions can be released by cells on such rough surfaces, facilitating the formation of osteoblastic differentiation-promoting adhesion components (e.g., calcium-phosphate deposits). On the other hand, hMSCs growing on polished surfaces with a surface roughness of ~0.67 μm exhibit a significantly lower rate of osteogenic differentiation. Likewise, a previous study on the effect of surface roughness for cells growing on a biodegradable polycaprolactone indicates that an average PCL roughness of 0.93 μm alone can serve as a compelling alternative to soluble osteogenic inducers for clinically relevant orthopedic applications.

In addition to micrometer-scale surface structures, in vivo extracellular matrix (ECM) enriched with specific nanotopographic features can affect different fates and functions of cells. Nanoscale structured adhesive ligands, including filamentous collagen, elastin, fibronectin, vitronectin, and laminin ranging in length from a few to hundreds of nanometers, have been observed to be present in in vivo ECM. Besides, nanotopographic structured adhesive molecules like integrin and the nanopodia of cells have demonstrated through extracellular nanotopographic features probing to have an effect on the enhancement of the cell membrane. Indeed, biomaterials with nanoscale topography have been shown to regulate both self-renewal and differentiation of pluripotent stem cells (PSCs) as well as mesenchymal, neural, and hematopoietic adult stem cells. Therefore, cell-surface interactions can be modulated by both the material type and surface roughness profiles.

9.2.6 Other Aspects Related to Biocompatibility

There are some other aspects related to biocompatibility, including the location of device adhesion, the effect of cell separation, and mechanical behaviors. Material degradation may cause the release of ions or molecules which can then trigger an immune response and cause system inflammation. Hence, inflammatory response is one of the most important issues of biocompatibility that biomaterials need to consider. Inflammation is a host's immune response to foreign objects, and implants are alien to the biological body. Improper implantation of a device can easily cause it to be attacked by the biological host's immune system, leading to inflammation. Inflammation is divided into acute inflammation and chronic inflammation. Severe acute inflammation can lead to implantation failure, and improper treatment can lead to widespread necrosis and life-threatening conditions. A mild inflammatory response may turn into chronic inflammation, which could result in local tissue fibrosis, further affecting the function of implants. Sometimes, biocompatibility requires long-term degradation, such as surgical sutures, whereas sometimes, biocompatibility requires long-term non-degradation, such as gold teeth. There are many ways for a material to degrade, from basic water degradation and enzyme degradation to advanced magnetic degradation, ultrasonic degradation, etc. By means of specific means, the degradation properties of biological materials can be precisely controlled through several methods, including changing material geometry, modifying the degree of crosslinking of polymers, and using hybrid materials. These are the areas in which biological material design needs to be considered.

As the body's main supplier of nutrients and oxygen, blood is a sensitive tissue. In the case of blood contact (artificial blood vessels, vascular stents, etc.), using inappropriate materials can lead to blood clots, and these clots are very serious and very difficult to manage. For blood compatibility, in vitro testing can be well evaluated, and prior animal testing is a must. Improved blood compatibility often involves improving the surface properties of the material. For incompatible materials, this can mean increasing the structure of a multilayer structure, simulating the surface structure of a vessel, or improving the anticoagulability of the material using a slow-release coating. Bone-related implants often provide mechanical structural support, such as artificial joints, bone cement, and so on. Bone compatibility involves promoting bone growth and connectivity to the bones. These properties directly determine whether the implant can withstand the weight of the body and will directly impact the time and degree of the patient's eventual recovery. Increased bone compatibility of a material, in addition to using bone compatible materials, such as bio-glass, hydroxyapatite, and collagen as an external coating, can still be done by changing the topology of the surface. In addition, the matching degree between material and bone mechanical properties should also be considered. The bone will gradually change depending on its environment, and the strength of the bone to material connection will decrease the strength of the bone itself. In addition to choosing suitable materials, the matching mechanical properties can be realized by combining materials, increasing the gap in the material, and reducing the relative density of the material.

9.3 Material Performance Index

9.3.1 Basic Concepts

The design of a mechanical component is specified by three factors: the functional requirements (the need to carry loads, transmit heat, store elastic or thermal energy, etc.), the geometry, and the properties of the material of which a component is made (including its cost). The performance of the component can be defined by aspects that are to be optimized, e.g., mass, volume, or cost. Optimum design can be considered to be selection of the material and geometry which maximize material performance in the final product. The optimization is subject to constraints, some of which are imposed by the material's properties. There are three groups of parameters which are said to be "separable": (1) functional requirements, (2) geometry, and (3) material properties. When these parameter groups are separable, the optimum choice of material becomes independent of the design details. The optimum material is the same for all geometries and all values of the functional requirements. Then, the optimum material can be identified without solving the complete design problem or even knowing all the details of the other separable parameters.

While we consider the mechanical properties of a biomedical device, we need to know the basic design relationships for strength, stiffness, and different types of leading and cross-sectional shapes. For example, for a solid bar/rod/cylinder with a length L in tension or compression along its length direction without any permanent deformation, the material yield strength σ_{yield} should be above the external tension/compression load F divided by the cross-sectional area A, and the material elasticity modulus E relates to the structural length variation x as described by $x = FL/(AE)$ as Eq. 2.3. For a bar/rod/cylinder bending under a load F exerted at the center along its length without permanent deformation, the material yield strength σ_{yield} must be sufficiently large:

$$\sigma_{\text{yield}} = I_c d_c M_c \tag{9.2}$$

where I_c is the moment of inertia of the cross section, d_c is the centroid to surface distance, and M_c is the bending moment. For a plate structure with length L, width W, and thickness T ($T < W$) as an example, the load is shared by both halves of the plate each with a length $L/2$, and thus, the bending moment $M_c = (F/2)(L/2)$. Clearly, $d_c = T/2$, and it can be checked that the moment of inertia $I_c = WT^3/12$. Therefore, for the plate structure, $\sigma_{\text{yield}} = 3FL/(2WT^2)$. Furthermore, the material elastic modulus E of the bending structure can relate to the distance at the central position along its length δ_c as

$$\delta_c = \frac{FL^3}{48EI_c}. \tag{9.3}$$

In order to further illustrate how to separate the functional requirements, geometry, and material properties in our design considerations, we may first look into two simple examples. Imagining that we are looking to select the optimum material for a light and stiff beam (Fig. 9.5) as a component in an implanted device, it is required that the beam has a fixed length of L where its width on the side b is adjustable to fit the design requirement. This beam is fixed with both ends after implantation. The mass of the beam is $m = \rho b^2 L$ where ρ is the density of the material. For a load F acting at the mid-length position, the central deflection δ_c is described as Eq. 9.3. As a stiffer beam would have a smaller δ_c upon the load P, we may consider the "structural stiffness" as P/δ_c.

Our job here is to choose the "best" material for a larger structural stiffness and a smaller mass. We may consider the following relation:

$$\delta_c = \frac{PL^3}{4Eb^4} = \frac{\rho^2 PL^5}{4E}\frac{1}{\rho^2 b^4 L^2} = P \times \frac{\rho^2 L^5}{4E}\frac{1}{m^2}, \quad \text{and thus} \tag{9.4}$$

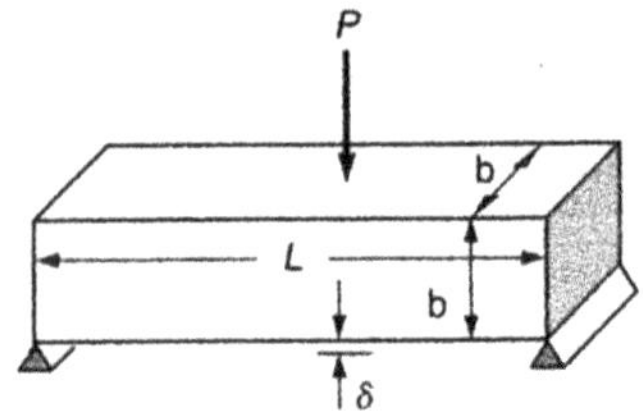

Fig. 9.5 Key parameters for a simple supported beam with a square cross-section

$$\left(\frac{P}{\delta_c}\times\frac{1}{m^2}\right)=\left(\frac{4}{L^5}\right)\left(\frac{E}{\rho^2}\right). \tag{9.5}$$

In the above equation, we can observe that the left term is related to the functional requirements of the design task, i.e., a larger value of this term implies a smaller mass or a larger structural stiffness. The first term on the right-hand side is a term related to the geometry requirement of L and the last term E/ρ^2 which only consists of material properties. For the required L, we should choose a material with a bigger value of E/ρ^2 for the better "performance," i.e., a larger value of the left term in the above equation. Therefore, we may define a material performance index as $E^{1/2}/\rho$.

Let's look at the other example. Imagine that we need to design a light and strong tubular beam with a fixed outer radius r as an intramedullary bone fixation nail. It is mounted by screws on its ends with a fixed separating distance L. Its function is to resist bending moments. The wall thickness t is fixed, whereas r is adjustable to fit the design requirement. Consider that the moment of inertia of a ring cross section with an outer radius of r and ring thickness t ($t \ll r$) $I_c = \pi r^3 t$. The intramedullary nail that can withstand with an infinite life cycle under a limit of bending moment M_b is

$$M_b=\frac{I_cE}{r},\quad \text{and hence}\quad r=\sqrt{\frac{M_b}{\pi tE}}, \tag{9.6}$$

where E is elasticity modulus of the material. The objective is to minimize the mass m. Now, we may express m by substituting Eq. 9.6:

$$m=2\rho\pi rtL=2\rho\pi tL\times\sqrt{\frac{M_b}{\pi tE}},\quad \text{and then} \tag{9.7}$$

$$\frac{\sqrt{M_b}}{m}=\sqrt{\frac{1}{4\pi tL}}\times\frac{E^{1/2}}{\rho}. \tag{9.8}$$

The better product performance should induce a smaller m and hence a larger value of the left term in the above equation. The first term on the right-hand side only depends on t and L which are the geometric requirements. The last term is related to the material properties. Thus, we may define a material performance index as $E^{1/2}/\rho$.

The above two examples demonstrate an enormous simplification: the overall performance for the product is achieved by maximizing the material properties, which is called the "performance index." Often, it is preferable to have lighter implant devices because they would become additional weight at a local site in the body, which may alter the body's balance and potentially cause risks during movements of the patient. Furthermore, often used material performance indices are listed in Table 9.3 based on different object shapes and exerted loadings.

9.3.2 *Ashby Charts of Biomaterials*

In the previous section, we attempted to specify some product device problems in three groups of quantities, i.e., functional requirements, geometry constraints, and material properties. Such quantities related to material properties induce the concept of "material performance index" and act as a guideline for us in choosing the material for "best" product performance. A straightforward approach for picking the "best" materials should be the computation of the material performance index for all material candidates and the selection of the materials with the maximum performance indices. For instance, the database of material properties can be found from the Internet or computer software, such as a widely used computer program called Cambridge Materials Selector (CMS) which summarizes a broad range of material properties and permits extensive evaluation of different materials. Also, a list of the material properties for commonly used biomaterials is also available in Table 9.4.

If we take a second look at the common material performance indices as listed in Table 9.3, we may realize that in many typical cases, these indices can be expressed as the ratios of different intrinsic material properties. Considering that a material has a material performance index relating to only two material properties, a value

Table 9.3 Some material performance indices and their corresponding cases for minimizing the product mass

	To maximize strength	To maximum stiffness
Bar in tension: load, stiffness, length are fixed; section area is variable	σ_s/ρ	E/ρ
Torsion bar: torque, stiffness, length are fixed; section area is variable	$\sigma_s^{2/3}/\rho$	$G^{1/2}/\rho$
Beam in bending: loaded with external forces or self-weight; stiffness; length fixed; section area free	$\sigma_s^{2/3}/\rho$	$E^{1/2}/\rho$
Plate in bending: loaded by external forces or self-weight; stiffness, length, width fixed, thickness free	$\sigma_s^{1/2}/\rho$	$E^{1/3}/\rho$
Cylindrical vessel with internal pressure: elastic distortion, pressure, and radius fixed; wall thickness free	σ_s/ρ	E/ρ

σ_s yield stress or strength, E module of elasticity, G shear modulus, ρ density

Table 9.4 List of physical properties of biomaterials

	Density (g/cm^3)	Elastic modulus (GPa)	Poisson's ratio	Yield strength (Mpa)	Strength (Mpa)	Elongation (%)	Wear coefficient	Hardness (kg/mm^2)
Cortical bone	1.85	15–30	0.3	30–70	70–150			
Metals/Alloys								
Stainless steel	7.5–8	190	0.3	221–1231	586–1351		1.7×10^{-5}	132
Cobalt -chromium alloys	8.5	210–235	0.31	448–1606	655–1896		1.87×10^{-6}	360
Titainium	4.5	110	0.32	485	760		1.1×10^{-5}	200
Titanium alloys	4.43	116	0.342	896–1034	965–1103		0.9×10^{-5}	320
Polymers								
UHMW-PE	0.93	1.1–2	0.45		30	300	5×10^{-5}	30
Polyethyloene	0.9–0.96	0.09–1.24	0.46		7.6–40	150–500	2.5×10^{-5}	50
Polypropylene	0.9	1.1–1.55	0.45		28–36	400–900	2×10^{-5}	70
Nylon	1.05–1.14	0.0012–0.0028	0.39		59–76	90–300	4×10^{-6}	105
Silicone rubber	1.12–1.23	<0.01	0.47		6–7	350–600	4.5×10^{-7}	50
PMMA	1.18	1.8–3.1	0.37	48	70	10	7×10^{-6}	92
PTFE (teflon)	2.16	0.4–1.8	0.46	30	43	400	N/A	58
Ceramics								
Graphite/carbon fiber	1.5–1.9	24	0.2		13.8–31		4×10^{-5}	60
Glass	1.5	24	0.17		27–62		8×10^{-6}	1550
Alumina	3.9	380	0.21		130–340		1.2×10^{-4}	2000
Hydroxyapatite	0.35–1.44	12–144	0.27		45–48		2×10^{-5}	5272

$P_m = X_{m1}^{1/\hbar}/X_{m2}$, where X_{m1} and X_{m2} based on two intrinsic material properties, and a constant $\hbar$, then we may convert the relation as

$$\ln(P_m) = K_{Pm}/\hbar = \ln(X_{m1})/\hbar - \ln(X_{m2}), \text{ and thus} \tag{9.9}$$

where K_{Pm} is another constant. Then,

$$\ln(X_{m1}) = \hbar \ln(X_{m2}) + K_{Pm}. \tag{9.10}$$

In this case, if we present the material properties X_{m1} and X_{m2} as a log-log scale plot, i.e., $\ln(X_{m1})$ against $\ln(X_{m2})$, materials having the value of P_m would lie along a straight line in such plot. The slope of such line $\hbar$ is determined by how P_m is related to X_{m1} and X_{m2}. The y-intercept of this plot K_{Pm} increases with P_m. Therefore, we can convert the material selection cases listed in Table 9.4 to two key log-log scale plots of different materials: strength against density (Fig. 9.6) and elasticity modulus against density (Fig. 9.7), which are called the Ashby charts as such representation has been proposed by Prof. Michael Ashby at the University of Cambridge. In particular, for the case of maximizing the structural stiffness of a bar in torsion, the corresponding material performance index ($=G^{1/2}/\rho$) should be applied with another plot of $\ln(G)$ against $\ln(\rho)$, and G is the modulus of rigidity which can be estimated by Eq. 2.15 using Poisson's ratios and elastic moduli listed in Table 9.4.

To compare material performance indices, we may draw the lines with the corresponding slopes of each index across each of the material points in the Ashby

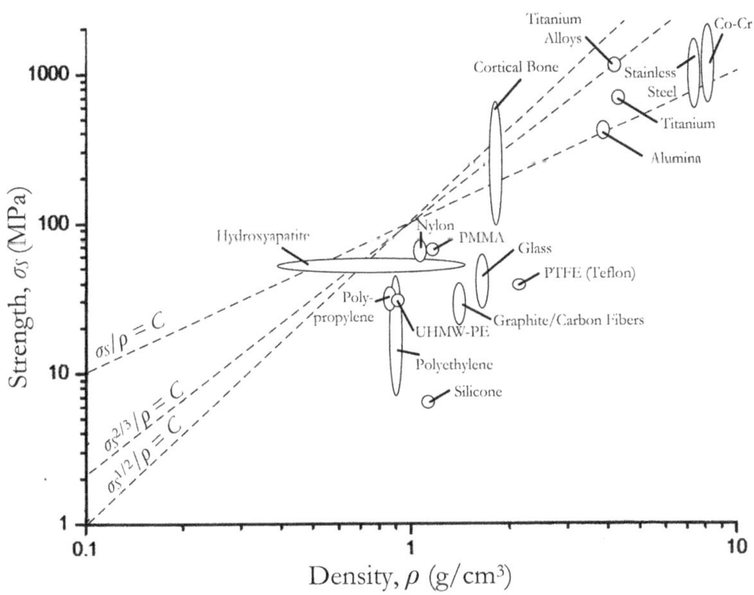

Fig. 9.6 Ashby chart of strength against density

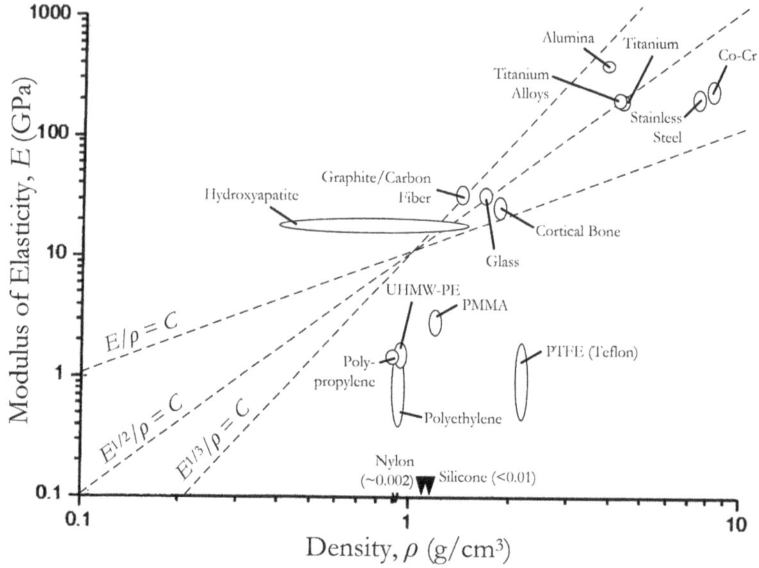

Fig. 9.7 Ashby chart of elastic modulus against density

chart and then compare their y-intercepts. The material with the largest y-intercept can be considered the "best" material. An easier way to search for the "best" material graphically is to move the line up and down on the Ashby chart, while the slope is maintained; the "best" material would be at the location intercepting with the moving line with a highest level. This graphical technique is particularly useful as we can search for the "best" materials for a product quickly and without any tedious computations.

Problems

Problem 9.1

An intramedullary nail with a circular cross section is designed to be loaded with force P at its free end to produce an allowable deflection of $\delta = 4PL^3/(3E\pi r^4)$. Explain why the material performance index based on the beam stiffness (P/δ) can be defined as $E^{1/2}/\rho$.

Problem 9.2

Are athletes likely to return to sports after tibial fractures? Sports fans have witnessed several horrific tibial fractures of athletes in games. In fact, 91.5% of patients with tibial shaft fractures treated surgically returned to sports. Typically, tibial shaft fractures require surgery, usually in the form of an intramedullary nail (a rod placed down the center of the bone with locking screws at the top and bottom of the rod). Returning to play requires the fracture to heal completely and in the proper position and alignment of the bone pieces. The athlete must regain full motion, strength, and function eventually.

In this question, we would like to think about issues on material selection and dimension configuration of an intramedullary nail used for the surgery of Luciano Almeida. The Brazilian soccer player suffered a severe injury during a match between Botafogo and Flamengo in 2007. He was out for 5 months to make a full recovery after surgery.

(a) Let's assume Luciano has a body mass of 72 kg. Throughout a soccer game, the legs of Luciano can be momentarily subjected to a force which is up to 20 times his own weight. Determine the maximum (i) normal and (ii) shear stresses developed in the tibia T of his leg at the midsection *a-a* as shown in Fig. 9.P1. The cross section can be assumed to be circular having an outer diameter of 4 cm and an inner diameter of 2 cm. Please note that a bone structure should be a hollow shape. Assume the fibula F does not support a load.

(b) Here, we have the responsibility of designing a light and strong tubular intramedullary nail for the required surgery of Luciano. The intramedullary nail should have a fixed outer radius R. The wall thickness T is adjustable, yet it should be sufficient to resist bending moment against permanent deformations. The cyclic bending moment can be considered as the product of the shearing force on the foot F and the distance between the fixing screws at both ends of the intramedullary nail (L). The bending moment (M_b) is given as $M_b = I\sigma_y L/(R \times \delta)$, where the moment of inertia $I = \pi R^3 T$, σ_y is the yield strength and $\delta =$ the nail deflection. Our goal is to choose a metal/alloy inducing a minimal mass m such that Luciano will carry a minimal extra load (i.e., the weight of the intramedullary nail) on his leg. What is the appropriate material performance index?

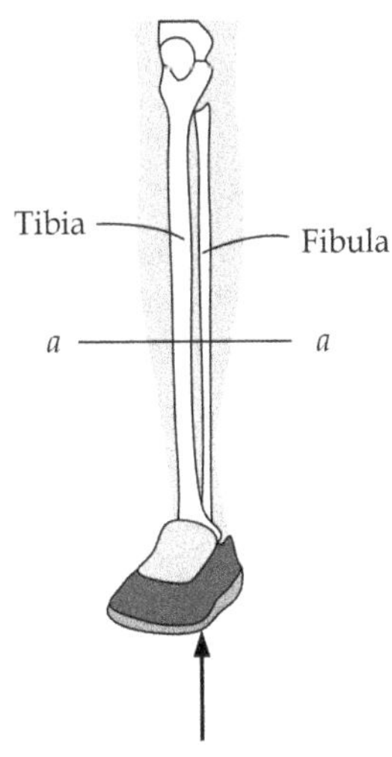

Fig. 9.P1 Momentary force acting on a leg

(c) Can you draw the Ashby chart as the yield strength against density using the material properties listed in Lecture 4 (page 12)? Which material would you prefer to be used for the intramedullary nail?

(d) Suppose L is 35 cm and R is 0.75 cm. What should the minimum mass of the intramedullary nail be?

Problem 9.3

External bone fixation plates are being considered for the upcoming surgery of a patient. Currently, you are asked by the doctor for advice on which type of materials are ideal as the structural materials. You are told that the bending of a fixed-fixed supported rectangular plate under a line force P exerted at the beam center has the maximum deflection $\delta = PL^3/(192EI)$ where E is Young's modulus and $I = bd^3/12$ is the moment of inertia. The length L and width b of the plate are fixed, but the thickness d may vary.

(a) Please use the list of material properties provided in the course webpage to draw the (i) Young's modulus-density and (ii) strength-density Ashby's charts.

(b) Please suggest a material performance index for the minimized mass. Explain your answer.

(c) Based on you own Ashby's charts, identify a short list of candidate materials and other possible selection considerations. Select the most appropriate material (s) with explanations.

(d) Please suggest and elaborate on another material performance index for the minimized material cost.

Problem 9.4

A rectangular bone fixation plate is planned to be implanted in a patient's body as shown in Fig. 9.P2. It is required to be able to have a displacement at its central length position of smaller than δ under a load P at the same position. Please consider that the plate cross section has width W and thickness d, where the W is fixed but d can be changed by choosing a different plate model provided by the manufacturer. The plate is fixed on the patient's bone by screws at its two ends with a separating distance of the screws of L. Determine the following using Table 9.2.

(a) The material that meets the design load and deflection limit at the lowest weight. Show all calculations.

(b) The material that meets the design load and deflection limit at the lowest volume. Show all calculations.

(c) If the design load is double, what is the effect on your calculations?

(d) If the deflection limit is halved, what is the effect on your calculations?

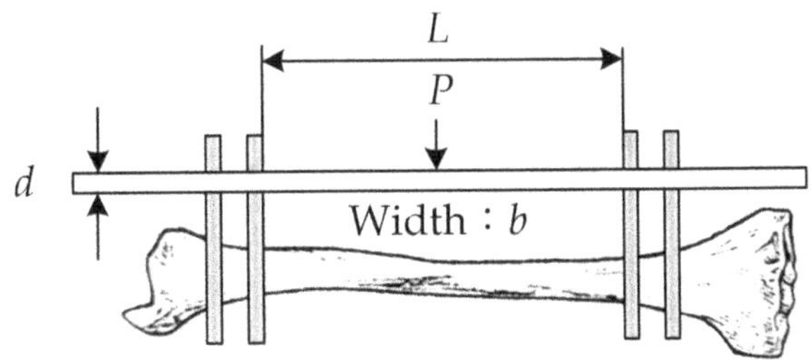

Fig. 9.P2 Configuration and key parameters of external bone fixation

Problem 9.5

A 70 kg patient has been arranged for an implantation surgery of an artificial knee joint after a severe car accident. The expected outcome after the surgery is described as Fig. 9.P3a. The lower tibial plate is confirmed to be covered by a polyethylene sheet, whereas the material for the upper femoral replacement component is yet undetermined. There are currently two material candidates under consideration for such femoral component: (1) stainless steel and (2) cobalt-chromium alloy.

Now, we may consider the femoral geometry simplified as shown in Fig. 9.P3b. Assume that the femoral plate is a semicylindrical plate with outer radius (Ro) 30 mm, inner radius (R_i) 25 mm, and width (W_{femur}) 50 mm and that the both ends are extended with a length (L_{ex}) 10 mm. Furthermore, each of the two extension plates contains two cylindrical hinge rods (filled with *dark gray* in the above figure; diameter, D_{hinge}; length, L_{hinge}) supporting the load during walking.

(a) Please suggest a material performance index for maximal strength upon bending of the hinge rods/beams with an explanation, and suggest the "better" material candidate (i.e., stainless steel or cobalt-chromium alloy).

Now, please consider that stainless steel has been chosen as the material for the femoral component. Please further assume $L_{\text{hinge}} = 10$ mm and $D_{\text{hinge}} = 5$ mm.

(b) While walking, the foot of a 70 kg man is momentarily subjected to a force (normal load on tibia) which is ~2 times of his weight. Please estimate the maximum daily weight loss (unit: μg/day) of polyethylene from the artificial joint caused by wear by considering that an individual should have ≤10,000 steps per day and that the "sliding distance" of the artificial joint is $\sim\pi R_o$ for every two steps (as one step for the left leg and the other for the right leg). Then, what is the thickness of the worn polyethylene after 10 years? Please use the material properties provided in Lecture notes.

(c) Please estimate the iron ion release rate (unit: μg/day) due to biochemical effects (e.g., interaction with blood) in the human body, based on info from Lecture notes. Does it exceed the patient's maximum allowable daily intake?

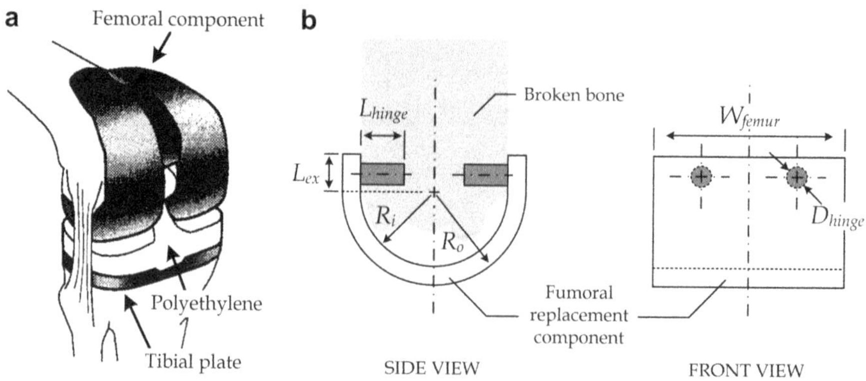

Fig. 9.P3 (**a**) Artificial knee joint, and (**b**) the key dimensional parameters for the femoral component

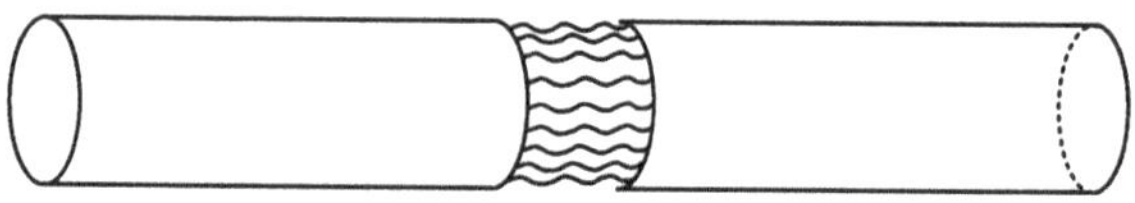

Fig. 9.P4 An artery grafted with a synthetic material

Problem 9.6

(a) An artery is grafted with a synthetic material for a length of 5 mm and a radius of 1 mm as shown in the Fig. 9.P4. If the synthetic material is not compliant, sketch how the vessel-graft combination will appear when pressure peaks in the blood circulatory system.

(b) Assume the radius of the artery is 1 mm, viscosity of blood is 3×10^{-2} Pa $\times$ s, and blood density is 1×10^{-3} kg/m^3. Can you estimate the gradients of pressure drop (G) for an overall blood velocity of 50 mm s^{-1} along the artery?

(c) Is the blood flow laminar or turbulent? Please explain.

(d) Consider the grafting material has been wrongly chosen so that blood cells will accumulate and form a layer over inner surfaces of the graft. Please sketch the plot as the blood pressure gradient along the graft against the thickness of the layer formed.

Problem 9.7

A cylindrical intramedullary nail is planned to be implanted inside the bone marrow of a patient's body as shown in Fig. 9.P5. It is required to be able to have a displacement at its central length position of smaller than δ under a load P at the same position. Please consider that the plate cross section has a diameter D, where D can be changed by choosing a different rod model provided by the manufacturer.

Fig. 9.P5 A rod fixed by screws on its both ends on a patient's bone

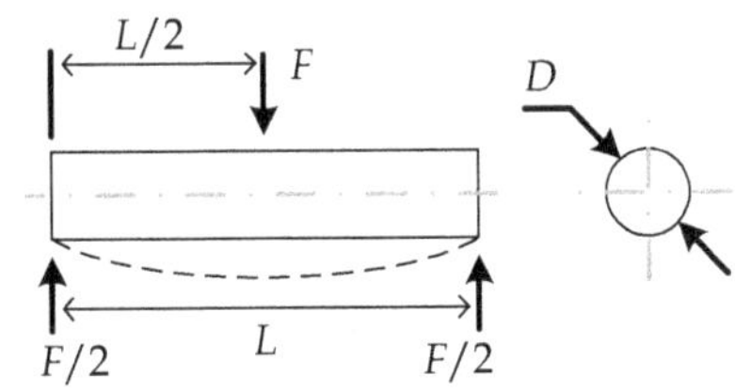

The rod is fixed on the patient's bone by screws at its two ends with a separating distance between the screws of L. Determine the following using Table 9.2 and given that the moment of inertia for a circular cross section is $\pi D^4/64$.

(a) The material that meets the design load and deflection limit at the lowest weight. Show all calculations.
(b) The material that meets the design load and deflection limit at the lowest volume. Show all calculations.
(c) If the design load is doubled, what is the effect on your calculations?
(d) If the deflection limit is halved, what is the effect on your calculations?

Problem 9.8

A hollow cylindrical intramedullary nail is planned to be implanted inside the bone marrow of a patient's body. It is required to be able to have a displacement at its central length position of smaller than δ under a load P at the same position. Please consider that the plate cross section has an outer diameter D and a shell thickness T, where D is fixed and T can be changed by choosing a different rod model provided by the manufacturer. The rod is fixed on the patient's bone by screws at its two ends with a separating distance between the screws of L. Determine the following using Table 9.2 and given that the moment of inertia for a ring cross section is $\pi D^3 T/8$.

(a) The material that meets the design load and deflection limit at the lowest weight. Show all calculations.
(b) The material that meets the design load and deflection limit at the lowest volume. Show all calculations.
(c) If the design load is doubled, what is the effect on your calculations?
(d) If the deflection limit is halved, what is the effect on your calculations?

Problem 9.9

We wish to select the optimum material for a light, stiff beam. The beam with square cross section is b cm on a side and L long. A is the area of the beam across section, and ρ and E are the density and Young's modulus of the material, respectively. For the beam being stretched with a force F, the beam extension δ is $\delta = FL/AE$.

(a) Explain why the material performance index can be defined as E/ρ. Please clearly describe the reasons and arguments in your explanation. (Minor marks will be deduced for unclear descriptions.)

(b) Consider now that we have three classes of materials: engineering metal alloys, engineering polymers, and engineering ceramics. With the help of the Ashby chart given above, can you select the "best" material class for the application described in this question? Please explain clearly for your answer with the Ashby chart.

Problem 9.10

Imagine that you need to design a light and strong tubular beam with a fixed outer radius r as an intramedullary bone fixation nail. It is mounted by screws on its ends with a fixed separating distance L. Its function is to resist bending moment. The wall thickness t is adjustable to fit the design requirement, whereas r is fixed. Consider that the moment of inertia of a ring cross section with an outer radius of r and ring thickness t ($t \ll r$) $I_c = \pi r^3 t$. The intramedullary nail can withstand with infinite life under a limit of bending moment $M_b = I_c E/r$, where E is elasticity modulus of the material. The objective is to minimize the mass m. Now, please suggest a material performance index.

References and Further Reading

1. Hill, D.: Design Engineering of Biomaterials for Medical Devices. Wiley-VCH, Chichester (1998)
2. Müller, U.: In vitro biocompatibility testing of biomaterials and medical devices. Med. Device Technol. **19**, 30–32 (2008)
3. Boutrand, J.-P. (ed.): Biocompatibility and Performance of Medical Devices. Elsevier, San Diego (2012)
4. Creese, R.: Introduction to Manufacturing Processes and Materials. CRC Press, Boca Raton (1999)
5. Musib, M.K.: Response to ultra-high molecular weight polyethylene particles. Am J Biomed Eng. **1**, 7–12 (2011)

Chapter 10
Design for Manufacturing

Abstract Design for manufacturing (DFM) aims to find the best pathway of producing the highest quality and reliability with the lowest amounts of cost and technical difficulties. In particular, machining is widely used in the production of biomedical devices because of its process flexibility. A computer numerical control (CNC) machining system further automates the machining process in three dimensions following preset paths of the machining tools. The machining process design is discussed in more detailed in this chapter.

10.1 Overview

When designing a biomedical product, we need to consider multiple aspects from the biomedical engineering point of view. We need to determine the material body shape, dimension, dimensional tolerance, and surface profile which will fulfill the function requirements. If the product contains electronic components, we need also to design the circuitry and power supply requirements. Sometimes, the choice of microprocessor as well as the corresponding programming and data communication issues must also be considered.

To mass-manufacture a product in the industry, we need to come up with a feasible manufacturing process flow with minimal cost and time that can be implemented as a production line. In this case, we also need to be concerned about tiny steps such as the tooling selection and fixture styles such that each component of the raw material can be inserted onto a primary manufacturing machine and then be transferred to other machines in the following manufacturing steps. Designs of molds/dies are also required in some processes such as casting, drawing, or injection molding. To achieve a better manufacturing system, we aim to adopt the most effective production approach which also involves scheduling, proper manufacturing process parameters, and facility layout.

*D*esign *f*or *m*anufacturing (DFM) is an activity that gathers and analyzes all elements (design and plan) of the manufacturing system. DFM aims to find the best pathway of producing the highest quality and reliability with the lowest amounts of cost and technical difficulties. Designers for manufacturing of biomedical

R. H. W. Lam, W. Chen, *Biomedical Devices*,
https://doi.org/10.1007/978-3-030-24237-4_10

products should consider a broad range of factors altogether, such as the number of components included, and how to finely tune a part design with the minimum cost of achieving every feature. They should also consider the capability of and interactions between materials, shapes, and manufacturing processes. Generally, they make the judgment on any product configurations needed according to some common principles such as:

1. Decreasing the total numbers of parts. This is the most direct way to reduce the cost and production duration as the demand of manufacturing machines and part assembly will decrease.
2. Developing a modular design. This can minimize cost by adding versatility during redesigning, allowing for partial changes of sub-parts for improving testing before assembling and using standardized parts.
3. Designing multifunctional components. Parts offering more than one function, such as electronic controllers and motor drivers, can be used in multiple products and shorten the development cycle for the future products.
4. Designing fabrication-friendly parts. Whenever possible, designers should choose shape features which are easier to manufacture. For instance, such feature selection may obtain compensation between the manufacture feasibility and the functionality of the final product. Nevertheless, designers may not reduce manufacturing efforts for some necessary shape features such as thin-walled and irregularly shaped components.
5. Standardizing components. Unlike custom-made components, standard parts can be purchased easily at a lower price and with higher availability and rcliability.

Additionally, it is not sufficient to consider only a subset of factors such as the material, geometric shape, and surface profile. An optimal manufacturing process demands a perfect balance between all different factors, including the production time and cost. For instance, if the manufacturing cost is too high, this will reduce overall income. The whole production schedule is affected by the influence of available time of different product components. The product supply rate is limited by the overall production schedule as some particularly difficult manufacturing procedures can drag the production cycle. Sometimes, the production may be affected by the unavailability of third-party components without other replacement component suppliers. Some manufacturing processes may impose harmful ions/molecules into the biomaterial body such that the chosen material is no longer biocompatible after fabrication. In some cases, a production process does not need to be extremely efficient if there is no intention to offer a large quantity because the process optimization would require efforts from biomedical engineers to analyze and estimate the optimal process parameters. For DFM, rules are suggested for processes such as casting, injection molding, forging, sheet metal forming, machining, and assembly.

For instance, improper cast design might lead to local voids of the final product. Gradual transition of thickness should be applied to change cross sections as exhibited in Fig. 10.1a. In case a step-alike feature has to be included in the product boundary, the aspect ratio (thickness-to-height) should not exceed 2. Otherwise, air

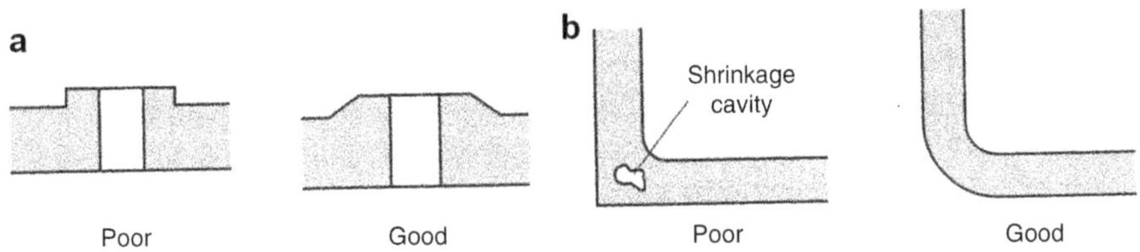

Fig. 10.1 Design for casting for (**a**) step and (**b**) turning features

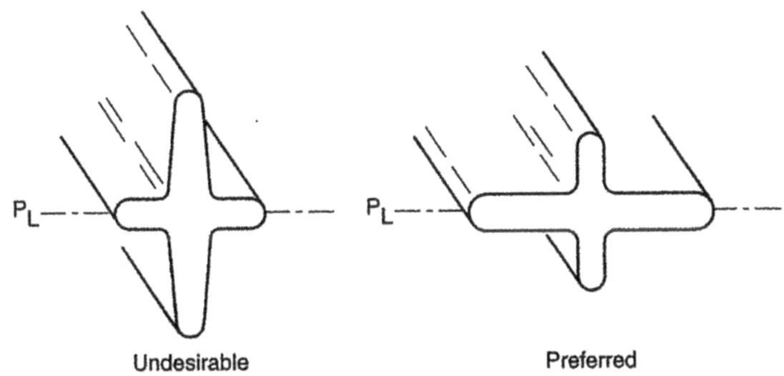

Fig. 10.2 Design for forging on the resultant aspect ratio

might be trapped at the corner of the step feature in the cast during processing, and this air pocket will then become a missing corner in the casted product. Cavities inside the casted body might also be created from shrinkage of cooled materials (Fig. 10.1b). Such shrinkage can be eliminated by reducing the number of walls and avoiding sharp convex/concave angles in the cast. That is, chamfer or fillet edges are preferred. For fillet edges, it is conservative to maintain the fillet radius to be between one-half and one-third of the section thickness to prevent the formation of shrinkage cavities. This criterion also facilitates smooth flow streamlines of molten metals during processing.

In design for forging, the fundamental principle is minimizing required deformation during processing. An example of this design consideration is shown in Fig. 10.2. Let's imagine that a rectangular block of material is forged to have a cross-shaped cross section with one direction longer than the perpendicular direction in the cross section. It is better to have the cavities on the forging dies to be shallower, rather than deeper. It is because we may consider the aspect ratio of cavity shape should roughly represent the required material strain for permanent deformation. The local spatial strain level should preferably be in the perfect plasticity region or the strain hardening region. If the strain is in the necking region or even beyond the fracture strain, cracks will form within the forged material body. On the other hand, if the strain is inadequate, the material would not be able to deform to the target shape.

In sheet metal forming, it is preferred to reduce the gaps between components being cut on the sheet so as to reduce material waste and increase efficiency. The number and length of edges should also be minimized so that the lifetime of tooling can be longer. One easy technique to achieve these requirements is to pack the components with shared cutting edges. As shown in Fig. 10.3, we may even modify the component outline to achieve edge sharing if the component outline is not critical to the product functions.

For the injection molding process, the basic principle is similar to casting. For example, chamfer or fillet edges are more favorable than sharp corners/edges, facilitating smother flow of polymer melts and preventing formation of voids (Fig. 10.4a). There is a more specific design rule for the inner fillet radius. It should be larger than 1.5 mm and at least 0.375 times of the wall thickness. The significant difference in thermal expansion coefficients of polymer melts versus metallic molds is an important concern. In this case, perfectly vertical wall structures should be avoided, but a draft angle should be added. Typically, this draft angle should be at least 2° such that the cooled and shrunk polymeric materials can remove themselves from the mold easier (Fig. 10.4b).

For better assembly of components, it is preferable to reduce the overall number of parts. A product consisting of many different parts consumes large amounts of cost and time of production, and the chances for error in the assembly process is higher, too. In case there is one defective part assembled in the product, the cost of examination and disassembling the product may be even higher than producing a new product. Therefore, it is better to modify the product design to one with fewer components. Another strategy is to assemble the product in, say, two stages, with the earlier stage assembling only components into modules and the later stage

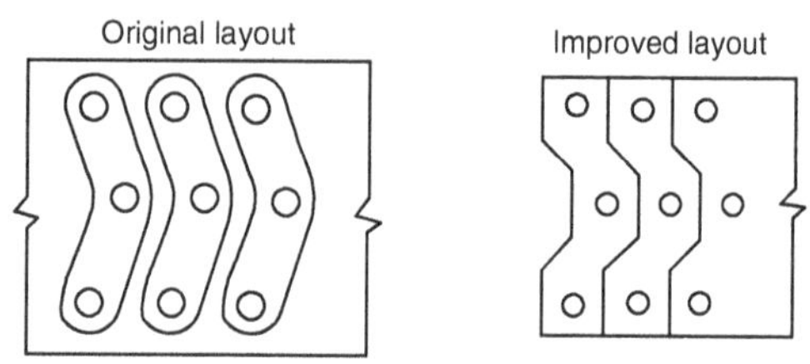

Fig. 10.3 Raw (left) and improved (right) design for components manufactured by sheet metal forming

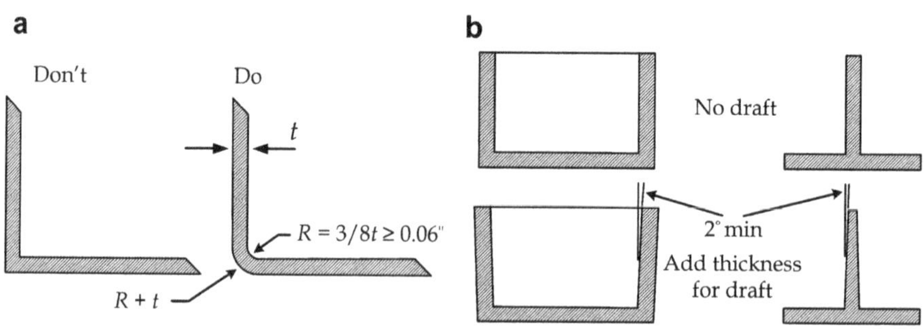

Fig. 10.4 Design for injection molding for (**a**) turning and (**b**) cavity structures

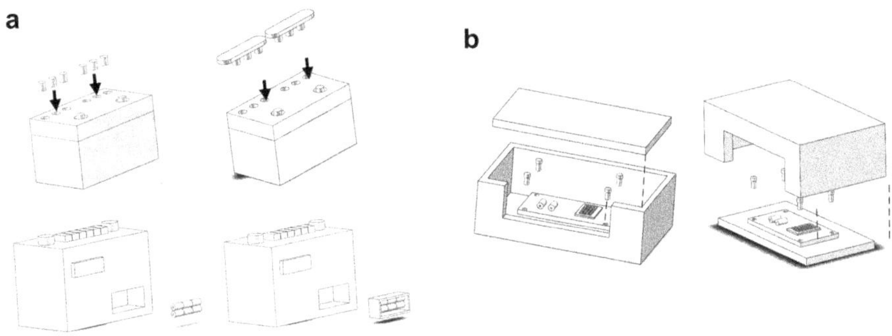

Fig. 10.5 Design for assembly: (**a**) modularizing multiple parts into single sub-assemblies and (**b**) designing open enclosures for higher accessibility of important components

assembling these modules into the final product as shown in Fig. 10.5a. Moreover, the important components should not be placed in a confined space because the higher accessibility of these components would enable replacement and maintenance if required in the future (Fig. 10.5b). Additionally, they should not be sealed with other parts of the product. If fasteners or screws are necessary, they should be located away from any obstructions with a wide opening region or on the product surface for higher accessibility.

In fact, nowadays, there are products (especially high-technology consumer electronics) designed with very high accessibility to key components included due to a strategic marketing concern. Manufacturers wish to prevent their products from being copied illegally. However, for medical device like a pacemaker, the functionality and the ease of maintenance/replacement is definitely the most important concern.

For machining, the production cost is determined by many factors such as material removal, number of setup steps for a machine tool, tools used on a machine, and specimen transfer between different machining machines. If there is a freedom of tuning the device design, it is favorable to modify the expected shape such that the time and cost of these factors can be minimized. One straightforward way to improve the design device for better manufacturability is to examine the "axis" of machining for every shape feature of the product. An example is given in Fig. 10.6. Considering the original design has more than one required milling directions (the related surfaces are highlighted by arrows), we may consider reorienting the shapes to match the cutting tool in the same direction/axis. In this case, this aims to reduce the required number of machining axes, and ideally, only one machining axis is required in the process. For concave angles, we should avoid generating sharp concave angles, or a fillet should be added at the tip of concave angles. In fact, machining of metals and alloys has a critical role in the production of custom-made bone fixation devices, artificial joints, and other prosthetic devices. Hence, in the next section, we will focus on operation parameters and their relation to machine cost and time for different machining processes. Furthermore, optimization of these parameters toward the best production cost and time is discussed in Chap. 12.

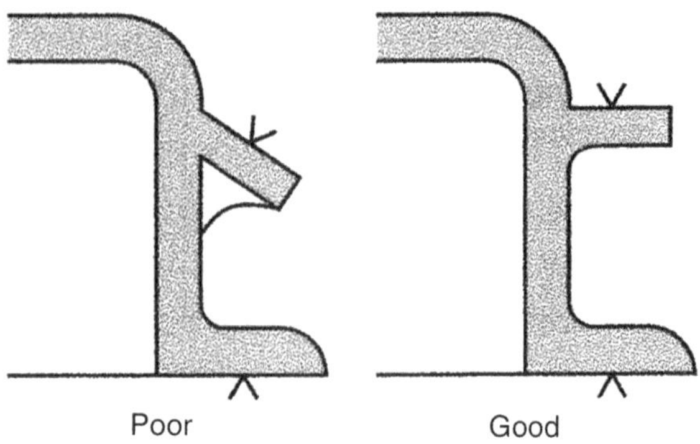

Fig. 10.6 Design for machining

On the other hand, *d*esign *f*or e*x*cellence (DFX) is another concept related to DFM. DFM brings together significant interests such as simplifying designs to maintain high reliability, reducing assembling and manufacturing costs, easing error diagnosis and recovery, improving quality, and shortening project duration. However, DFX is not an independent system. Instead, it is a mixture of different objective-oriented criteria, and "X" can be replaced by different kinds of measures such as "manufacturing and assembly," "reliability," and "serviceability." Meanwhile, the concept of environmental friendliness has risen up in the past few decades and has also become elements of concern. For biomedical devices, "biocompatibility" must be an essential element.

Before production, there are still many other factors which need to be considered, including tooling, operation step sequence, operation conditions, etc. Briefly, we should first confirm if the production volume can be supported by the manufacturing machine, whether or not to decide if the product should be divided into multiple parts, and if each of the part sizes can be supported by the manufacturing machines. In each product part, we should identify the shape characteristics, such as planar shapes, curved plates, holes, interior or exterior undercuts, uniform walls, uniform cross sections, rotational symmetry, and captured cavities. Based on their shape capabilities from some manufacturing processes such as those discussed in Chap. 6, we may select the appropriate manufacturing process for each of the product parts; a sample mapping is summarized in Table 10.1. The manufacturing processes may have different characteristics dependent on the shape features in aspects including the resultant material consolidation condition, process repeatability, and flexibility of product shapes (as a shape feature still includes a larger variety of shape configurations). Roughly speaking, the solidification processes induce perfect material conditions and have a very high process repeatability. However, since they rely on the cast/mold to define the resultant shapes, there is no process flexibility. The bulk deformation processes change the material shapes by stressing the raw material beyond the yield stresses which causes negative effects in terms of the material properties, so the process repeatability is not guaranteed. Typically, the methods of bulk deformation utilize dies/molds to define the resultant shape, and no process flexibility can be offered, unless a new die/mold is separately designed and

Table 10.1 Compatibility between manufacturing process and shape feature

Process capabilities for shape generation														
Processes		Shape attributes									Process performance (0–5)			
		Depressions ID	Depressions >ID	Uniform-Wall	Uniform-section	Axis of rotation	Regular cross section	Captured cavities	Enclosed	Draft free surfaces	Material consolidation	Repeatability	Design for Assembly Index	Process flexibility
Solidification	Investment casting	Y	Y	U	Y	Y	Y	Y	N	N	5	5	4	0
	Die casting	Y	Y	U	Y	Y	Y	N	N	N	4	5	4	0
	Injection molding	Y	Y	U	Y	Y	Y	N	N	N	5	5	5	0
	Blow molding	Y	Y	G	N	Y	Y	G	Y	N	3	4	0	0
Bulk	Forging	Y	Y	Y	Y	Y	Y	N	N	N	3	2	2	0
Deformation	Hot extrusion	Y	N	Y	G	Y	Y	N	N	Y	2	2	0	0
	Transfer molding	Y	Y	Y	Y	Y	Y	N	N	N	2	2	2	0
Metal	Machined from stock	Y	Y	Y	Y	Y	Y	Y	N	Y	2	3	3	5
Removal sheet	Sheet metal bending	Y	Y	G	Y	Y	Y	N	N	N	4	3	0	3
Forming	Thermoforming	Y	Y	G	Y	Y	Y	N	N	N	3	3	0	1

U yes when the tools are in good conditions, *G* reasonably good, but not ideal manufactured for a new process. Sheet metal processes locally deform the raw sheet materials, and hence, the negative effects on the material properties are limited. These processes can provide a certain extent of process flexibility, but the repeatability is not very well controlled. In particular, we can realize that machining provides excellent process flexibility and supports custom designs of product shape and dimensions and makes it highly applicable in manufacturing biomedical devices. We discuss more details on the machining process design in the next section.

10.2 Machining Process Design

10.2.1 Overview

Machining is a process in which raw material is removed in order to form the desired shape. However, before the removal process, designers need to conduct several preparation steps in order to build up an optimum machining process:

1. Analyze the part requirements. Designers should thoroughly analyze design specifications of the part such as the shape, size, tolerance, and surface roughness such that the correct machining method(s) can then be applied. Designers should then choose an appropriate process mode/tooling. For example, if a planar surface requires being smooth, then face milling would be the better option over regular milling.
2. Choose the raw material. In the manufacturing industry, there are more than 2000 materials available, including metals, ceramics, polymers, and even composites. It is essential to decide the proper material through different factors such as strength, thermal expansion, biocompatibility, etc. Considering design requirements and material characteristics together, designers can pick the best material (e.g., the one with a higher material performance index).
3. Choose machine tools. Different machining modes, e.g., milling and turning, can generate materials with different shape features. Designers should pick the feasible machine tools to cover all the shape features of the product.
4. Choose a machining sequence. Designers should determine both the machining tools required and the proper sequence of applying these tools. It is important to select the machine mode and tool with a fast enough material removal speed for the highest process efficiency and lowest cost. Yet, for straighter product requirements with lower tolerance and finer surface finish, a slower material removal speed might be required instead. In this case, a better approach would be to first remove the unwanted material outside the product geometry with a higher speed, followed by the slower machining process over the expected product surface. Besides, in case a large hole is required to be generated, a more feasible approach is to drill a hole with a smaller diameter and then enlarge the hole by reaming. The actual removed volume by machining tools is less and therefore the tools can survive with more manufacturing operations.

5. Select cutters. There are more than 10,000 different cutters on the market, including nearly all combinations of materials and dimensions. Designers should pick the proper cutters according to their process requirements.
6. Choose and/or design fixtures. Designers should choose proper fixtures to mount the raw materials on the stage of the machine. If a standard fixture cannot be used, the designers then need to design and manufacture a fixture to support the materials during the required machining process.
7. Determine cutting conditions, and calculate the machining time and cost for each planned operation. Designers should be knowledgeable enough to determine the process parameters based on the key machining relations. (Such relations for different machining processes are discussed in the next subsection.) These parameters may include the cut speed, feed rate, and cutting time. These parameters should be set with values which meet the design demands, such as the shortest process time.

A process plan can help manufacturers estimate production time and cost to better process management. The manufacturing designers should come up with an optimized plan by considering all the related factors together, including process type, tooling type, process sequence, fixture design, and process parameters. For biomedical devices, there is sometimes an urgent need for producing a custom-made device in a necessary time frame, and therefore, process planning is particularly important in biomedical/clinical applications.

10.2.2 Machine Time and Material Removal Rate

In order to determine the most feasible machine operation parameters to achieve a material removal process with minimal manufacturing cost and time, we should understand the relations between the different operation parameters and the manufacturing concerns, such as machine time and machine removal rate (MRR).

Drilling

The typical configuration of drilling is shown in Fig. 10.7. The drilling head with a diameter of D rotates at a machine speed (N) with a unit of revolution per minute (rpm), and therefore, the cutting speed v (unit: mm per min) at the tip of cutting edge is $v = \pi ND$. In other words, if a process requires a cutting speed v, we need to set the drilling speed accordingly. In practice, we might wish to apply a higher drilling speed such that an ideal process is guaranteed:

$$N = \frac{kv}{\pi D} \tag{10.1}$$

where k is the unit constant, whose suggested value can be checked from a manufacturing handbook. $k = 1$ when considering an ideal case and $k > 1$ for a higher hole quality.

Fig. 10.7 Configuration of drilling

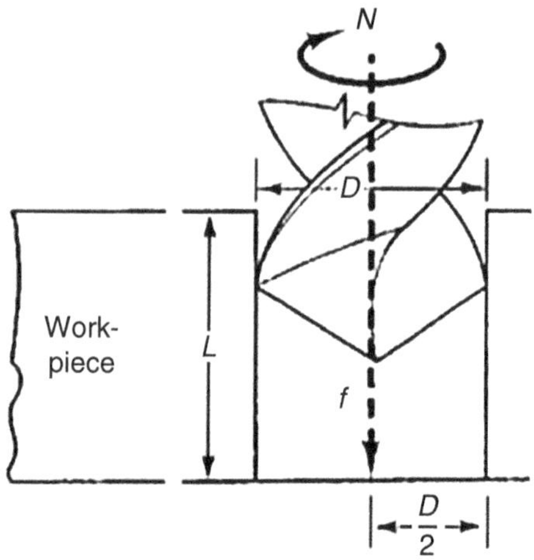

While the drilling head is rotating, it is moving into the material body with a feed rate of the spindle f with a unit of millimeter per minute; and the feed distance per revolution (L_{feed}) is equal to f/N. The total moving distance of the drilling head includes the generated hole depth (L) and an offset distance of the drilling head placed away from the material surface at the process beginning (L_{A}). The cutting time of the drilling process (T_{cut}) can be approximated by

$$T_{\text{cut}} = \frac{L + L_{\text{A}}}{f}. \tag{10.2}$$

The offset distance is usually stated as $D/2$ as the tip angle of the drilling head is generally >90°.

Furthermore, the MRR can be considered as the removed material volume throughout the process divided by T_{cut}. For the case $L \gg L_{\text{A}}$:

$$\text{MRR} \approx \frac{\pi D^2 f}{4} = \frac{kDL_{\text{feed}}v}{4} \tag{10.3}$$

Turning

For turning, the material sample with a diameter D is mounted on a lathing machine which rotates with N revolutions per minute, whereas the cutting tool approaches the outer material layer gradually with a feed distance per revolution (L_{feed}) or a feed rate f along the sample axial direction and the total cutting length along the axial direction L. as shown in Fig. 10.8a. The maximum cutting speed v is then equal to πND. There is an offset distance L_{A} of the cutting head position away from the sample surface when the process starts. Here, though the symbol definitions are different from the drilling process, the cutting speed v and the cutting time T_{cut} are still Eqs. 10.1 and 10.2, respectively.

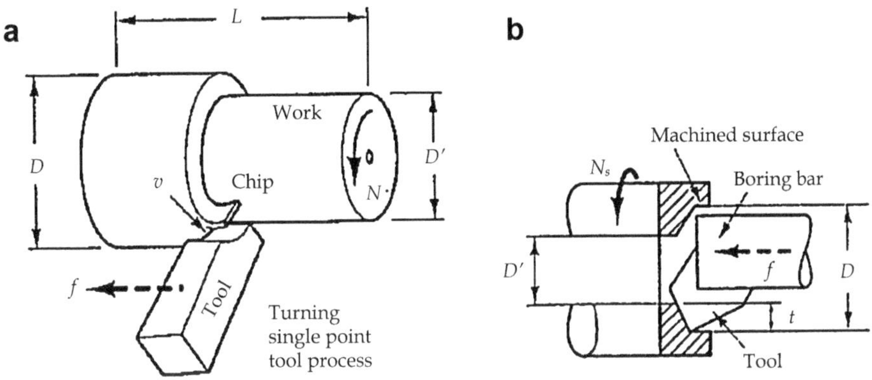

Fig. 10.8 Configuration of (**a**) turning and (**b**) boring

In the turning process, it is assumed that the thickness of the material layer being chipped off is t, which is often much thinner than material diameter D. If we further let the material diameter after the process equal D' $(= D - 2t)$, then MRR can be expressed by

$$\mathrm{MRR} = \frac{\pi\left(D^2 - D'^2\right)f}{4} \approx ktL_{\mathrm{feed}}v \tag{10.4}$$

Boring

Boring is another machining process with the lathing machine as configured in Fig. 10.8b. A basic boring operation enlarges the inner diameter D' of a pipe-shaped structure to a larger diameter D by moving a cutting tool into the pipe cavity with a feed speed f. Under a rotational speed N with a unit of rpm, the maximum required cutting speed (v) is then $v = \pi ND$, the feed distance per sample rotation (L_{feed}) is f/N, and the thickness of cut t is $D' - D$, which should be much thinner than the inner wall diameter. The length of cut is assumed to be L. The cutting tool should have an initial offset distance of L_A. In this case, the machine speed N, cutting time T_{cut}, and MRR can be expressed again by Eqs. 10.1, 10.2, and 10.4, respectively.

Facing and Slot Cutting

In the processes of facing (Fig. 10.9a) and slot cutting (Fig. 10.9b), the cutting tool moves into the sample along the radial direction of the sample rotation on a lathing machine. We define the initial sample diameter as D and the sample rotation speed in rpm as N, which can be configured as Eq. 10.1 given the cutting speed (v) is $v - \pi DN$. Given a constant machine speed N, the spindle/machine speed only decreases during these processes as the radial position reduces, and therefore, the machine speed as considered in Eq. 10.1 is the maximum level. The tool should be placed with an offset of L_A from the sample centerline when the process starts, and the length of cut (L) creates the final diameter of the cutting site with a diameter of D'

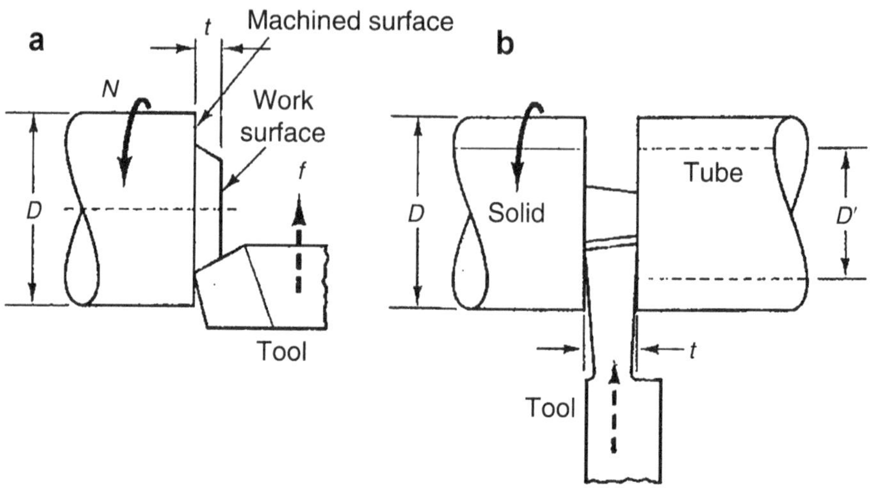

Fig. 10.9 Configuration of (**a**) facing and (**b**) and slot cutting

($= D - L$). If we further set the feed distance per sample revolution as L_{feed} with a feed rate $f = NL_{\text{feed}}$, the cutting time T_{cut} is Eq. 10.2, accordingly. For instance, if L is set as $D/2$ in the slot cutting process, the outer part of sample after cut would come off, and hence, we may call this particular case the cutoff operation.

The width of cut is further defined as t. In the facing operation, t can be determined by the tool position relative to the workpiece surface (i.e., the overlapping length of the tool and the workpiece part). In the slot cutting operation, as the tool is cutting a middle part of the workpiece, the overlapping length or t is basically the tool width. The MRR for either process can then be calculated as

$$\text{MRR} = \frac{\pi\left(D^2 - D\prime^2\right)tf}{4(D - D')} \approx \frac{ktL_{\text{feed}}(D + D')v}{4D} \tag{10.5}$$

Milling

The multiple (n) teeth on the milling head with a diameter of D rotating with a speed of N in rpm perform multiple cuts over the material surface with a thickness of B during the milling operation. As illustrated in Fig. 10.10 (*left*), the direction of the rotational axis is perpendicular to the feed direction, and these two axes are parallel to the material surface. The material surface area being removed is defined by the length of cut L and the width of cut (i.e., the tooth length) as indicated in Fig. 10.10 (*right*). The cutting speed v is then determined by Eq. 10.1. Meanwhile, the material sample mounted on a movable stage moves toward the milling head with a feed rate f (millimeter per minute), and the feed distance per single tooth cut ($L_{\text{feed\_tooth}}$) is $f/N/n$ (or the feed distance per single revolution $L_{\text{feed}} = f/N$).

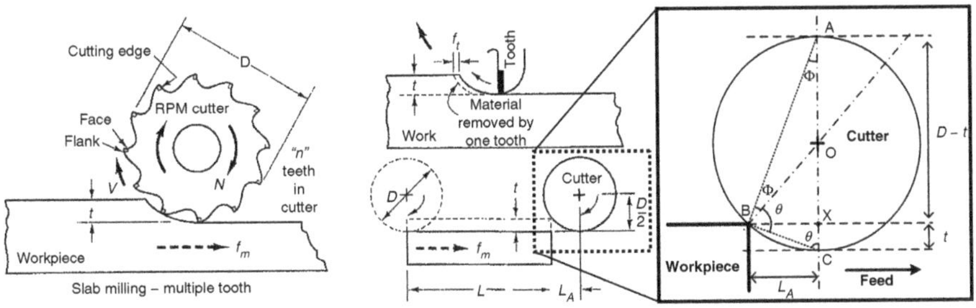

Fig. 10.10 Configuration and key parameters of milling

The cutting time T_{cut} is calculated by Eq. 10.2, whose length of initial offset position (L_A) can be further estimated by this simple proof provided below[1]:

$$L_A = \sqrt{t(D - t)} \tag{10.6}$$

where t is the width/depth of cut. On the other hand, the MRR with $L_A \ll L$ is

$$\text{MRR} \approx \frac{LWB}{T_{\text{cut}}} = fWB = \frac{kv}{\pi D} WBL_{\text{feed}} = WBNnL_{\text{feed\_tooth}} \tag{10.7}$$

Face Milling

For face milling (Fig. 10.11a), the rotational axis of a milling head with a diameter D is normal to the machined surface. With the same definitions of most process parameters, the machine speed and feed rate are computed as the previous section. The cutting time T_{cut} should consider both the offset at the beginning (L_A) and ending (L_O) of the process:

$$T_{\text{cut}} = \frac{L_A + L + L_E}{f} \tag{10.8}$$

where L is length of cut on the sample.

In more detailed considerations, the actual values of L_A and L_E are determined by different situations of the cutting width W. For example (Fig. 10.11b):

- If the cutter is not fully engaged and W is larger than $D/2$, then $L_A = L_E = D/2$.
- If the cutter is not fully engaged and W is smaller than $D/2$, then $L_A = L_E = \sqrt{W(D - W)}$.

---

[1]As shown in the right inset of Fig. 10.11, since the lengths OA = OB = OC = $D/2$ (radius of mill head), angle BAO = angle ABO, and angle OBC = angle OCB. Considering triangle ABC, 2 $(\theta + \Phi) = 180°$, and thus angle ABC = 90°. Considering triangle ABX, $\tan\Phi = L_A/(D - t)$. Considering triangle BCX, angle XBC = Φ, and thus $\tan\Phi = t/L_A$. Therefore, $\tan\Phi = L_A/(D - t) = t/L_A$.

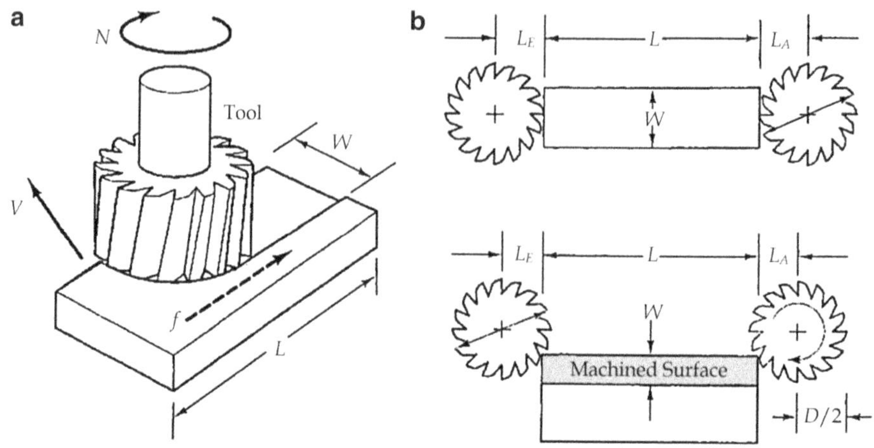

Fig. 10.11 Configuration of (**a**) isometric and (**b**) top views of face milling

The MRR can then be approximated as Eq. 10.7, ignoring the offsets of engagement (L_A) and over-traveling (L_E).

10.2.3 Other Machining Process Design Parameters

Machine Cost

One major concern of machining processes is the cost of manufacturing. When considering the cost, there are many factors which can be separated into two categories: "fixed" and "variable" costs. The fixed costs include factors that must consume resources and cannot be ignored such as heating, lighting, product research and development (R&D), production overheads, administration, etc. Variable costs include factors that can be adjusted such as material price, material volume, labor wages, and working hours. In general, people will analyze these factors and calculate the machine cost in order to get the optimum combination that minimizes machine time while maximizing machine removal rate with a processing path that can be achieved.

Machine Power

In most of the machining processes, the point of cut can be simplified as the condition described in Fig. 10.12. The cutting action with a cutting speed V should support a sufficient cutting force F_C in order to chip off the material locally. Given that the fore surface of the cutting tool has an angle of α with the material surface normal vector, such that the folding line of the material layer being chipped off has an angle ϕ from the cutting direction, the shearing force along the material folding line F_S is $F_C\cos(\phi)$. The essential F_S can be estimated based on the shear strength (σ_S), $F_S = \sigma_S A_{chip}/\sin\phi$, where A_{chip} is the cross-sectional area of the material layer

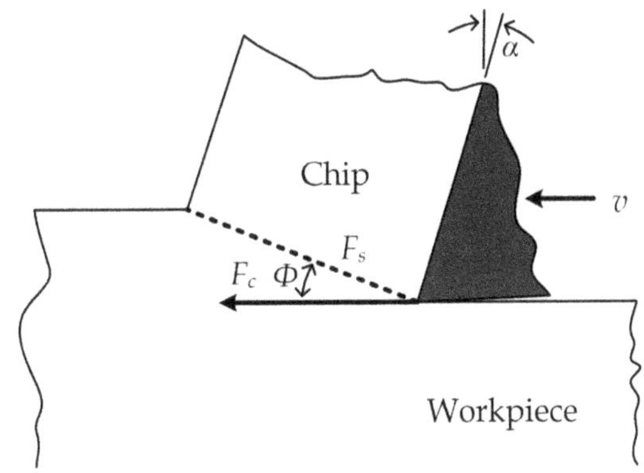

Fig. 10.12 Force exerted at the point of cutting

Table 10.2 Specific cutting energy for different materials

Material	Specific energy (W s/mm^3)
Aluminum alloys	0.4–1
Cast irons	1.1–5.4
Copper alloys	1.4–3.2
High-temperature alloys	3.2–8
Magnesium alloys	0.3–0.6
Nickel alloys	4.8–6.7
Refractory alloys	3–9
Stainless steels	2–5
Steels	2–9
Titanium alloys	2–5

being chipped (e.g., $L_{\text{feed}}t$ for lathing and Wt for milling). The deformed chip thickness is equal to $t/\sin\phi$. The minimal required F_C should be larger than $2\sigma_S A_{\text{chip}}$.

The required cutting power P_c is $F_C v$, which can be further related to the cutting energy for the product material per material removal or the specific cutting energy (E^*_{cut}) and the material removal rate: $E^*_{\text{cut}} \approx P_c/\text{MRR}$. Some suggested ranges of E^*_{cut} for different materials are listed in Table 10.2. In fact, MMR is proportional to v, and thus, the suggested E^*_{cut} and other machining process parameters can be used to determine F_C. Notably, these suggested ranges of E^*_{cut} are related to the minimal required F_C as the angle of folding interface of the material layer chipped off ϕ may not be within the optimal angle range. Tools after multiple runs of machining process will become dull, and therefore, the required specific cutting energy will then be higher. A safe estimate in such cases is to multiply the specific energy by a factor of 1.25 for dull tools, or F_C for dull tools can be scaled up by a factor of $k_{\text{machine}} = 1.25$.

$$F_C = \frac{k_{\text{machine}} E^*_{\text{cut}}}{(\text{MRR}/v)} \tag{10.9}$$

Such estimated value of F_C can help to decide requirements on machine specifications. For instance, a machine offering adequate torque or cutting force should be chosen. Although we can always select machines that provide a very large cutting power and support the target processes, we should avoid using a machine which provides largely excessive cutting power in order to reduce the unnecessary cost of using such machines.

Furthermore, the machining process induces temperature of the machined site. Such temperature increment could be approximated based on the following assumptions:

- No heat escapes from the shear plane into the workpiece.
- Heat conduction can be neglected as compared to heat mass transfer in the direction of the chip motion.
- It is acceptable to use average values, equal for all materials, for shear-plane angle, rake-rake friction coefficient, and chip/tool contact length, and to consider them to be unaffected by cutting speed.
- Thermal parameters of materials are assumed as they apply at room temperature, and their change with temperature is neglected.

Accordingly, the maximum temperature increment (T_C) at a local material site along the tool-material interface should be proportional to the cutting power P_C. More specifically,

$$P_C \approx \Phi_C \frac{F_C v}{\alpha \rho C} T_C \tag{10.10}$$

where Φ_C is roughly a constant, ρ is the density, C is the specific heat capacity, and α is the thermal diffusivity.

Although the scaling approximation of Eq. 10.10 has largely simplified from the real case, the estimated T_C is not significantly different from the one obtained by much more complex computations with an error of <20%.

Surface Roughness by Milling

In the milling process, multiple cuts by the teeth feeding along the material surface generate a roughness profile connected by multiple arcs. The angle of a tooth cutting into the material surface can be approximated by the depth of cut z and the feed distance per tooth $L_{\text{feed\_tooth}}$ as $\tan^{-1}(2z/L_{\text{feed\_tooth}})$, which is also similar to the orientation of the tooth tip with respect to the milling head center with a diameter of D, i.e., $\tan^{-1}(L_{\text{feed\_tooth}}/D)$, and hence, $z \approx L_{\text{feed\_tooth}}^2/(2D)$. A quick estimation of the surface roughness (δ_S), which is the variation of the surface height, can simply be the average z over every single-tooth feed distance, i.e., $\delta_S \approx L_{\text{feed\_tooth}}^2/(4D)$. In fact, a more accurate estimate of the surface roughness should also consider the tooth feed and sample feed directions. Figure 10.10 shows the up-milling case in which both the directions are the same. On the other hand, if the milling head is rotating counterclockwise, in this case such that its direction is opposite to the sample feed direction, this is called down-milling. To consider the roughness, it is then expressed as

$$\delta_S \approx \frac{L_{\text{feed\_tooth}}^2}{4[D \pm L_{\text{feed\_tooth}} n/(2\pi)]} \tag{10.11}$$

where n is the number of teeth over the milling head. For the "±" sign mentioned here, the up-milling adopts the "+" sign, whereas the down-milling adopts the "–" sign.

10.3 Computer Numerical Control (CNC) Machining Process Planning

10.3.1 Hardware

Typically, a computer numerical control (CNC) machining system (Fig. 10.13) can control the machining process in three dimensions following the paths preset with a digital script loaded in a motion control unit (MCU) of the system. These programming scripts can automatically choose the proper machine tools based on different machining parameters, such as depth and width of cut, cutting speed, feed rate, usage of coolant, etc. CNC can calibrate the machine tool conditions and update the machining parameters for the best manufacturing outcome if necessary. More advanced CNC systems can support both milling and lathing with automatic machine tool switching, coordinate measurement machines, and multi-axis positioning machinery. For example, a turning process can also be extended with the rotation of cutting tools for supporting more complicated shapes. Vertical milling can be upgraded by adding the rotation of the platform for better surface profile conditions. Other system parameters such as temperature and material residual removal frequency can also be set.

The development of CNC systems has gone through multiple generations. In the 1940s, when the system was still called the "numerical control" system, where computers could only support basic computation, operators used punched tape as data storage systems for defined machine paths and parameters. In the 1950s, changeable tapes or even floppy disks were integrated into the systems, allowing for multi-machining modes and very flexible machining paths. However, the programing scripts could not be edited directly inside the CNC machines. To date, CNC machines include a high-end computer installed with computer-aided design software and computer-aided manufacturing algorithms and are connected to the Internet. The system architecture of CNC machines includes a microprocessor, memory, bus, display interface, Programmable Logic Controller (PLC) interface, spindle

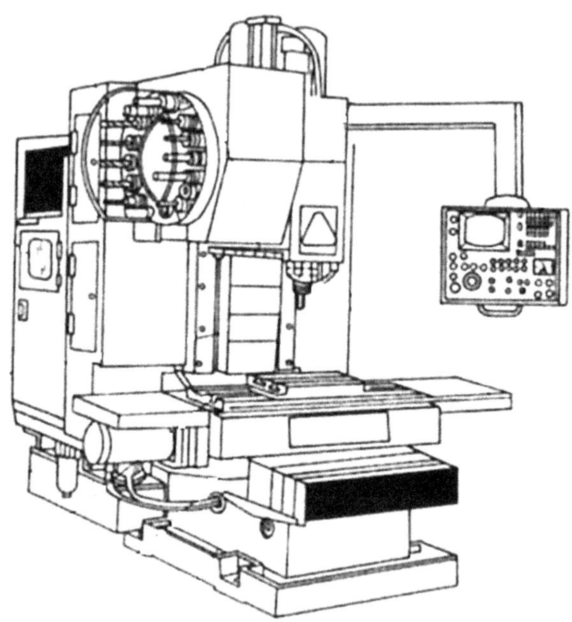

Fig. 10.13 A CNC machine

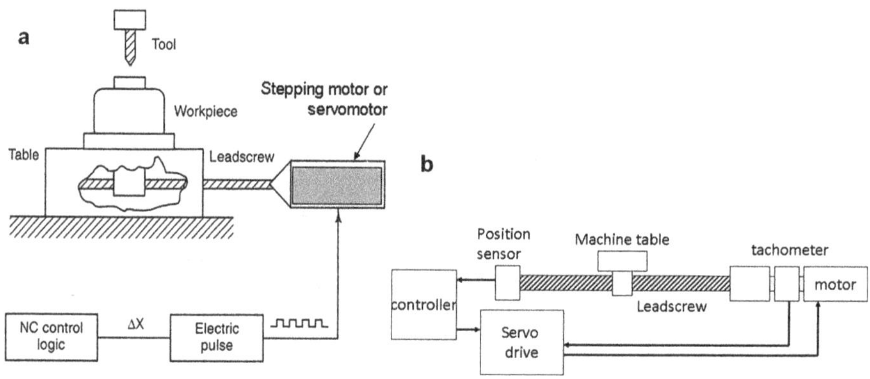

Fig. 10.14 (**a**) Motion control scheme of a CNC machine. (**b**) Motion control of a servomotor

control, tape reader interface, communication interface, etc. Operators can now perform any real-time adjustment and optimization of the machining process.

In particular, the motion control unit (MCU) as shown in Fig. 10.14 is important in that it controls machining operations through reading the preset program scripts and sending and receiving motion-related signals. The MCU consists of electronic circuits (hardware) that are useful in reading and interpreting the instructions (program scripts) and converting such instructions into mechanical actions of machine tools. An MCU knows how to read the code description, decode encoding instructions, produce axis motion commands, receive feedback signals of the tool position and speed, and adjust the driving signals in real-time.

The motor is the key component for driving movements of the stage and machining tools. Without a proper control scheme for the motor movement, the CNC machine no longer functions well. Servomotors are widely used in the CNC machines for modulating position and velocity. The word "servo" originates from "*servus*" in Latin, meaning "slave." In a servomotor device, an optical-electric rotary encoder is mounted on a motor and detects its rotation. The detected value will be adjusted and fed back to the servo amplifier, and the value will be compared with the command/target value based on the proportional-integral-differential (PID) control scheme. Thereby, the control system reduces the error between its detected position/speed and target values.

10.3.2 Programming Language

Since implementation of CNC relies on digital computers, it must use a programming system to execute the machining procedure. For instance, "M- and G-code" and "Automatic Programming language (APT)" are the coding systems that CNC machines commonly use. "M- and G-code" are accepted as the standard of machine tool controlling language with its syntax following RD-274. It was developed for low memory computer; therefore, the words used are compact. This is why it will use letter plus number to represent specific operations. In addition, as most of the commands start with G and M, people named it "G-code," "M-code," and "G- and M-code." Generally, an integral CNC program consists of three parts: the title, the subject, and the epilogue. The program is scripted with several sections that act in the accordance of different machining stages, including warming up and shutdown steps. These script sections are made of several instruction characters. The initiative character in a program should be "%" or "O," and the program title should be "Oxxxx" as a single row. The number is ranged from 1 to 9999, known as "O0001"–"O9999." Moreover, the program epilogue must be a "%."

The letter "G" refers to machine preparation such as position, scale, selection, unit, etc. G is sometimes referred to as the preparatory code since it starts with the letter G in the CNC machining program. G-code is a language that designers use to specify the CNC machine on "How" to manufacture a product step by step. These steps defined by the G-code instruct the controller and accordingly tell the CNC machine which part to move, where to move, how fast to move, and which way to go. The most common case is that the machining tool and stage move according to these instructions along a preset path. The same concept can further extend to other non-cutting manufacturing machines for processes like forming, polishing, light drawing, and even calibration/dimension measurement.

The letter "M" refers to miscellaneous machine functions such as switching coolant, tool changing, spindle control, etc. In all numerical control programs, M-code is essential to ensure proper operation of the G-code. By combining different letters and numbers, this system has at least hundreds of commands despite the fact that only around 30 commands are commonly used as listed below.

Code	Description	Code	Description
M00	Compulsory stop	M22	Mirror, y-axis
M01	Optional stop	M22	Tailstock backward
M02	End of program	M23	Mirror OFF
M03	Spindle on (clockwise rotation)	M23	Thread gradual pullout ON
M04	Spindle on (counterclockwise rotation)	M24	Thread gradual pullout OFF
M05	Spindle stop	M30	End of program, with return to program top
M06	Automatic tool change (ATC)	M41	Gear select – gear 1
M07	Coolant on (mist)	M42	Gear select – gear 2
M08	Coolant on (flood)	M43	Gear select – gear 3
M09	Coolant off	M44	Gear select – gear 4
M10	Pallet clamp on	M48	Feed rate override allowed
M11	Pallet clamp off	M49	Feed rate override NOT allowed
M13	Spindle on (clockwise) and "flood" on	M52	Unload Last tool from spindle
M19	Spindle orientation	M60	Automatic pallet change (APC)
M21	Mirror, x-axis	M98	Subprogram call
M21	Tailstock forward	M99	Subprogram end

Moreover, although most of the commands are the same between machines, some may be different because of their unique machine-based functions, configurations, and options. Each line in the script can only offer one command, and each line can be inserted with only one movement/configuration called "Block." Finally, those blocks will arrange to form a program and trigger the complete machine operation.

A G-Code Example

A 2.0″ × 2.0″ square is to be milled using a 1/2″ end mill as shown in Fig. 10.15 on the *right*. The G-code program is

```
O0001
N0010 G41 S1000 F5 M03    Begin compensation, set feed and speed, and turn on spindle
N0020 G00 X6.00 Y6.00     Move to lower left corner
N0030 G01 Z-1.00          Plunge down the cutter
N0040 Y8.50               Cut to upper left corner
N0050 X8.50               Cut to upper right corner
N0060 Y6.00               Cut to lower right corner
N0070 X6.00               Cut to lower left corner
N0080 Z1.00               Lift the cutter
N0090 G40 M30             End compensation, and stop the machine
```

APT is not merely a programming language but a program compiler which calculates and generates the machine commands and detailed operation sequences. APT adopts highly readable English-like syntax as its coding scheme, and the compiler will translate the code to machine executable commands. Its coding

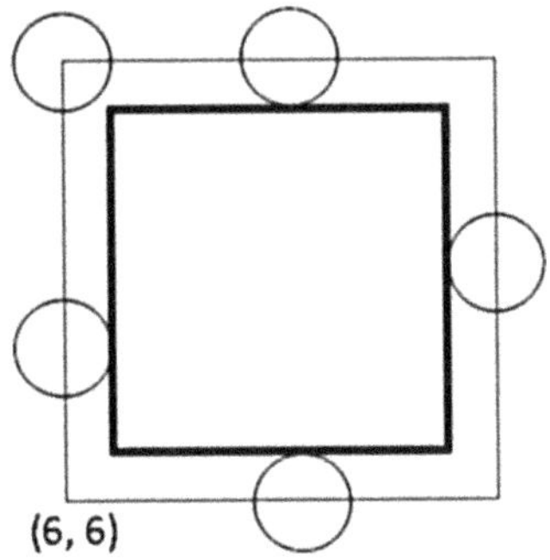

Fig. 10.15 Sample milling path

concept is similar to "automatic programming," which refers to some mechanisms of generating low-level machine codes that allow a programmer to write code at a higher level of abstraction with fewer technical hurdles and shorter code development cycles.

10.3.3 Automatic Tool Path Generation

Development of "G-code" and "APT" manually requires a long time for writing and debugging. Taking milling as an example, writing the scripts for removal of unwanted material volume outside the target geometry is tedious. As this operation includes a larger number of back and forth moving of the mill cutter to remove those volumes layer by layer, many line of paths have to be generated, and mistakes may be made if the scripting is done manually. Also, for objects with curved outlines or surfaces, computing the working milling path can be exceptionally tedious and difficult. In fact, the conversion from object shapes to machining paths is mostly a routine task of geometric calculations, and hence, such conversion can be executed by some preset algorithms without errors/mistakes. Automatic tool path generation is the key feature of computer-aided manufacturing (CAM). Algorithms for path generation are computer programs for improved implementation of manufacturing processes. In many cases, these algorithms can realize an optimum design as a set of key locations along the machining path dependent on the cutter size, cutter shape, and other process configurations.

The path generation is typically based on a sequence of computations. The first set of paths is generated around the outline of the final surfaces and planes, and the tool orientation is aligned with the principal axis of the coordinate system. For curves, they are first broken down into multiple arcs linking up together, followed by generating multiple lines combining as an approximation for each of the arcs. The precision of such an approximation is defined by the number of lines along the arc. For curved or inclined surfaces, a polyhedral machining scheme can be used. Next, isoperimetric lines are generated for the remaining volume outside the final shape.

Importantly, the actual tool paths are generated after importing the machine parameters, such as the type of tool, base, speed, position, etc., into the coding system as the geometry statement in APT, for example.

The position values of each key point along the tool path are often described in the Cartesian coordinate system. The origin of the machine is fixed, and it is at physical location which the machine builder defines. It differs from the origin of the path coordinate system that the machine origin uses to calibrate and control the relative motion system of the machine table, slide, and tool, also known as machine zero. Instead, the workpiece origin should be considered as the origin of the CNC scripts. This should also be the origin in the computer-aided drawing in the process design stage. The product dimensions are determined in the "workpiece" coordinate system in the drawing software. The coordinate axes of the workpiece coordinate system are parallel to the corresponding coordinate axes of the machine coordinate system. After a workpiece is installed on the machine tool with the fixture, the coordinate distance between the origin of the workpiece and the origin of the machine tool is measured. This distance is called the workpiece origin offset. During machining, the preset value of the workpiece origin can be automatically added to the workpiece coordinate system so that the numerical control system can determine the coordinate value during machining according to the machine coordinate system.

The tool path is first generated for the rough machining (Fig. 10.16). It is designed to remove material as quickly as possible. However, in order to prevent excessive removal, it is preferred to separate the material into layers and conduct the process per layer. This involves part contour calculation, machining pattern selection, width of cut consideration, machine method determination (e.g., pre-drill or ramping), and cutting condition consideration (e.g., depth, speed, and coolant). In particular, for the machining pattern selection, there are some basic rules: (1) If the cutting area is larger than half of the material, a zigzag pattern is preferred. Otherwise, a spiral pattern is preferred for curved outlines. (2) If the shape of the part is complex and contains multiple islands or cavities, the whole area should be split into different zones which are then machined one by one. (3) The offset should be calculated based on both the tool size and the required tolerance so that material can be removed at a maximum rate.

The machining style is configured with the different surface finishing stages, e.g., roughing, semifinishing, or finishing. Unlike the rough machining, the finish machining step requires high accuracy to fulfill the final product specifications. The corresponding tool path is aimed at removing the remaining material over the expected product surface after the rough machining. In other words, the finish machining performs a few additional tasks like smoothing surfaces and edges, fixing angles, or removing excessive teeth. Taking milling as an example, it is common to apply the machining mode with five axes or with three axes equipped with a ball-end mill (Fig. 10.17) as these configurations can generate curved surfaces, fillet edges, and inclined surfaces without the tooth-alike surface profiles.

Once the machining configuration has been set, the tool path can then be computed. For instance, the length of each line of machine tool movement is flexible. Such length can be adjusted based on the required tolerance and the final shape

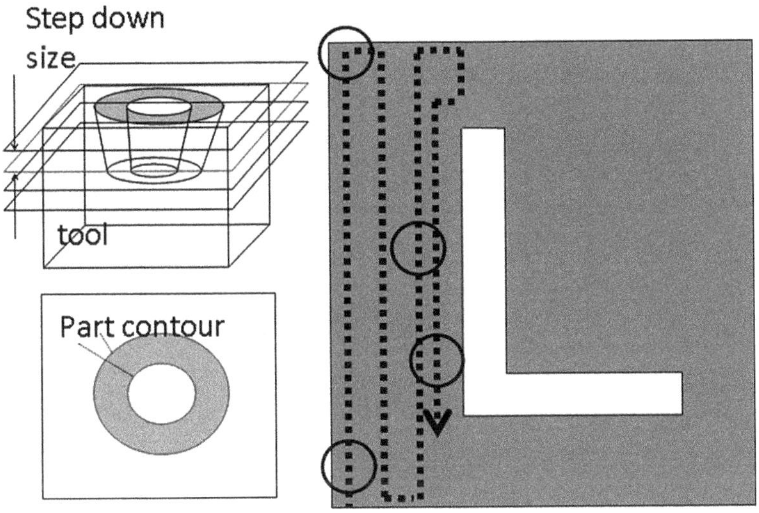

Fig. 10.16 Example of machine tool path generation

Fig. 10.17 Milling schemes for finish milling using either a five-axis ball-end mill operation mode (left) or a five-axis flat-end mill operation mode (right)

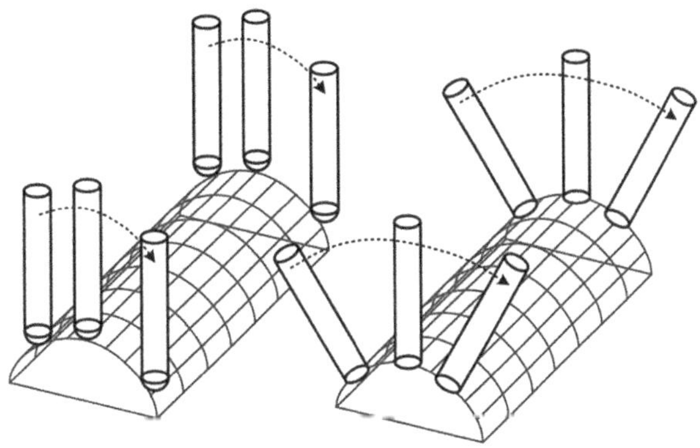

features, e.g., inclined angle of side surfaces and the radius of fillet edges, as exemplified in Fig. 10.18a. If the final shape has a curved feature having a changing radius of curvature along the shape outline, this outline can be broken down into multiple piecewise straight lines, each with a different length to meet the required tolerance with the minimum number of lines (or key points) as shown in Fig. 10.18b.

After the automatic tool path generation, the path can then be verified using a computer simulation on the machining process. The simulated resultant product shape can then be examined on whether or not it fulfills the product specifications. Such simulation verification can save time and cost in the product development cycle as this step does not require the actual material and manufacturing implementation in that we can obtain the resultant product shape instantly. Designers can then revise the tool path if any error is found in the verification step before the tool path is downloaded to the machine tool for the finalized machining process.

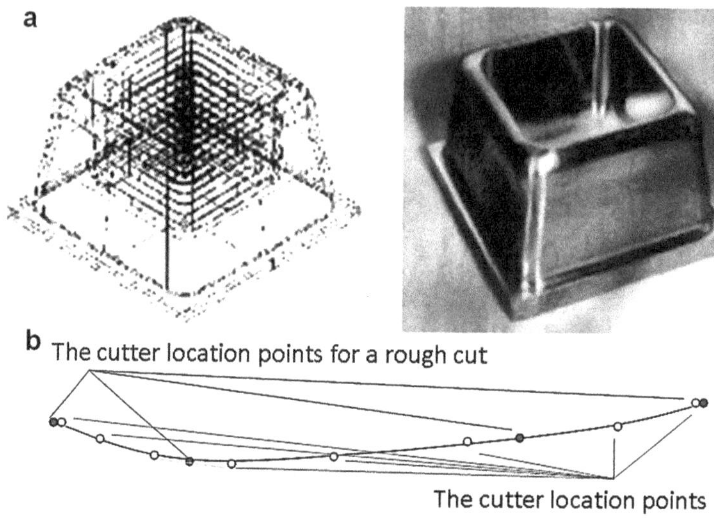

Fig. 10.18 (**a**) Example of an object with rounded edges. (**b**) Path planning for curves compromised with the required tolerance

Problems

Problem 10.1

Here we focus on the manufacturing of a compartment of knee prosthesis as the figure shown in Fig. 10.P1. The front shiny part is made of titanium alloy, while the back screw structure is made of cobalt-chromium alloys. Please design the required manufacturing processes (in sequence) based on the manufacturing methods discussed in Chap. 6. The considerations may include the material type, feature shape, material attachment, and surface conditioning.

Problem 10.2

As shown in Fig. 10.P2, we plan to apply machining to process the given material (*left*) to be the final product (*right*).

(a) Consider one slice of structure as shown in Fig. 10.P3 (*left*). Now we apply a milling process for this slide of structure during the machining process, and the mill end has diameter *D*. Now, please design on how the milling tool can manufacture the structure by briefly drawing the milling path and the structure on the rectangle below (*right*). The path need not to be quantitative, but important labels for the mill positions should be mentioned.

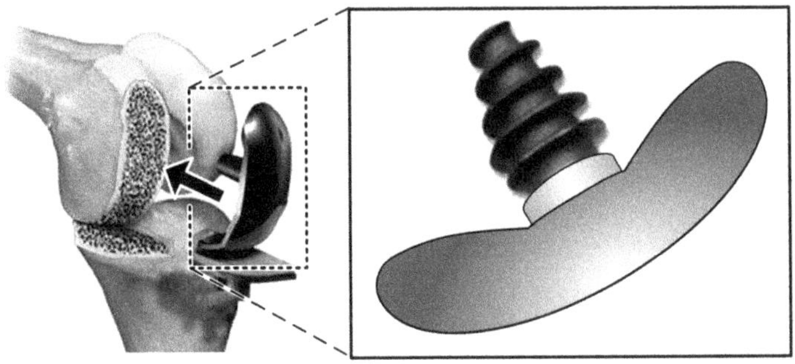

Fig. 10.P1 Knee prosthesis made of cobalt-chromium alloys

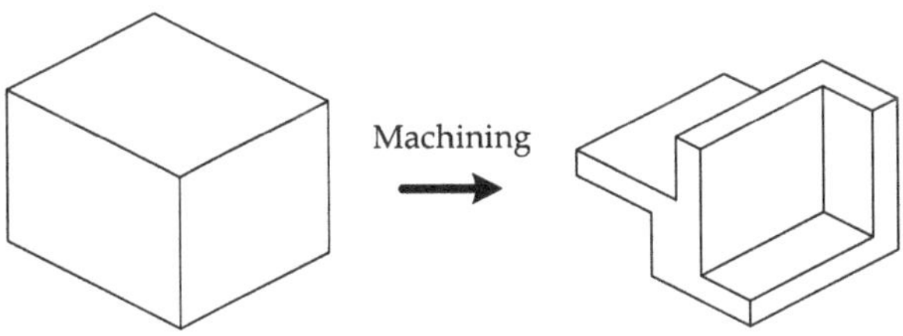

Fig. 10.P2 Change of material shape before and after milling

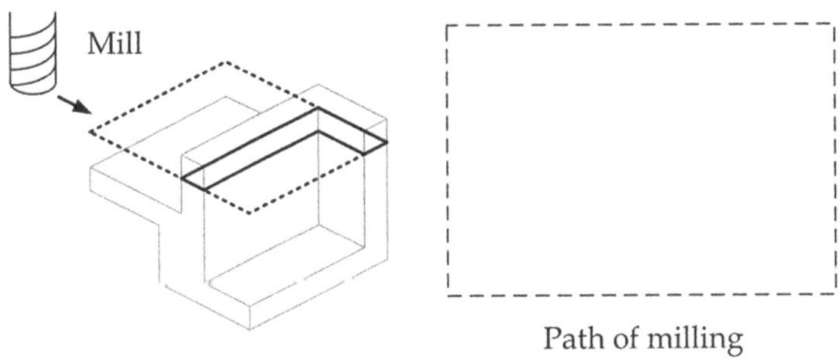

Fig. 10.P3 Path planning of a cross-section for milling

(b) Are there any features in part (a) difficult to be manufactured by milling, and why? If so, please identify the feature(s) directly in the path in part (a).

Problem 10.3

A turning operation has the following operation parameters: outer diameter $D_1 = 75$ mm, inner diameter $D_2 = 0$ mm, length $= 80$ mm, cutting speed

(v) = 150 m/min, feed rate (f) = 0.3 mm, depth of cut = 5 mm, and specific force = 2000 N/mm^2. Determine:

(a) Tangential force (N)
(b) Metal removal rate (cm^3/min)
(c) Cutting power (kW)
(d) Machining time (min)

Problem 10.4

A shaft of stainless steel is turned down to a diameter of 80 mm from 100 mm. The cutting speed v is 120 m/min with a feed rate f of 0.25 mm. Determine:

(a) Chip area (mm^2)
(b) Metal removal rate (cm^3/min)
(c) Cutting power (kW) using the specific force listed in Table 10.2
(d) Machining time (min)

Problem 10.5

A face-milling operation is being performed. A 300-m-diameter cutter with 16 teeth is cutting a 200-mm-diameter wide strip of stainless steel, which is centrally located under the cutter axis. The operating parameters are depth of cut = 5 mm, cutting speed v = 150 m/min, and feed per tooth f_t = 0.25 mm. Determine:

(a) Machine speed (rev/min)
(b) Feed rate (mm/min)
(c) Metal removal rate (cm^3/min)
(d) Average chip thickness (mm)
(e) Cutting power (kW) using the specific force listed in Table 10.2

Problem 10.6

Please prove the relation on the resultant surface roughness from a milling process as described in Eq. 10.11. Please kindly mention the assumptions you have made. In case you can find an even better relationship than Eq. 10.11, please kindly discuss about the accuracy of Eq. 10.11.

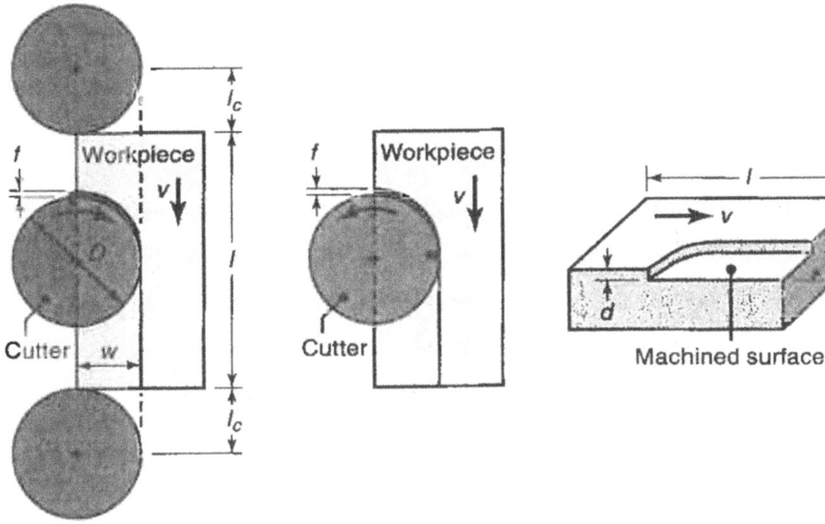

Fig. 10.P4 Face mill and the key parameters

Problem 10.7

Refer to Fig. 10.P4 and assume that $D = 150$ mm, $w = 60$ mm, $l = 500$ mm, $d = 3$ mm, $v = 0.6$ m/min, and $N = 100$ rpm. The cutter has 10 tooth inserts, and the workpiece material is UHMW-PE. For a face mill through the material surface with a thickness of $D/2$, calculate the material removal rate, cutting time, and feed per tooth, and estimate the power required.

Problem 10.8

Investigate machining titanium alloys with cutting speed $v = 0.5$ m/s. Depth of cut $t = 0.2$ mm, width of cut $b = 10$ mm, and chip angle $\phi = 23°$, implying that $A_{\mathrm{chip}} = tb/\sin\phi$. The cutting force F_{C} can be converted from the shearing force F_{S}, which itself is obtained from the shear flow stress $\tau_{\mathrm{S}} = 800$ N/mm^2, such that $F_{\mathrm{S}} = \tau_{\mathrm{S}} tb/\sin\phi$.

(a) Please find out an expression for the maximum temperature increment T_{C} at the machined site.
(b) Find out the relative value of T_{C} comparing to your answer in (a), if now the chip angle $\phi = 35°$.
(c) Adding the next change of $t = 0.05$ mm, find out again the relative value of T_{C} comparing to your answer in (a).

Fig. 10.P5 Artificial hip joint and its components

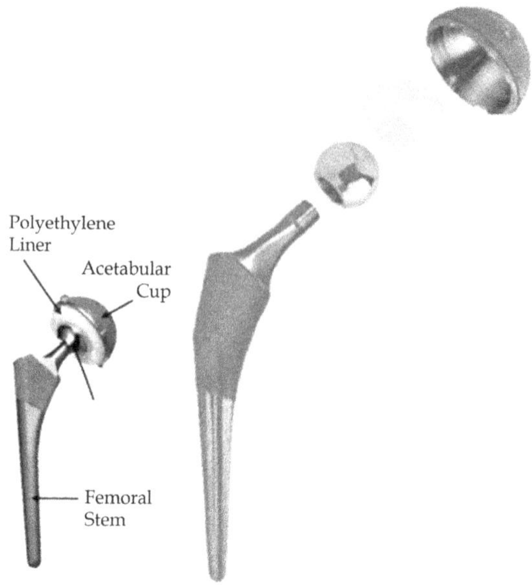

(d) Now, see if you could afford to triple the speed to $v = 1.5$ m/s (with the changes in (b) and (c) maintained) and still keep the peak temperature below the value you calculated in (a).

Problem 10.9

Please design the manufacturing process for an artificial hip joint shown in Fig. 10.P5.

Problem 10.10

A patient has been arranged for the implantation surgery of the artificial knee joint after a severe car accident. The expected outcome after the surgery is described as Fig 10.P6a. The lower tibial stainless steel plate is confirmed to be covered by a polyethylene sheet, whereas the material for the upper femoral replacement component is also stainless steel. Here, we may consider the tibial plate (with side length $L_{\text{plate}} = 60$ mm) simplified as shown in the following figure:

The gray region in Fig. 10.P6b shows the surface needed to be milled in order to generate an appropriated surface roughness for better bone cell attachment. The required surface roughness for cell attachment is between 2 μm and 3 μm. The

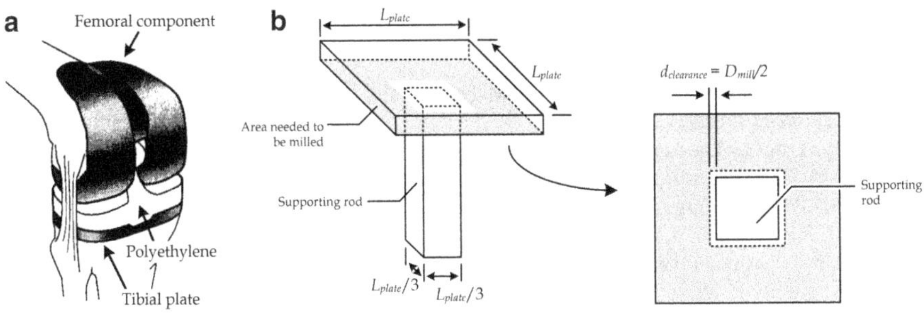

Fig. 10.P6 Artificial knee joint and simplified geometry of the tibial plate

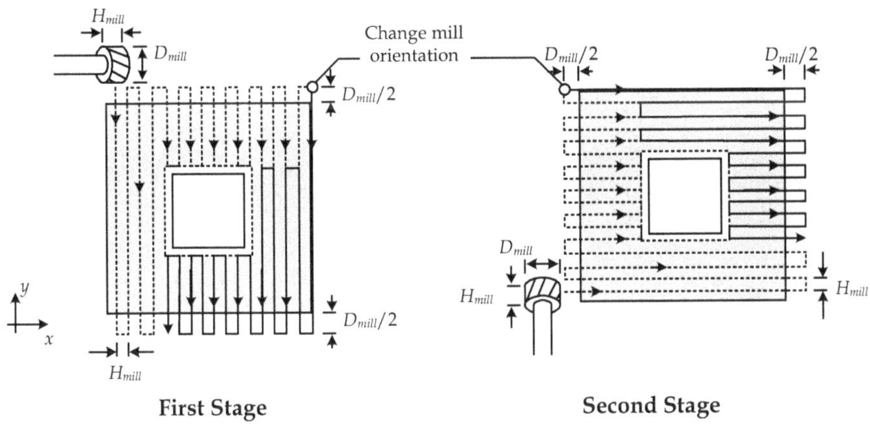

Fig. 10.P7 Milling of the lower surface on the tibial plate for bone attachment

milling cutter has a diameter D_{mill} (= 5 mm) and a height H_{mill} (= 3 mm), with the number of teeth on the cutter n (= 2). The milling path includes two stages of slab milling as shown in Fig. 10.P7. The moving velocity (feed rate) is to be set the same throughout the milling stages. Between the two stages, the milling head would move from the end position of stage 1 to start position of stage 2.

The maximum moving speed/feed rate of milling head ($F_{\max}$) is 50 mm/s. The maximum rotational speed of the milling cutter is 6000 rpm, yet the suggested rotational speed for this milling operation (N) is only 300 rpm.

(a) Please estimate (1) the required distance of moving path for either stage 1 or stage 2 of the milling process and (2) the moving distance between the two stages. (*Note:* Use symbols only, but not the given values, for this part.)
(b) Please find out the range of feasible moving speed/feed rate such that the resultant roughness is within the required range.

References and Further Reading

1. Walsh, R.A., Walsh, R.A.: Handbook of Machining and Metalworking Calculations. McGraw-Hill, New York (2001)
2. El-Hofy, H.A.G.: Fundamentals of Machining Processes: Conventional and Nonconventional Processes. CRC Press, Boca Raton (2013)
3. Davim, J.P.: Machining of Hard Materials. Springer, London (2011)
4. Arias, E.R.: Analysis of Surface Roughness for End Milling Operations. Texas Technological University, Lubbock (1983)
5. Samuel, R.: Formulas and Calculations for Drilling Operations. Wiley, Somerset (2011)
6. Davim, J.P.: Machining of Titanium Alloys. Springer, Berlin, Heidelberg (2014)
7. Overby, A.: CNC Machining Handbook: Building, Programming, and Implementation. McGraw-Hill, New York (2010)
8. Kief, H.B., Roschiwal, H.A.: CNC Handbook. McGraw-Hill, New York (2013)
9. US Office of Technology Assessment: Computerized Manufacturing Automation-Employment, Education, and the Workplace. US Office of Technology Assessment, Washington, DC (1984)

Chapter 11
Scaffold Design

Abstract The aim of tissue engineering is to create a working tissue for a replacement body part. Scaffold is the key biomedical device in tissue engineering applications, and in this chapter, we focus on scaffolds, including the key design parameters. In particular, a more detailed discussion on electrospinning is provided.

11.1 Tissue Engineering

Tissue engineering is a multidisciplinary field that applies the principles of engineering and life sciences toward the development of artificial tissues for implantation into a living organism in order to repair or replace existing biological tissues. It combines cells, biological macromolecules, and scaffolds to create complex organs and structures from relatively simple yet biomimetic parts. This growth of complex systems from simpler "seeds" has fascinated the human mind for generations. The generation of Prometheus and the birth of Eve from Adam's rib are some of the examples of early literary imaginations of the concept of tissue engineering. Artificial and non-native implantation of devices is used to heal human conditions. For example, going back as far as the time of the Romans, dental implants can be seen as the first step in the evolution of regenerative medicine. Research into tissue engineering has grown in parallel with research into artificial biomaterials. An example of a milestone in this regard was the use of synthetic fibers in artificial skin grafts as a form of burn treatment during the 1960s. A decade later, heparin, an artificial scaffold made up of organic polymers and biomaterials, was used as a means to aid in blood coagulation. Later, rapid research into tissue engineering led to the development of artificial skin made out of silicone which covers porous, crosslinked collagen with chondroitin as a skin treatment for burn victims. Nowadays, tissue engineering is used to develop artificial livers made from hollow fiber reactor and pig liver cells and neurotransmitter secreting organs. The next step in tissue engineering is regenerative medicine which is the implantation of foreign biomaterials and tissues to regrow cells and tissues [1].

The aim of tissue engineering is to create a working tissue for a replacement body part. From an engineering perspective, tissue engineering is, in a nutshell, the

R. H. W. Lam, W. Chen, *Biomedical Devices*,
https://doi.org/10.1007/978-3-030-24237-4_11

creation of desired biological organs or body parts by controlled and directed synthesis of constitutive cells by using a base structure so that cells conform to the part being synthesized, i.e., a "scaffold." This can take place either in vivo or in vitro, meaning that the scaffold materials are either directly implanted into the patient's body site or pre-cultured with "seeds" of cells, i.e., small amount of cells, before the implantation. Scaffolding provides a way to replace malfunctioning tissues inside our bodies. In particular, the in vitro tissue formation process can be broadly split into three steps: (1) cell selection, involving the sourcing of cell and their isolation and seeding; (2) scaffolding, providing the structure that houses the cells; and (3) implantation of the cell scaffold into the patient body.

Scaffold is the key biomedical device in tissue engineering applications, and we will focus on the discussion of scaffolds in this chapter. Scaffolds are required to have a structure which can support three-dimensional tissue formation and mimic the structures and functions of extracellular matrices. It has been demonstrated that drug testing of cells in conventional two-dimensional culture platforms (culture flasks or well plates) disagrees with subsequent animal trials, whereas drug tests with three-dimensional scaffolds provide more representative outcomes compared to animal and clinical trials. Very likely, cell monolayer culture does not convey a comparable microenvironment which sufficiently mimics that in the human body. The structural material should be carefully selected and constructed with ideal physical and biochemical properties for the target cell types to achieve successful regeneration of the damaged/abnormal tissues. The scaffold materials must be biocompatible, biodegradable, and preferably bioresorbable while also being able to be functionalized with biomolecules. One pivotal consideration is ensuring that there is an appropriate microenvironment for cell growth and tissue regeneration. Ideally, cells seeded in the scaffold should have the same signaling and interactions as in the in vivo tissue microenvironment to perform proper behaviors including apoptosis, differentiation, growth, proliferation, growth, and invasion. In human tissues, cell signaling biomolecules such as soluble growth factors, transcription factors, and ligands to cell receptors create a complex meshwork inside the extracellular matrix (ECM). Further, the cellular microenvironment in scaffolds should allow for dynamic cell regulations and interactions that guide cells into proper morphology, physiology, and other behaviors. Taking a wound healing device as an example, successful procedures of applying scaffolds should be able to imitate the different microenvironments in the cascaded stages of wound healing in order to guide skin regeneration and ideally speed up the overall healing process.

There are a broad range of technologies developed for tissue engineering which have been developed in the past few decades. For example, microcarrier technologies produce many functional cells with specific phenotypes and proper biocompatibility. Microtissues from natural microcarriers can also be transplanted into a bioreactor to produce large-scale tissues and organs as illustrated in Fig. 11.1 which provide many new opportunities for both tissue engineering and regenerative medicine.

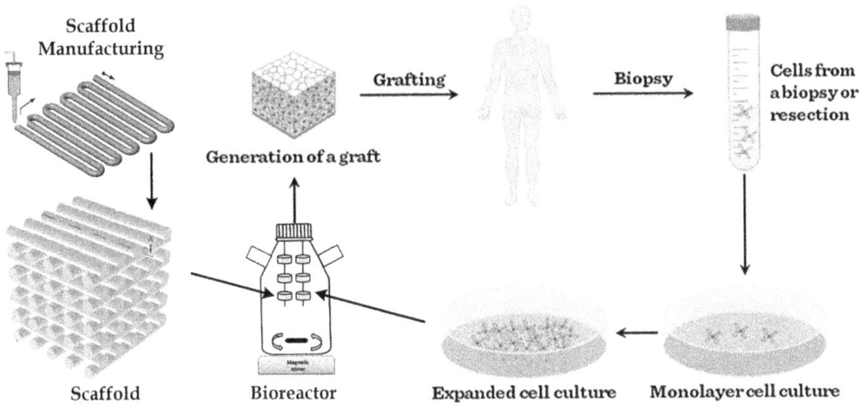

Fig. 11.1 Basic workflow of tissue engineering implementation

11.2 Scaffold Design Factors

11.2.1 Basic Requirements

Scaffolds are structures that imitate the form of the tissue/organ that is to be replaced. They provide the extracellular matrix environment in which cells can grow to form tissues. Before a scaffold is used, it needs to be assessed whether a scaffold-based, a cell-based, or a cell-loaded scaffold is required. Synthetic scaffolds, loaded with proper growth factors and proteins, have been used as physical supports to capture cells, as localized hormone delivery therapies, as induction agents for growth of artificial and functional tissues, and as protective gels to help in wound healing. However, a big challenge remains for tissue engineering in searching for a reliable delivery mechanism which mimics the in vivo extracellular matrix of tissues.

The design specifications of scaffolds should include the types of viable cells and biomolecules on or being released by the scaffold surfaces. The scaffold supports cell migration, growth, and differentiation while also guiding tissue development and organization. Engineering of scaffolds is a complicated task since there are many design parameters to take into consideration, including micro-scale architecture, structural mechanics, surface conditions, degradation by-products, biological components, etc., and these parameters change over time. Any scaffold must provide sufficient mechanical strength and stiffness to compensate for the loss of the mechanical functions of the targeted diseased or damaged tissue. A good scaffold has surface properties that allow for the attachment and migration of target cells. Scaffolds can be generated by manufacturing porous architectures of biocompatible materials, decellularizing extracellular matrix from live tissues, phase separation, or controlled crosslinking of hydrogels. In summary, a scaffold used in tissue engineering has to meet three primary requirements:

- Biocompatibility: it cannot be toxic to cells nor elicit severe inflammatory responses. It also must provide biomimetic niche signaling for cells.
- Biodegradability: after usage, it can be easily disposed as the new cells produce their own extracellular matrices. Also, the biodegraded product cannot be toxic to cells.
- Mechanical properties: properties of the scaffold should be consistent with the implantation anatomical site and must be sturdy enough to allow for surgical handling.

11.2.2 Extracellular Matrix (ECM)

Cell growth and differentiation, especially in three-dimensional tissue cultivation, require the presence of a structured environment to allow for interactions among cells. These interactions occur in the ECM, a polymeric network of several types of macromolecules in combination with smaller molecules and ions. It should be reiterated that such a cellular microenvironment gives rise to the complicated dynamic regulations that dictate cell morphology and physiology. By anchoring cells, the ECM functions through enzymes, effector molecules, growth factors, diffusion gradients, and chemotaxis. ECM is composed of fibrous proteins such as collagens, elastin, fibrillin, fibronectin, laminin, and hydrophilic proteoglycans. It is assembled by cells and can be modified by them as they proliferate, differentiate, and migrate. For example, the molecular structure of collagen fibers is described in Fig. 11.2. Through protein, particularly motifs, the receptor-to-ligand binding initiates a biochemical pathway that can change cell morphology or cause a chemical reaction to take place. The stiffness of the matrix creates a mechanical signal which has an overall effect on the activation of chemical pathways. This reaction to stimuli can be classified under a current studying field of mechanotransduction. The key variables associated with ECM in scaffold design are dimensions, pore size, stiffness, and permeability.

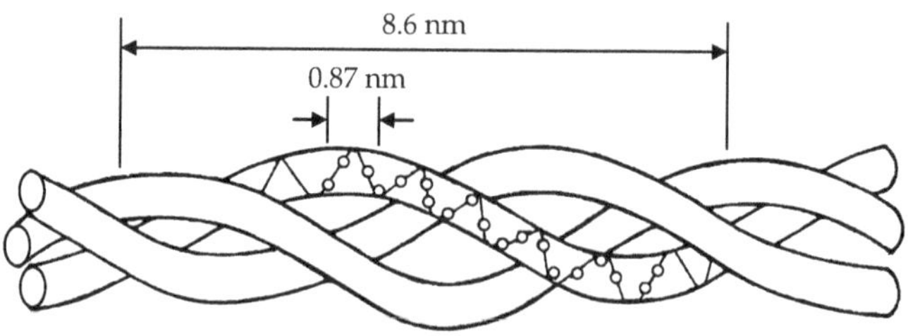

Fig. 11.2 Basic molecular structure of collagen fibers

11.2.3 Design Parameters

The ideal scaffold should have an interconnected pore structure and high porosity to allow cells to permeate freely inside the structure and allow for diffusion of nutrients. The porous structure should also diffuse waste by-products without any chemical reactions interfering with the surrounding environment. Pore size needs to be small enough to deter large macromolecules but large enough to let cells migrate altogether. There should ideally be a minimal ligand density to establish an effective binding of large numbers of cells to scaffolds [10].

Pore Size

The size of pores in scaffolds can be classified as three groups. Macro-pores are defined as above 50 μm whose scale influences tissue function. Pores greater than 300 μm are used for bone growth as well as vascularization. On the other hand, micropores, or those below 50 μm, influence cell function such as cell attachment. Mammalian cells are approximately 10–20 μm in diameter. Nano-porosity refers to the surface textures and/or pore architectures on a nanoscale or those between 1 and 1000 nm. The macro- and micro-porosities allow for the presence of cells, irrespective of seeding methods used. However, pore size can regulate some cell behaviors. Researchers have applied fused deposition modeling (FDM) to engineer three-dimensional scaffolds containing a homogeneous 600 μm pore/gap size (Fig. 11.3a) or an anisotropic pore size gradient (Fig. 11.3b) with pore/gap sizes from 200 μm, 600 μm, and 1600 μm from the superficial layer to the inner layer of the scaffold body. After seeding human chondrocytes on the top surface, the positions of the majority of cells have been shown to be different for the two scaffold types with a short migration distance and a smaller migration rate for the scaffold with a porosity gradient as its surface layer has a smaller pore size.

Pore Interconnectivity and Accessibility

A scaffold may be porous and should be interconnecting. Otherwise, it will not provide any accessibility to cell migration and rapid molecular transport other than the relatively slow molecular diffusion via the scaffold materials. The

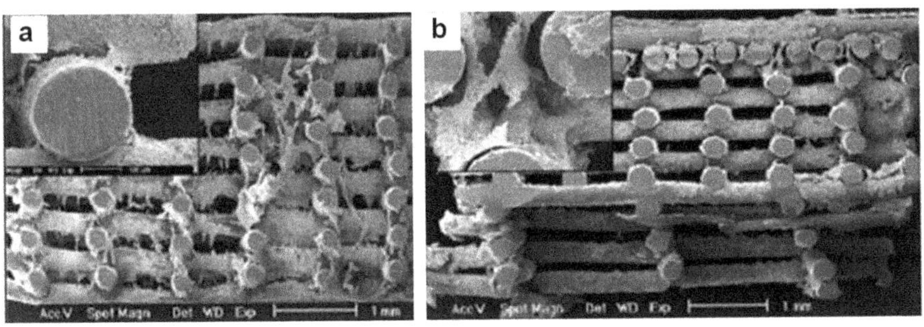

Fig. 11.3 Cell seeds in artificial scaffolds with (**a**) uniform and (**b**) gradient porosities from the surface to the inner scaffold body. (Woodfield [13], Tissue Engineering. Adapted with permission)

interconnecting pores need to be large enough to allow cell migration and proliferation. Ideally, scaffolds should all be interconnecting in their pore volume so that they can maximize the diffusion gradient and chemotaxis. Here, we introduce the concept of interconnectivity to represent the proportion of pores which can always offer a vacant path to connect to all other pores within the interconnectivity region. The level of interconnectivity can be considered as the volume of the largest interconnected region relative to the entire vacant pore volume of the scaffold. In general, a scaffold should have a significant number of pores over its surface, implying that a 100% interconnectivity would ensure molecules/particulars from outside have a chance to reach all pores in the scaffold via diffusion or other transport means. We further consider these pores as the "accessible" pores and consider the volume of accessible pores related to the total pore volume as the level of accessibility. For instance, we may consider an ideal scaffold with 100% accessibility and 100% interconnectivity as shown in Fig. 11.4a, as we can always find a path connecting any two pores in the vacant regions in the scaffolds and always find a path connecting any pore to the environment outside. In case there is a blocking from the interconnection between two pores (Fig. 11.4b, highlighted with a square), the corresponding interconnectivity should be <100%. However, as long as all pores are accessible to the exterior environment, the accessibility can be still 100%, even though the interconnectivity is not. In some cases, if there is an isolated pore region inside a scaffold (Fig. 11.4c, outlined with dotted lines), neither the accessibility nor interconnectivity is ideal.

Mechanical Properties: Random Versus Designed Architectures

Most scaffold designs need a high degree of porosity to allow for mass transfer and tissue development. In other words, the volume fraction of the scaffold material should be low. Equally as important, scaffolds should possess adequate stiffness and strength for physical support in some applications, meaning that the volume fraction should be sufficiently high. Hence, there is a need to evaluate the relationship between porosity and the mechanical properties of the scaffold. Additionally, a scaffold must offer the maximal transport of cells, nutrients, and secreted molecules during the target tissue regeneration process. As a very basic prediction of the bulk

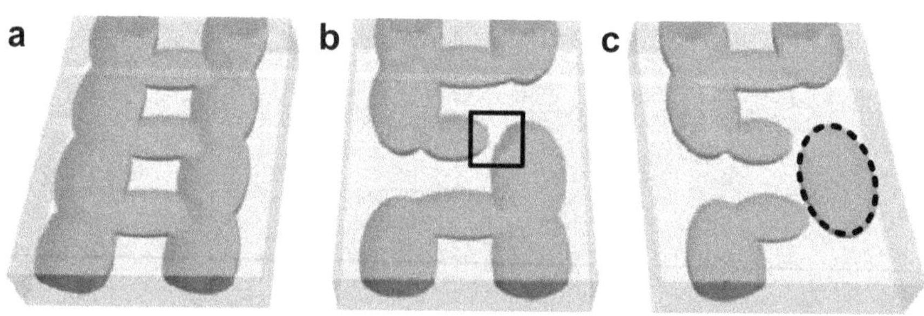

Fig. 11.4 Examples of pore structures with different pore interconnectivities. (**a**) 100% accessibility and 100% interconnectivity. (**b**) 100% accessibility and <100% interconnectivity. (**c**) <100% accessibility and <100% interconnectivity

stiffness and strength of a scaffold, we may consider compressive stiffness (E) and yield strength in compression (σ_S) to be each related to the porosity β_p (between 0 and 1) with the following relations:

$$E = E_o\left(1 - \beta_p\right)^{n_p} \tag{11.1}$$

$$\sigma_S = \sigma_{So}\left(1 - \beta_p\right)^{n_p} \tag{11.2}$$

where E_o and σ_{So} are the modulus of elasticity and strength of the scaffold material (i.e., for $\beta_p = 0$), respectively, n_p is a constant, and $(1 - \beta_p)$ is the volume fraction of material relative to the bulk scaffold volume. Importantly, n_p is dependent on architecture of micro-/nanostructures inside the scaffold; very rough estimates of the exponent values are $n_p \approx 2$ and $n_p \approx 3$ for scaffolds with "designed" and "random" architectures, respectively.

A "designed" architecture of scaffolds represents microstructures inside the scaffold set repeatedly inside the scaffold body. For example, the structure shown in Fig. 11.3 is a designed architecture. On the other hand, "random" architecture shows no predicted geometrical or structural features in the scaffold, i.e., these features are defined randomly during the scaffold formation process. Generally, we can maintain a higher level of accessibility and interconnectivity for designed architectures as we can precisely define the microstructures inside the scaffolds facilitating cell migration with a higher effectiveness, as illustrated in Fig. 11.5. Taking the porosity $\beta_p = 0.8$, for example, the designed scaffold architecture could then have a stiffness and a strength around five times compared to those of the random architecture. Anyhow, the above two relations are just very rough estimates, and more precise modeling should be done through very careful theoretical analysis, solid free-form fabrication techniques, or computational topology design techniques. In recent years, there have been more and more successful application cases of applying biodegradable, highly porous scaffolds, and the corresponding variation in structural stiffness and strength of the scaffold-cell structure during the tissue regeneration process has continued to be a challenging topic.

Biochemical Properties: Growth Factor and Molecule Release

Growth factor (GF) simulates cellular growth, proliferation, wound and tissue regeneration, and cellular differentiation. Current research suggests that spatiotemporal control on the location and bioactivity of growth factors after being introduced into the body can lead to tangible therapeutic effects. The goal of regenerative medicine is to reinstall the developmental processes, restore tissue integrity, and maintain functionality. Thus, GFs are imperative to regenerative medicine due to their role in signaling along with other molecules in cell development. GFs have a variety of effects on cells, such as chemotaxis (onset of direct migration), mitogenesis (simulation of cell division), morphogenesis (induction of cellular differentiation), apoptosis (initiation of programmed cell death), and metabolism (modulation of metabolic activity). Some widely used growth factors are listed in Table 11.1.

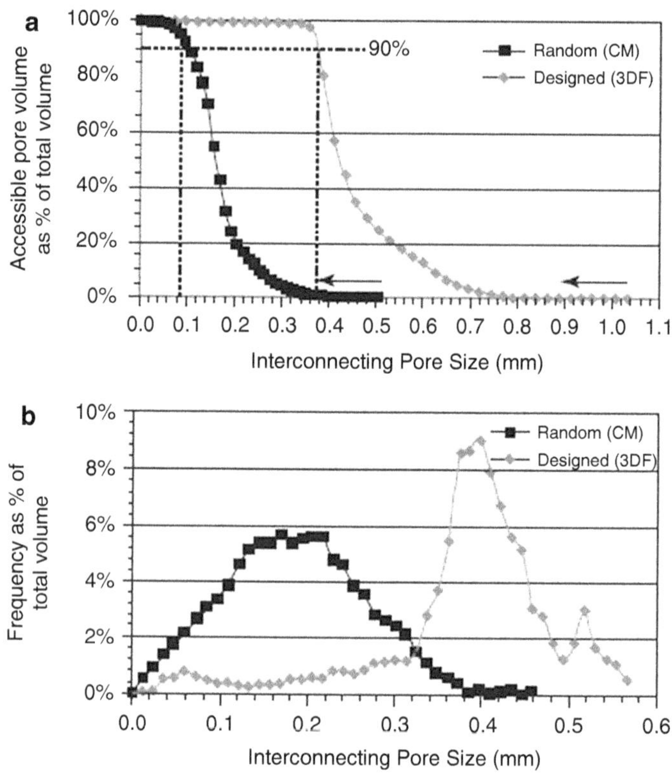

Fig. 11.5 (**a**) Accessibility and (**b**) occurrence of interconnecting pores in scaffolds as functions of the pore size. (Woodfield [13], Tissue Engineering. Adapted with permission)

The ideal carrier for GFs should be non-cytotoxic and biodegradable and prepared through a feasible process that does not inhibit protein activity. A sufficiently robust protein coat can restrict the protein's conformational mobility, but it can also protect it from physical and chemical degradation. GF carriers should also have a high loading efficiency and a controlled release profile. The carrier should also be able to target the desired site and excrete in a time manner that mimics the temporal profile of the usual healing process. GF carriers need to be made from elastic materials that conform to the size and shape of wounds while also possessing mechanical support to withstand in vivo internal forces. Bioactivity for cellular attachment and migration should also be taken into consideration. The elastic material has to be non-immunogenic and readily absorbed after the tissue regenerates.

Growth factors and other molecules can be embedded in a polymeric gel as the structural scaffold material. The gel is a chemically crosslinked molecular network swollen in a solvent. Some polymer solutions, such as solid proteins and animal tissues, will increase in viscosity when forming gels. Gels cannot be dissolved or melted. They can be considered both concentrated polymer solutions and also high

Table 11.1 Commonly used growth factors

Abbreviation	Tissues treated	Representative function
Ang-1	Blood vessel, heart, muscle	Blood vessel maturation and stability
Ang-2	Blood vessel	Destabilize, regress, and disassociate endothelial cells from surrounding tissues
Fgf-2	Blood vessel, bone, skin, nerve, spine, muscle	Migration, proliferation, and survival of endothelial cells, inhibition of differentiation of embryonic stem cells
Bmp-2	Bone, cartilage	Differentiation and migration of osteoblasts
Bmp-7	Bone, cartilage, kidney	Differentiation and migration of osteoblasts, renal development
Egf	Skin, nerve	Regulation of epithelial cell growth, proliferation, and differentiation
Epo	Nerve, spine, wound healing	Promoting the survival of red blood cells and development of precursors to red blood cells.
Hgf	Bone, liver, muscle	Proliferation, migration, and differentiation of mesenchymal stem cells
Igf-1	Muscle, bone, cartilage, bone, liver, lung, kidney, nerve, skin	Cell proliferation and inhibition of cell apoptosis
Ngf	Nerve, spine, brain	Survival and proliferation of neural cells
PDGF-AB (or -BB)	Blood vessel, muscle, bone, cartilage, skin	Embryonic development, proliferation, migration, growth of endothelial cells
TGF-α	Brain, skin	Proliferation of basal cells or neural cells
TGF-β	Bone, cartilage	Proliferation and differentiation of bone-forming cells, anti-proliferative factor for epithelial cells
Vegf	Blood vessel	Migration, proliferation, and survival of endothelial cells

elastic solids with small molecules such as water permeating and diffusing in them. Although the crosslinked polymer cannot be dissolved by a solvent, it can absorb some solvents to swell and then form a gel. In particular, when a scaffold made of gel containing growth factor swells, a swelling-controlled release process occurs where diffusion of molecules from the gel is faster than the gel swelling.

During the swelling process, solvent molecules permeate into the polymer network and expand its volume. At the same time, this expansion of volume leads to the extension of polymer chains in three dimensions. The expanded molecular network endures stress and produces elastic shrinkage energy to shrink the network. When these two opposite energies counteract each other, the swelling reaches equilibrium. The swelling ratio for a crosslinked polymer Q is defined as the volume at swelling equilibrium to that before swelling. Clearly,

$$\lambda_x = \lambda_y = \lambda_z = Q^{1/3} \tag{11.3}$$

where λ_x, λ_y, and λ_z are the extension ratio along one of the three-dimensional directions. The value of Q depends on temperature, pressure, degree of crosslinking, and the properties of the solvent and solute. The quantitative relationship of Q can be derived from quasi-lattice solution theory and statistic theory for rubber elasticity.

The change in free energy of polymeric molecules during swelling ΔG includes two parts: (1) the change in free energy in the mixing of polymer and solvent ΔG_{M} and (2) the elastic free energy of the molecular network ΔG_{E}, i.e., $\Delta G = \Delta G_{\mathrm{M}} + \Delta G_{\mathrm{E}}$. Recall Eq. 2.24 on the elastic modulus E as a function of temperature T and Eq. 4.25 for the strain energy of rubbery elastic materials:

$$\begin{aligned} \Delta G_{\mathrm{E}} &= -\Delta U_{\mathrm{n}} \approx (-\Delta E/6)\left(\lambda_x^{\,2} + \lambda_y^{\,2} + \lambda_z^{\,2} - 3\right) \\ &= (\eta_{\mathrm{T}} T/6)\left(\lambda_x^{\,2} + \lambda_y^{\,2} + \lambda_z^{\,2} - 3\right) \end{aligned} \tag{11.4}$$

where η_{T} is the temperature coefficient. Note that there should be no elastic free energy at a temperature of 0 K. Although the above relation is obtained from thermodynamic concepts, a brief understanding is that a higher elastic free energy represents less stretched polymer chains and hence the lower strain energy. Substituting Eq. 11.3,

$$\Delta G_{\mathrm{E}} \approx (\eta_{\mathrm{T}} T/2)\left(Q^{2/3} - 1\right) \tag{11.5}$$

A more precise description of this relation for polymers would be

$$\Delta G_{\mathrm{E}} = \frac{3\rho_{\mathrm{P}} R_{\mathrm{g}} T}{2M_{\mathrm{P}}}\left(Q^{2/3} - 1\right), \tag{11.6}$$

where ρ_{P} is the density of the polymer, R_{g} (= 8.31 J K^{-1} mol^{-1}) is the gas constant, and M_{P} is the mean molecular weight of the effective chains.

Besides those, ΔG_{M} can be described by the Flory-Huggins theory derived from the statistical thermodynamic relations which are out of this book's scope. It can be expected that ΔG_{M} of free-moving solvent molecules is proportional to the temperature T; and the actual relation is

$$\Delta G_{\mathrm{M}} = R_{\mathrm{g}} T(n_{\mathrm{S}} \ln \varphi_{\mathrm{S}} + n_{\mathrm{P}} \ln \varphi_{\mathrm{P}} + \chi_{\mathrm{S}} n_{\mathrm{S}} \varphi_{\mathrm{P}}) \tag{11.7}$$

where n_{S} and n_{P} are the mole fractions of solvent and polymer, respectively; φ_{S} and φ_{P} represent the volume fractions of solvent and polymer, respectively, with the relation $\varphi_{\mathrm{S}} + \varphi_{\mathrm{P}} = 1$; and χ_{S} is the dimensionless Flory-Huggins parameter, dependent on the polymer-solvent interactions, with a typical value ranging from −0.5 to 0.5. For example, χ_{S} has the values 0.37 for polystyrene in toluene, 0.40 for polyisoprene in benzene, 0.14 for cellulose in acetone, 0.19 for poly(ethylene oxide) in benzene, 0.45 for poly(dimethyl siloxane in toluene) in toluene, and 0.40 for poly (ethylene oxide) in water. Furthermore, for a unit cubic polymeric material after swelling, $n_{\mathrm{P}}\underline{V}_{\mathrm{P}} = 1$, where $\underline{V}_{\mathrm{P}}$ is the molar volume of polymer; and φ_{P} is the ratio of initial unit polymer volume. The resultant volume after swelling, as a function of the molar volume of solvent $\underline{V}_S$, is

$$\varphi_P = 1/(1 + n_S\underline{V}_S), \text{ implying } \varphi_S = n_S\underline{V}_S/(1 + n_S\underline{V}_S). \tag{11.8}$$

Recalling that the swelling ratio Q is the ratio of volume after and before swelling, clearly we can obtain

$$Q = (\varphi_P + \varphi_S)/\varphi_P = {\varphi_P}^{-1}. \tag{11.9}$$

Now, we may consider that the swelling process of a polymer has reached a stabilized maximum level. Then, the total free energy of the polymeric molecules ΔG is no longer dependent on n_S. Substituting Eq. 11.8, this means

$$\frac{\partial \Delta G}{\partial n_S} = \frac{\partial \Delta G_M}{\partial n_S} + \frac{\partial \Delta G_E}{\partial n_S} = 0, \text{ and thus} \tag{11.10}$$

$$R_gT\left\{\ln(1-\varphi_P) + \varphi_P - \frac{\underline{V}_S}{\underline{V}_P}{\varphi_P}^2 + \chi_S{\varphi_P}^2\right\} + \left[R_gT\left(-\frac{\rho_P}{M_P}\right){\varphi_P}^{-5/3}\right](-\underline{V}_S{\varphi_P}^2) = 0 \tag{11.11}$$

Considering the molecular sizes of solvent and polymer molecules, $V_p \gg V_S$,

$$\ln(1-\varphi_P) + \varphi_P + \chi_S{\varphi_P}^2 \approx -\left(\frac{\rho_P\underline{V}_S}{M_P}\right){\varphi_P}^{1/3}. \tag{11.12}$$

Typically, the swelling ratio is large, i.e., $Q \gg 1$, such that $\varphi_P \ll 1$. We can then approximate $\ln(1 - \varphi_P)$ by the Taylor expansion "$\ln(1 - x) = -x - x^2/2 - x^3/3 - \ldots$," with the expression

$$\left(-\frac{M_P}{\rho_P\underline{V}_S}\right)\left[\frac{\ln(1-\varphi_P) + \varphi_P}{{\varphi_P}^2} + \chi_S\right] = {\varphi_P}^{-5/3} \tag{11.13}$$

$$\frac{M_P}{\rho_P\underline{V}_S}\left(\frac{1}{2} - \chi_S\right) \approx Q^{5/3} \tag{11.14}$$

Notably, the volume and pore size of gel scaffolds after applying the cell culture are different from when they are first prepared in dry conditions. In case a scaffold is implanted in the human body, the swelling and molecule release take place at the implantation site. Though the swelling ratio of gel is high, the scaffold porosity can offer sufficient vacancy for the volume increase during swelling. When the implanted site has only limited space and constraint for the gel swelling, the swollen scaffold has to have a reduced size and thus negative strains. Residual stresses would then establish in the scaffold material.

11.3 Scaffold Fabrication: Electrospinning

11.3.1 Working Principle

For designing the architecture of scaffolds, we may manufacture the required microstructure using the rapid prototyping-related technologies. However, the manufacturing speed can be slow. Alternatively, if we can sacrifice a level of structural precision, we may be able to find manufacturing processes for random porous architectures of biocompatible materials with a higher yield.

Electrospinning is a process used to create fibrous scaffolds for tissue culture. These fibrous scaffolds mimic the size and functionality of the ECM that they replicate and can also be used to dictate cellular behavior, including, but not limited, to bone repair and stem cell differentiation. This process was first investigated by C.V. Boys in 1897. The first successful use of electrospinning to make fibers was performed by Formhals in 1934 [12]. However, it was only in the early twenty-first century that electrospinning would be explored for widespread use as a polymer processing method for applications in tissue engineering and drug delivery. In general, electrospinning relies on the application of electrostatic force using the phenomenon of electrospray which conceptualizes that a droplet with high charge will be torn into small droplets when passing through a voltage gradient. With the help of electrostatic force, electrospinning can produce fibers with extremely small diameters (upper range of natural 50–500 nm) which lead to a higher surface area when compared with other methods. A basic electrospinning device as shown in Fig. 11.6 contains three key components: a feeding unit, a power supply, and a collection unit. A feeding unit is usually a syringe or pipette with an extremely small output needle in which the continuous solution flow is driven with a pump. A power supply is connected to the needle and the collection unit. Normally, a voltage of 15–25 kV is provided with voltages of up to 50 kV in some processes. This is used to charge the solution in order to force it to travel through and tear from the needle. A

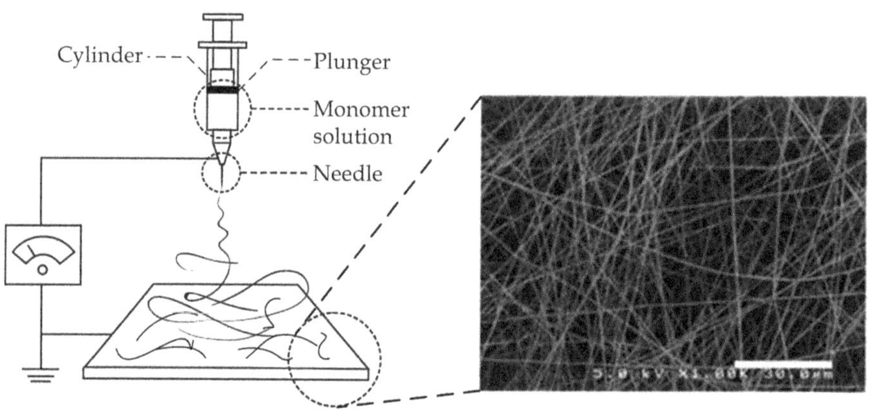

Fig. 11.6 Basic setup of electrospinning

collection unit usually connects to the ground of the power supply. This unit is used to collect the fibrous products. The connector can be configured with different shapes and charged modes for different expected nanofiber architectures.

In electrospinning (Fig. 11.7), the material used, whether it is a polymer, a material dissolved in proper solvent, or a material melted into liquid form, is ejected from a syringe under a continuous pumping. The solution is gradually sent to the feeding unit. An interesting phenomenon takes place due to the interaction of the application of voltage on an electroconductive medium. Initially, the shape of the solution at the tip is dictated by the surface tension, taking on a hemispherical shape. However, as the electrical field at the tip increases, Coulombic forces lead to the separation of ions, elongating the hemispherical solution into a Taylor cone. When the voltage exceeds the critical value, the solution's surface tension force is overcome by the repulsive electric forces. Hence, the charged solution is discharged from the surface of the Taylor cone and travels to the grounded surface as a jet. Beyond the conical base and immediately at the end of the capillary tip, the jet continues to become thinner. This jetting mode is known as the electrohydrodynamic cone jet. The jet will initially travel in a straight line toward the collector but will eventually become unstable. The splaying is actually one single fiber rapidly bending or whipping, causing the fiber to make lateral exclusions that grow into spiraling loops. As the jet travels further, it becomes thinner and separates out because of the Coulombic repulsion, leading to the random deposition and formation of a nonwoven nano-fibrous mat.

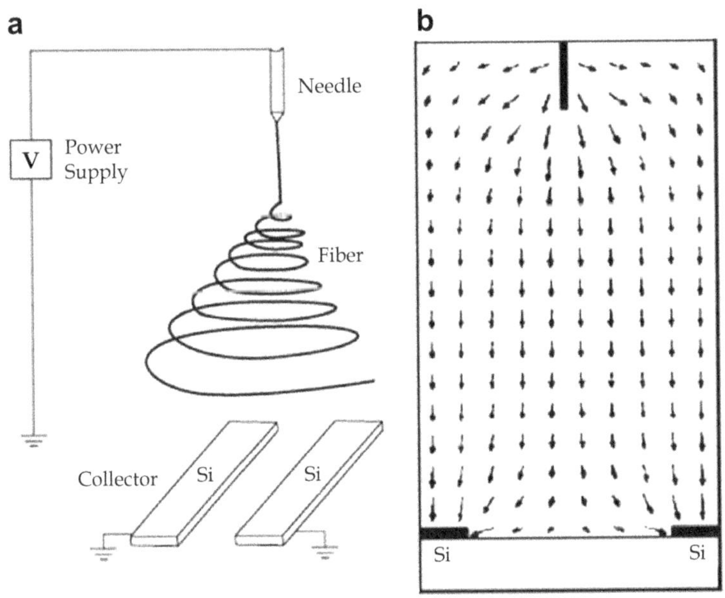

Fig. 11.7 Sample (**a**) configuration and (**b**) electric field in electrospinning

In tissue engineering, electrospinning is used for the fabrication of nanofibric scaffolds from a variety of natural molecules, such as collagen, chitosan, silk fibroin, and fibrinogen. Further synthetic polymers such as poly L-lactic acid, polylactide-co-glycolide, poly(ethylene terephthalate) (PET), and poly-ε-caprolactone (PCL) are also used in the scaffold fabrication. Electrospun three-dimensional nano-fibrous structures meet the essential design criteria of an ideal tissue-engineered scaffold based upon their unique action in supporting and guiding cell growth. Electrospun nano-fibrous structures are capable of supporting cell attachment and proliferation. These structures feature a morphological similarity to the extracellular matrix of natural tissues, which is characterized by a wide range of pore diameter distribution, high porosity, and effective mechanical properties.

Electrospun nano-fibrous membranes can also be used as novel wound dressings. Histological examinations indicate that the dermis becomes well organized when wounds are covered with electrospun nano-fibrous membrane. There, membranes are particularly important because of their favorable properties, such as high specific surface area combined with antibacterial and drug release functionality if degradable polymeric materials are used. Nano-fibrous dressings could promote hemostasis; have better absorptivity, semipermeability, and conformability; and allow for scar-free healing. Nano-fibrous materials also allow for controlled evaporative water loss, excellent oxygen permeability, and promoted fluid drainage ability, but they can still inhibit exogenous microorganism invasion due to their ultra-fine pores.

11.3.2 Process Parameters

The fiber morphology formed in electrospinning has been shown to be dependent on solution properties, process conditions, and ambient conditions. The choice of solvents directly affects a wide range of properties, including the resultant fiber morphology. The polymer solution must have a concentration high enough to cause polymer entanglements, yet not so high that viscosity prevents polymer motion induced by the electric field. The resulting fibers' diameters usually increase with the concentration of the solution according to a power-law relationship. Concentration, surface tension, and charge of the liquid jet do affect the formation of beads along nanofibers which is a common problem in nanoscale fiber fabrication by electrospinning. Decreasing the polymer concentration in the solution produces thinner fibers. However, a concentration below a threshold value causes the formation of beads instead of fibers. The main factors affecting the formation of beads have been shown to be solution viscosity, surface tension, and the net charge density carried by the electrospinning jet. High viscosity and net charge density with low surface tension would favor fiber formation and fewer beads. Surface tension can be decreased by adding a surfactant to the solution. The reduction in fiber thickness is due to the solution conductivity which reflects the charge density of the jet and thus the elongation level. The net charge density can be induced by adding salt in the solution.

The parameters in the process are spinning voltage, distance between the tip of the capillary and the collector, solution flow rate (feed rate), needle diameter, and motion of the capillary tip during process. Voltage and feed rate show different tendencies and are less effective in controlling fiber morphology as compared to the solution properties. A high voltage might result in splaying and irregularities in the fibers. A bead structure is formed when the voltage is either too high or too low. A high voltage also leads to a high evaporation rate of the solvent which in turn might further lead to solidification at the tip and instability in the jet. Morphological changes in the nanofibers can also occur upon changing the distance between the syringe needle and the substrate. Increasing the distance or decreasing the electrical field reduces the bead density, regardless of the concentration of the polymer in the solution.

Ambient conditions include humidity, temperature, air velocity in the spinning chamber, and atmospheric pressure. For example, the humidity and temperature of the chamber in which the electrospinning is carried out are important factors that affect the diameter of synthesized nanofibers. Humidity primarily controls the formation of pores on the surface of the fibers. The precise mechanism behind the formation of pores and texturing on the surface is complex and is dependent on a combination of molecular aggregation and phase separation.

Here, we look into the relations of a few process parameters during electrospinning such as supplied voltage, solution viscosity, and temperature. The spinning velocity can be estimated by the concept of power balance:

$$I_e V_e = Q_s {v_s}^2/2 \tag{11.15}$$

where I_e and V_e are the current and voltage between the syringe needle and the collector plate, Q_s is the liquid flow rate in the syringe, and v_s is the spinning velocity. The spinning velocity can actually be close to the velocity of sound in air. In some electrospinning processes, the spinning velocity of the fibers reaches 140–160 ms^{-1}.

Additionally, a reduced viscosity has been reported to be a major factor in crosslink formation along the fiber length. The fiber diameter depends on how fast and how far individual polymer chains can move among each other along with the solution flow. Liquids of small molecules and most dilute polymer solutions act as Newtonian fluids, whereas the viscosity of polymer melts or concentrated solutions may not flow according to the law of Newton viscosity. These fluids are known as non-Newtonian fluids. Only when the shear stress is very low do these fluids behave as Newtonian fluids.

After dissolution of a polymer in a solvent, the solution viscosity is often higher than that of the pure solvent as viscosity increases with concentration. In order to predict viscosity of polymeric solution, we first consider an "intrinsic viscosity," denoted by $[\mu]$, to represent the contribution of polymer to the solution viscosity

$$[\mu] \approx \frac{\mu - \mu_0}{c_P \mu_0} \tag{11.16}$$

where μ_0 is the solvent viscosity, μ is viscosity of the resultant solution, and c_P is the polymer concentration in the solution, with a small value of c_P.

As a basic understanding of the viscosity increment for high-molecular-weight polymers dissolved in a solvent, we may first consider the classical Einstein's viscosity relationship by further assuming polymer molecules as spherical, impenetrable particles at a very low density:

$$[\mu] = \frac{5}{2}\frac{N_P}{c_P}\frac{V_P}{M_P} \tag{11.17}$$

where N_P is the number of polymer particles in the solvent, V_P is the volume of a polymer particle, and M_P is the mean molecular weight. We may further express V_P as the effective radius of polymeric molecules R_P, i.e., $V_P = (4/3)\pi {R_P}^3$. Then,

$$[\mu] = \frac{10\pi}{3}\frac{N_P}{c_P}\frac{{R_P}^3}{M_P} \tag{11.18}$$

This equation is still highly representative for dissolved polymers if the hard sphere radius is replaced by the hydrodynamic radius of the polymer coils. Now, we further consider the case that the solvent does not affect the intramolecular forces and molecular structure of the polymer, such that R_P can be maintained as the original molecular radius R_{P0} in its dry condition. We call this type of solvent the θ-solvent. The corresponding intrinsic viscosity $[\mu]_\theta$ is

$$[\mu]_\theta = \frac{10\pi}{3}\frac{N_P}{c_P}\frac{{R_{P0}}^3}{M_P} = \Phi_\theta \frac{{R_{P0}}^3}{M_P} \tag{11.19}$$

It has been measured that the Φ_θ has a constant value of about 4.2×10^{24}. However, many solvents would increase the molecular radius of dissolved polymer, and thus, we can define the expansion factor of the molecular radius α_P as

$$ {\alpha_P}^3 = [\mu]/[\mu]_\theta \tag{11.20}$$

The value of α_P has a relationship with the degree of polymerization (DP), calculated by M_P/M_0 where M_0 is the molecular weight of the corresponding monomer:

$$ {\alpha_P}^3 = \Gamma_P (M_P/M_0)^{(a_P+1)} \tag{11.21}$$

where Γ_P is a constant depending on the polymer type and environment conditions and a_P depends on the solvent-polymer interactions. Typically, $a_P = 0$ for θ-solvent; and $a_P \approx 0.1$ or smaller for good solvent. By substituting Eqs. 11.21 and 11.19 into Eq. 11.18, the intrinsic viscosity becomes

$$[\mu] = \Phi_\theta \frac{\Gamma_{\mathrm{P}} R_{\mathrm{P0}}{}^3}{M_0{}^{(a_{\mathrm{P}}+1)}} M_{\mathrm{P}}{}^{a_{\mathrm{P}}} = K_{\mathrm{P}} M_{\mathrm{P}}{}^{a_{\mathrm{P}}} \tag{11.22}$$

where $K_{\mathrm{P}} = \Phi_\theta \Gamma_{\mathrm{P}} R_{\mathrm{P0}}{}^3 / M_0{}^{(a_{\mathrm{P}}+1)}$. The above equation is known as the Mark-Houwink or Mark-Houwink-Staudinger equation. The empirical values of K_{P} and a_{P} can be obtained from some handbooks of polymer materials.

In addition, the ambient temperature in which the spinning is performed has a substantial effect on the fibers produced as the solvent evaporation rate increases. In a warmer environment, individual molecules of solvent will have more energy to jump from the liquid to gaseous form. On the other hand, viscosity decreases with an increase in temperature. This is caused by the fact that a higher temperature induces softening of polymers. The polymer chains have a greater degree of freedom to move due to the increase in energy of breaking or weakening crosslinks, van der Waals forces, and hydrogen bonding between the chains. The viscosity μ upon change of temperature can be modeled using Barr's approximation of Vogel's original equation for the viscosity of hydrocarbon-based liquids over a range of temperature T (unit: K):

$$\mu = K_{\mathrm{T}} e^{b_T / ^T (T+\theta_{\mathrm{T}})} \tag{11.23}$$

where K_{T}, b_{T}, and θ_{T} are empirical constants.

11.3.3 Process Configurations

As mentioned in the previous section, the morphology and the diameter of the nanofibers formed can be affected by many factors: the solution factors (surface tension, viscosity, polymer structure, etc.), process factors (power supply, travel distance, feeding rate, electric field, etc.), and ambient factors (humidity temperature). The strength and configuration of the electric field (with collector shape) also have a considerable effect on the final product. This is because changes in the electric field and collector shape affect the travel and deposition format of the polymer. For example, in the vertical case, the configuration of two earth electrodes as shown in Fig. 11.7a can collect a suspension of fibers, while a single collector will lead to a homogeneous electric field and complete even (random) deposition. In addition, if a series of copper rings are added (spiral), the electric field and deposition will group inside the rings. Therefore, the distinct combination of collector and electric field can result in different formats of polymer deposition.

Some common schemes are applied in the industry as shown in Fig. 11.8. The collector can be a single ground plate (Fig. 11.8a), causing the collected electrospun fibers to be randomly distributed and the area of collection to be limited. This ground plate can be modified as a rotating ground cylinder for a larger collection area (Fig. 11.8b). The ground plate can also be modified as multiple ground bars. If

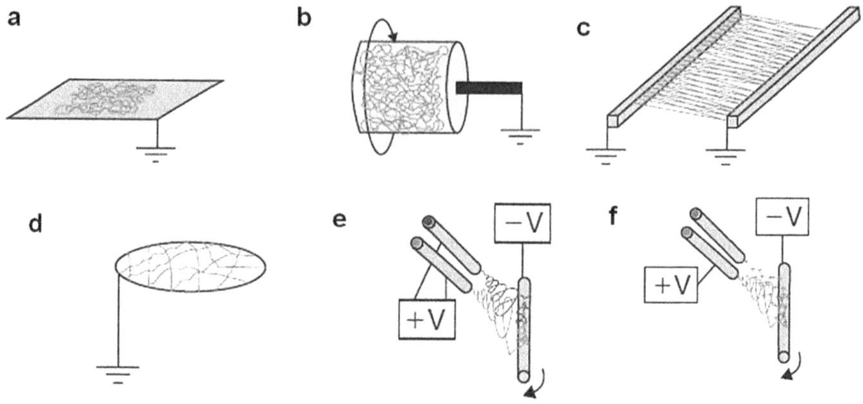

Fig. 11.8 Different fiber generation and collection schemes: (**a**) single ground plate, (**b**) rotating ground plate, (**c**) bar collector, (**d**) ring collector, (**e**) parallel spinnerets, and (**f**) electrospinning simultaneously with deposition of other particles

there are two bars connected to the ground, the collected fibers will tend to align perpendicularly to the ground bars (Fig. 11.8c). In case a fibrous mesh is desired, a single horizontal ring and the ground port can be used (Fig. 11.8d). Other than the collector side, multiple spinnerets can be used for simultaneous generation of fibers of different materials (Fig. 11.8e). The electrospinning can also be performed together with deposition of particles, including cells or powders of other materials, such that these particles can be embedded inside the electrospun scaffold (Fig. 11.8f).

11.3.4 Electrospinning of Multiple Components

The electrospinning process is highly versatile and allows not only for the processing of many different polymers into polymeric nanofibers but also for the co-processing of a polymer with other materials (either another polymer or a low-molecular-weight nonvolatile material). This is done simply by including the components in the solution for electrospinning to achieve extended nanofiber functionalities. The goals of this process include adjustment of the fiber morphology and characteristics. Polymer blends, core-shell structures, and side-by-side biocomponent electrospinning are some application examples. Another interesting aspect of nanofiber processing is that it is feasible to modify surface structures in order to create various chemically reactive functionalities.

The electrospinning technique also provides the capacity to lace together a variety of types of nanoparticles or nano-fillers to be encapsulated into an electrospun nanofiber matrix. Several functional components (e.g., nanometer-sized particles, nano-fillers, carbon nanotubes, drugs, enzymes, and DNA) can be dispersed in the initial polymer solutions which are then electrospun to form composites in the form of continuous nanofibers and nano-fibrous assembles. When the nano-fibrous materials are implanted into a human body, the polymeric component will swell and release the secondary components such as drugs, enzymes, and DNA to provide further treatment capability of the implanted device. Electrospun micro- or nanofibers support rapid reactions of enzymes and catalysts due to their ultra-thin sizes and large surface areas.

Electrospun fiber mats have also been explored as drug delivery vehicles with promising results. The application of electrostatic spinning in pharmaceutical applications has resulted in dosage forms with useful and controllable dissolution properties. For instance, hydroxy propoxy methylcellulose, a cellulose derivative commonly used in pharmaceutical preparations, together with drugs has also been tested. Poly-(L-lactic acid) (PLLA) and poly(D,L-lactide-co-glycolide) (DLPLGA) nanofibers are other polymers that have been electrospun with an encapsulated drug and have shown promising drug release properties. Incorporation of an antibiotic in fibers developed for scaffold applications has also been reported. The combination of mechanical barriers based on nonwoven, nano-fibrous biodegradable scaffolds and their capability for local delivery of antibiotics makes them desirable for the prevention of postsurgical adhesions and infections.

In addition, unlike common electrospinning which produces random structure nanofibers, self-assembling styles involve designing the polymeric characteristics before the construction process. Molecular self-assembly has been utilized to produce supermolecular structures. Materials fabricated by this method are frequently referred to as hydrogels. Despite the self-assembly taking a longer time for production, a smaller diameter and pore sizes with higher-quality product can be generated by this method. When this style applied, scientists design the polymer structure in order to let the polymer carry out different functions, such as storing other materials, emitting materials when reacting with specific elements/chemicals, or binding to specific spots. For this purpose, scientists also design the polymeric 3D structure. Therefore, it is common to use non-covalent interactions to achieve the designed 3D structure such as ionic bonds, hydrogen bonds, van der Waals forces, etc. Since some of the self-assembling electrospun products will have contact with organisms in areas for reconstruction such as blood vessels, nerves, or bones and need a filter for environmental and chemical factors, most of the scaffold materials use biopolymers. This is because those materials can minimize the probability of triggering immune responses. Also, they can contain materials which promote bioactivity, helping to trigger growing factors.

In addition, an extended model of electrospinning is to contain "dual spinnerets" which has mainly two formats: sequential and simultaneous spinnerets. Sequential spinneret technique is similar to normal electrospinning where a layer of polymer is sprayed onto the collector. After the first layer solidifies, another layer with a different polymer is sprayed on top of the existing layer, resulting in multilayers of nanofibers with different contents. For simultaneous spinnerets, both polymers are deposited on the collector at simultaneously; meanwhile, the collector moves at high speeds such that the two different fibers tangle in the scaffold.

11.4 Characteristics of Fabrication Techniques

Scaffolds can be made by different types of fabrication techniques which can create either designed or random architectures. These fabrication techniques can determine both bulk and surface properties of the resultant scaffold. Selection of the appropriate approach is exceptionally critical to generate scaffolds fulfilling the requirements for the specific tissue regeneration site with the desired characteristics, including porosity, pore size, pore interconnectivity, cell attachment, cell distribution, molecule release rate, etc.

Up to this point, we have already discussed a number of scaffold fabrication techniques, especially for the ones with designed architectures. Rapid prototyping techniques can be applied to generate porous designed architectures by controlling the path of polymer solidification as previously described in Fig. 11.3. A few rapid prototyping techniques are discussed in Chap. 7. Fused deposition modeling (FDM) can manufacture scaffolds by printing fibers of pure polymers like polycaprolactone (PCL) and polylactic acid (PLA), which are biodegradable in human body. The printed scaffolds typically have fibers of a diameter of >200 μm, a pore size of ~150–700 μm, and a porosity of <80%. FDM can generate scaffolds with adjustable mechanical properties in the design-architectural manner (e.g., as described in Eqs. 11.1 and 11.2) for bone tissue engineering applications. In case the nozzle is replaced by a syringe needle in the FDM configuration as another 3D printing technology, polymer solvents can then be printed to form gels. As much thinner fibers can be generated after printing and solidification, the fiber diameter can be pushed down to 10–30 μm. The fibers are generally softer and weaker such that the higher porosity limitation of <60% is required. The polymer solvent can be PLA or poly(lactic-co-glycolic acid) (PLGA), and additive particles such as hydroxyapatite (HA), tricalcium phosphate (TCP), or other supplemental molecules can be added into the solvent. This 3D printing technology can generate scaffolds for bone tissue engineering and in vitro soft tissue regeneration of cells, e.g., fibroblast, epithelial cells, and vascular smooth muscle cells. The solid ground curing (SGC) methods can induce scaffolds with a pore size of 100–500 μm. This method is particularly suitable for generating thin tissue layers. It has been demonstrated that poly(ethylene glycol) diacrylate (PEG-DA), gelatin, or polysaccharides (with the photo-crosslinkers) can be used as scaffold materials for soft tissue regeneration for nerves and blood vessels. Stereolithography (SL) can also be applied to fabricate scaffolds with materials similar to those for SGC as the solidification process of SL is also based on photo-crosslinking as SGC. Pore sizes ranging from 150 to 400 μm can be generated. The typical application of SL is bone tissue regeneration. Powder-based rapid prototyping techniques can be applied by sintering microspheres of poly(lactic-co-glycolic acid) (PLGA or PLG) or other polymers. The resultant porosity is only <40%, and the pore size is ~50 to 300 μm. Moreover, a couple manufacturing techniques capable of generating porous scaffolds with random architectures are also discussed in Chap. 7, i.e., solvent casting, gas foaming, freeze-drying, and fiber bonding. Generally, these methods induce randomly aligned polymers and unpredictable size and distribution of pores. Future directions of tissue engineering include application of engineered scaffolds with clinical implementation and coupling scaffolds with cells, genes, and controlled release of drugs and molecules for promoting specific stem cell differentiation and tissue regeneration.

11.5 Implementation of In Vitro Tissue Regeneration

11.5.1 Bioreactors

Understanding the implementation procedures of using scaffolds provides important insights on how to come up with better scaffold design and proper selection of manufacturing process in the development of biomedical device. Bioreactors are chambers where cells, growth factors, and scaffolding material are placed together to incubate and develop an artificial tissue prior to implantation. Bioreactors support cell attachment and proliferation on material surfaces of the scaffold. Bioreactors also allow cell maturation and secretion of molecules, including extracellular matrices which are essential in the subsequent tissue regeneration process.

Incubator-based bioreactors can regulate nutrient flow, gaseous ions, and signaling molecule levels. Multi-pass filtration seeding can be applied to place cells in three-dimensional scaffolds. Apparently, this approach produces more uniformly distributed cells inside scaffolds than static seeding, i.e., cells which are just pipetted onto the scaffold body inside culture media. Furthermore, the flow transport of oxygen and soluble nutrients can be enhanced through culturing the construct in a stirring spinner flask (Fig. 11.9a). The spinner generates semi-controlled fluid shear which produces detrimental turbulent eddies which mix oxygen and nutrients, thereby reducing the boundary layer at the construct surface from a reduction of the diffusional limits through the generation of a dynamic laminar flow with minimal levels of shear. Other than agitation using a spinner, the chamber can be configured with a rotating wall (Fig. 11.9b) to induce low shear stresses and high mass transfer rate. The rotating wall has an advantage of balancing forces to stimulate "zero gravity" conditions, meaning that the cells, nutrients, and biowastes do not accumulate at the chamber base due to gravity.

In addition to the incubator chambers, perfusion techniques can be integrated into a bioreactor system. Culture media are continuously perfused through the scaffolds placed inside the perfusion chamber (Fig. 11.9c). The culture media that exited the chamber may then recirculate at the inlet of the chamber, or alternatively, fresh culture media can always be supplied from the inlet. In this case, the continuous flow ensures nutrients and dissolved oxygen are continuously supplied to cells in the scaffold and the biowastes from cells do not accumulate around cells. Besides, the perfusion bioreactors can be inserted with the semipermeable hollow fibers (Fig. 11.9d) which contain both cells and scaffolds with the fiber direction perpendicular to the media flow. This approach maintains the cells in the scaffolds and culture regions in the hollow fibers while also allowing for the required mass

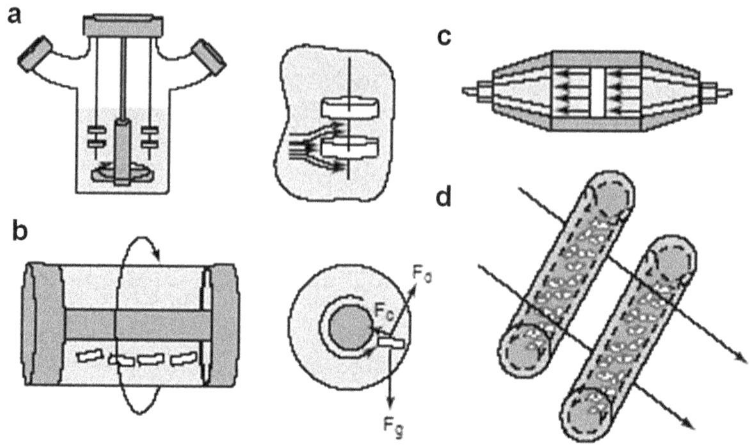

Fig. 11.9 Different configurations of bioreactors. (**a**) Incubator with spinner. (**b**) Incubator with rotary chamber wall. (**c**) Perfusion chamber. (**d**) Hollow tubes with culture media perfusion

transport of oxygen and nutrients to the highly metabolic cells. Furthermore, these bioreactors can also be applied with mechanical conditioning as the scaffold is physically contacted with the fibers. Cyclical mechanical stretch on the fibers can induce direct stresses on the scaffolds and the cells inside. Cyclical mechanical stretching can also improve the mechanical properties of some human cell types including myocytes and tenocytes. Mechanical stimulus can also enhance cell proliferation, organize the matrix structure, and stimulate expression of cell proteins like elastin and collagen.

11.5.2 Cell Preparation

In vitro tissue regeneration starts from the selection of cells. Most tissue engineering approaches are based on seeding and extensive in vitro culturing of cells within a scaffold prior to implantation. An important consideration is the choice of the cell source and the ability to control cell proliferation and differentiation in the case of stem cells. Primary cells derived from the patient's own healthy tissues (i.e., autogenic cells) could be the first obvious choice since this avoids many of the problems associated with immune rejection of foreign tissues. Primary cells derived from normal donors of the same (i.e., allogeneic cells) or different species (i.e., xenogeneic cells) are, at least in principle, readily available in sufficient quantities due to the number of potential donors and cryopreservation possibilities. However, potential rejection by the host's immune system and the possibility of disease transmission are serious risks to be considered. It is safer to use autologous cells as they are inherently more biocompatible. A more feasible approach would be to isolate the so-called progenitor cells from tissues at a healthy/undamaged site of the patient's own body (i.e., syngeneic cells). These stem cells are considered to be only partly differentiated, and therefore, their conditions for specific downstream differentiations are well known and manageable. Still, these cells are not yet fully differentiated. They stay flexible enough to provide several different cell types [3]. Bone marrow-derived mesenchymal stem cells are an example of such progenitor cells and can provide stem cells for bone and cartilage repair. These mesenchymal stem cells can differentiate so that the generic marrow cell population evolves into a specific osteogenic lineage. A number of questions have to be taken into account while devising scaffold design and selection of proper fabrication strategies, including (1) what cell source, type, population, or density should be used, (2) what type of scaffold could be used, (3) how to create the scaffold, and (4) how the regenerated cell including scaffold should be implanted in the patient's body.

The seeding of cells onto biocompatible scaffolds is a determinant step in the attainment of functionality of engineered tissues. Passive seeding (static seeding and gravitational seeding) is the simplest and most widely used seeding method, but it is the least efficient method. "Dynamic seeding" strategies have increased efficiency, uniformity, and penetration of cells in the scaffold. For example, sheet-based cell seeding creates congruous sheets of cells. Sheet-based cell seeding allows for the manipulation and removal of an intact sheet of cells without any usage of digestive enzymes to break cellular attachments. Electrostatic cell seeding manipulates electrostatic properties of vascular scaffolds. However, electrostatic cells run the risk of morphological maturation before implantation and cell retention after implantation. Magnetic cell seeding uses magnetic forces to increase cell seeding efficiency. Photo-polymerized hydrogels for cell seeding mimic the ECM properties through engineering the hydrogel in the scaffolds. Hybrid systems combine different seeding systems such as rotational systems, pressure differential system, vacuums, and bioreactor systems [5].

11.6 Application Examples

11.6.1 Artificial Skin

Deep dermal wounds caused by physical damages or burns cannot heal quickly or effectively. Despite scar formation, the sensory or movement capability of the damaged site may not be fully restored if it is not treated well. Skin autografts can be used to assist the natural healing process for better recovery of the skin's functions. Using skin cells isolated from another site on the patient's own body is ideal as there is no immune rejection. Part of the epidermis with a thin layer of dermis, also called split-thickness skin grafts, can be harvested from the patient's uninjured skin. During a split-thickness graft process, a sharp blade shaves a thin layer of skin (either epidermis or dermis) parallel to the skin surface. If the wound is large and deep, a mesh of multiple grafts might be harvested and applied over different areas to cover the entire wound. In more severe cases in which there are inadequate amounts of skin allografts or xenografts tissues available, the wound could be temporarily covered by biocompatible materials during the stage of dermal healing. This action reduces water loss from the wound and prevents bacterial and pathogen infections through the wound. However, the disadvantages include a new wound created after removal of the biomaterial for the later epidermis healing process, scarring with skin pigments, or, in the worst case, permanently losing the dermis layer.

The idea of tissue engineering can be applied in skin engineering. "Artificial skin" can be applied to regenerate skin with the help of a collagen scaffold. The engineered scaffold adheres to the substrate while maintaining its durability while also becoming elastic enough to not be totally deformed. It acts like natural skin by preventing water loss through evaporation and provides a barrier against microbes. The new skin should also promote hemostasis and regeneration without an immune response resulting in inflammation.

The concept of artificial skin has already been successfully applied for severe (third-degree) skin injury in practice. Taking the artificial skin called "Integra" as an example, a scaffold material consists of collagen, glycosaminoglycan, growth factors, and cell adhesion molecules with a silicone cover that can be first applied to the damage site (Fig. 11.10a). Meanwhile, a small amount of epidermal skin cells are extracted from the patient and are expanded using the traditional tissue/cell culture facilities. The following steps are performed only after at least 2 weeks once the epidermal cells reach an adequate quantity. This waiting period offers time for the wound healing process, including dermal regeneration guided by the implanted scaffold (Fig. 11.10b). With the physical support of the scaffold and biochemical supply via gradual release of growth factors and cell adhesion molecules from the swelling process, dermal formation is more rapid, and the uniformity of the newly formed vessels is improved. Furthermore, as the scaffold is covered by the silicone membrane, such regenerated dermis should have a flat exterior surface profile. Afterward, the doctor removes the silicone cover (Fig. 11.10c) and replaces the outermost layer with the cultured epidermal skin cells. These cells are plated on the regenerated dermis as a mesh-like layer (Fig. 11.10d). These cells might have phenotypic changes after proliferating outside the human body environment, but this mesh cell layer can function as an autograft for the following regeneration of epidermis.

11.6.2 Cartilage and Bone Repair

Both bones and cartilage are made from specialized forms of connective tissue which contain a network of cells inside the extracellular matrix. Cartilage's mechanical properties include high flexibility and resistance to compressive forces and the physical properties of being very thin and avascular. Healthy cartilage facilitates joint movements. It allows bones to glide over each other with minimal friction. In contrast to cartilage, bones are much more vascularized because of their calcified matrix.

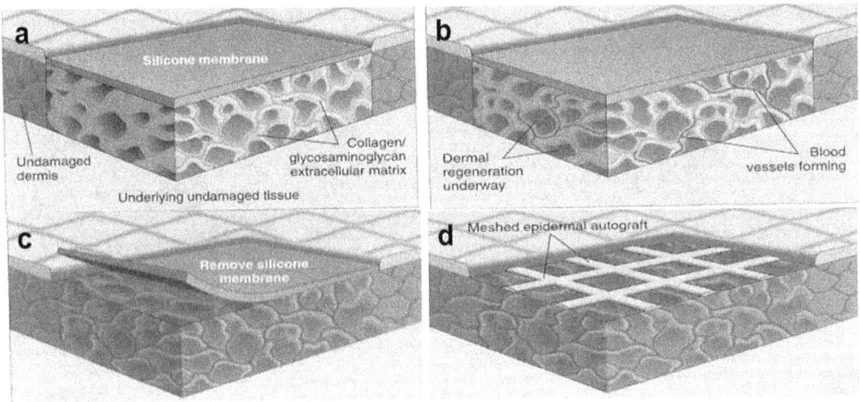

Fig. 11.10 Implantation procedures of an artificial skin with two layers of wound matrices: (**a**) scaffold placement, (**b**) dermal regeneration (about the 7th day), (**c**) covering membrane removal (about the 14th day), and (**d**) epidermal cell seeding and healing (after 2 weeks of application)

Cartilage Repair

Articular cartilage is a smooth tissue that covers the ends of bones where they meet to form joints (Fig. 11.11a). Articular cartilage is composed of chondrocyte cells inside a dense extracellular matrix. After injury, articular cartilage is limited in self-regeneration because of the limited intrinsic biological factors available at its sites. On the other hand, articular cartilage under chronic shear from knee joint movements can cause defects on contacting surfaces. These focal defects can be either chondral or osteochondral. The chondral lesions are on the cartilage and not in the subchondral bone, while osteochondral defects are cuts in the vascularized subchondral. Chondral lesions do not heal instantaneously since the subchondral vessels are not affected by the cut. Osteochondral lesions have some instant repair because the cut causes mesenchymal chondroprogenitor cells to form cartilage spontaneously. Yet, a full recovery is not instant as the newly formed tissues are fibrous but lack the functioning mechanical properties of the hyaline cartilage. On the other hand, shear forces from trauma can separate articular surface among the radial cartilage and the calcified cartilage and bone layers. Full-thickness defect of articular cartilage is severely limited in its ability to heal and be repaired. Since cartilage is avascular, the injury will most likely result in necrosis since there are no networks or veins for cells to replace and heal the damage. In the situation of large osteochondral defects, a joint replacement utilizing artificial prostheses can be performed for the healing process, but this would create another issue on material wear of the artificial knee prostheses.

Alternatively, scaffolds with seeded chondrocytes are used for cartilage regeneration because the chondrocytes can secrete type II collagen and contain the native ECM over the healing duration. These chondrocytes can be harvested from other healthy articular cartilage cells in the patient. Chondrocytes can also be differentiated from mesenchymal stem cells through chondrogenesis. Synovium content can be extracted arthroscopically from donors without harming the donor due to their

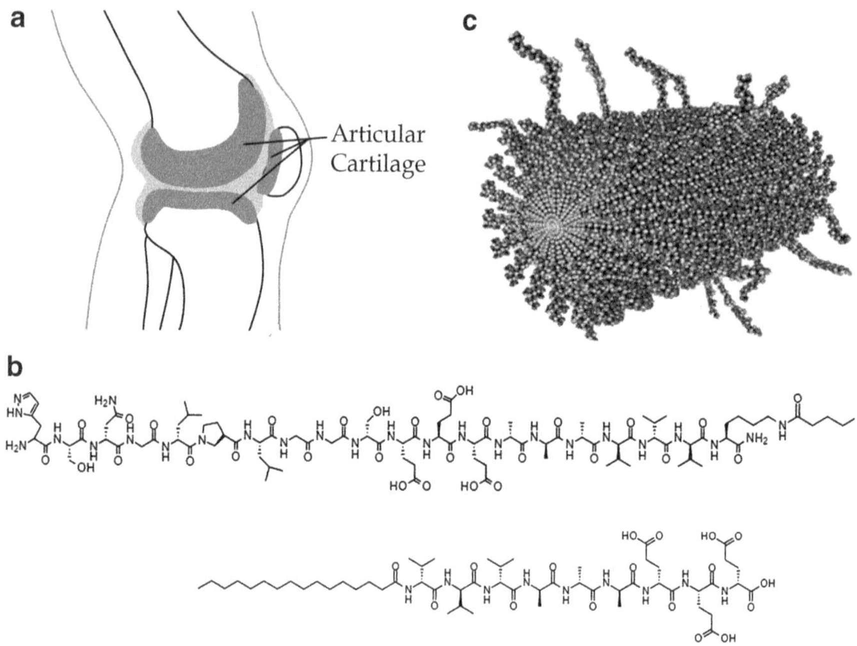

Fig. 11.11 (**a**) Articular cartilage. (**b**) Molecular structure of transforming growth factor beta 1 binding peptide amphiphiles (PA). (**c**) Protein fiber formed by the PA molecules

regenerative nature. The engineered scaffold must be mechanically stable to support the continuous growth of ECM (including collagen II) and then be biodegradable after the ECM network fully matures. Furthermore, molecular release from scaffolds can guide and promote the regeneration process. Hydrogels are an excellent scaffold material for this purpose. As hydrogels are networks of hydrophilic polymers which absorb water or other liquids, components embedded in hydrogels can be released as the hydrogel swells. For instance, bone morphogenetic proteins can be released to continuously simulate mitosis of chondrocytes and synthesize the ECM. If human mesenchymal stem cells are used, such proteins can trigger MSCs to differentiate into chondrocytes. Besides, transforming growth factor beta 1 (TGFβ-1) regulates cartilage regeneration and its attachment with adjacent tissues.

Regeneration of cartilage tissues involves the self-assembly of binding peptide amphiphiles (PA). In practice, this self-assembled PA material can be fabricated by electrospinning by using it as the scaffold material. After implantation into the damaged tissue site, PA can become part of the recovering cartilage. PA also offers swelling-controlled molecular release of any embedded molecules, and thus, growing factors such as TGFβ-1 can be bound with PA in the electrospinning process. As shown in Fig. 11.11b, PA contains a TGFβ-1 binding peptide site (*red*) separated by a glycine spacer from polar domain (*purple*) at one end and a filler PA that does not contain an epitope on the other end. Such TGFβ-1-binding PA molecules assemble together via the binding of TGFβ-1 within themselves, and this self-assembly

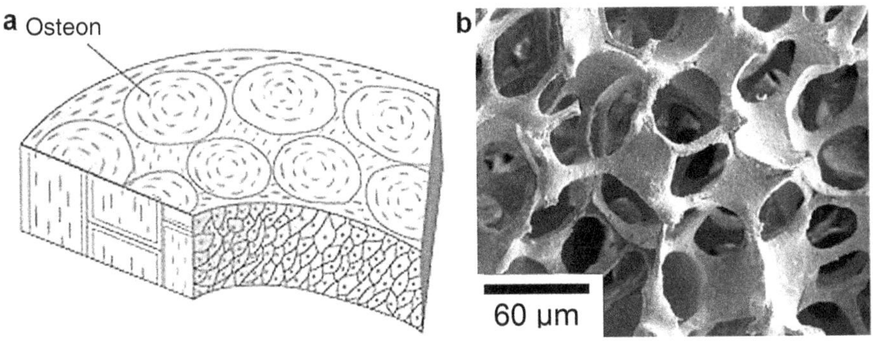

Fig. 11.12 (**a**) Bone tissue structure. (**b**) Example of a hydroxyapatite scaffold with random architecture to promote osteogenesis. (Teixeira [14], Journal of Materials Science: Materials in Medicine. Adapted with permission from Springer)

process induces fibrin structures at the molecular level (Fig. 11.11c) and a soft and porous scaffold material at a larger scale. TGFβ-1 molecules can then be released at injured sites and promote tissue regeneration.

Bone Repair

In contrast to cartilage, bone is continuously remodeled in the human body. Bone remodeling is the result of the balance between the activities of two different cell populations: osteoclasts and the osteoblasts (Fig. 11.12a) which are responsible for bone resorption and deposition, respectively. Bone lesions/defects occur in a wide variety of clinical situations, and their reconstruction to provide mechanical and functional integrity is a necessary step in the patient's rehabilitation. Because of the potential of bone to spontaneously regenerate, most bone lesions, such as fractures, heal well with conventional conservative therapy, surgery, or bone fixation implants. Nevertheless, a bone graft or a bone substitute is often required in orthopedic and maxillofacial surgery to assist healing of large traumatic or postsurgical defects and for congenital osseous deformities.

To fix bones, autologous cell transplantation of mesenchymal stromal cells combined with other biodegradable scaffolds (Fig. 11.12b) can be used. Other cell-based tissue therapies for bone repair use targeted recruitment of specific cells at the lesion site on a degradable scaffold with the native matrix. Cell delivery into the scaffold can be done through injection of osteocyte-containing liquids. This scaffold allows for cell infiltration, proliferation, differentiation, and integration from the in vitro cells with the existing cells onto the lesion [14].

Problems

Problem 11.1

Consider the configuration of electrospinning for a highly porous biomaterial as a scaffold for tissue engineering applications. A polyvinylpyrrolidone (PVP)

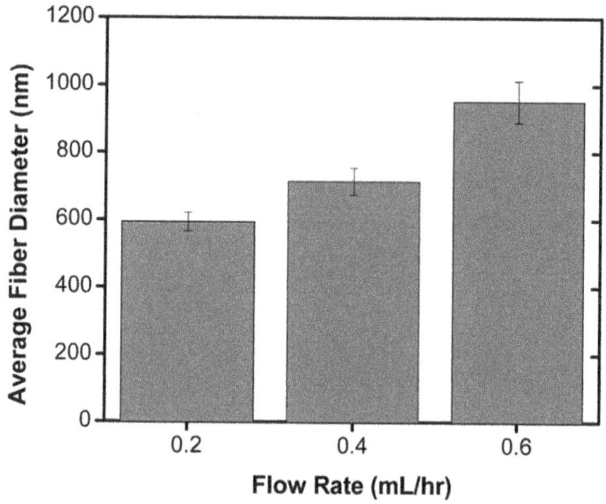

Fig. 11.P1 Average fiber diameter against flow rate in the electrospinning process

monomer solution of 10 ml (V_D) needs to be deposited on the substrate material. The solution has a viscosity of 10×10^{-3} kg/m s and a density of 1.2 g/cm^3. This solution should flow through a needle with a length of 1.5 cm (L) and an inner radius (R_b) of 0.5 mm. The radius of the plunger (R_p), where the force for fluid motion is obtained through pressure exerted by the thumb, is 5 mm. A calibration experiment was performed previously for the relation between the flow rate and the average fiber diameter produced by electrospinning as shown in Fig. 11.P1.

In this question, we consider for manufacturing of a scaffold with a target fiber diameter of 800 nm.

(a) What are the *Reynolds* number of the flow along (i) the cylinder and (ii) the needle?
(b) What are the fluidic resistance ($\Delta P/Q$) along (i) the cylinder and (ii) the needle?
(c) Can you compare between the pressure drop in the cylinder and that in the needle?
(d) What is the minimum required force applied to support the required rate of the liquid injection?

The polymerization process of PVP can be described as Fig. 11.P2 below:

Consider also the molar weight of the PVP polymer (M_w) is 43,000 g/mole and the atomic weights of the atoms listed below (Table 11.P1).

(e) What type of polymerization should this process be and why?
(f) What is the degree of polymerization (DP)?

Problem 11.2

A sample for 3D printing of a scaffold is prepared by dissolving 0.10 g of polystyrene (molecular weight: 4.26×10^6 g/mol) in 100 ml of butanone. The sample has a

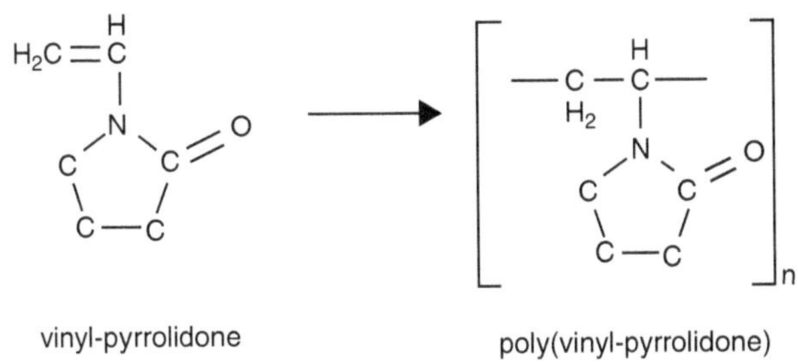

Fig. 11.P2 Polymerization of poly(vinyl-pyrrolidone)

Table 11.P1 Atomic weights of the atoms found in poly(vinyl-pyrrolidone)

	Carbon (C)	Hydrogen (H)	Nitrogen (N)	Oxygen (O)
Atomic weight (g/mole)	12	1	14	16

viscosity 1.273 times of the pure butanone. Given the empirical values of K_P and a_P in the Mark-Houwink-Staudinger relation have been previously obtained as 3.9×10^{-3} and 0.58, respectively, what is the sample viscosity?

Problem 11.3

The extension ratios ($= \lambda$) for four different molar masses (M_P) of polyisobutylene in cyclohexane obtained at 30 °C are listed in Table 11.P2. Check that these data are consistent with Eq. 11.14.

Problem 11.4

Imagine that you are responsible for manufacturing an artificial bone for replacement of a bone fragment missing in the skull of a patient who had a serious injury in a car accident. A surgery was performed previously to remove a broken piece of bone. During the primary bone healing of the patient, your role is to manufacture a scaffold for the subsequent bone cell seeding and culture in order to develop the in vitro artificial bone. In the next surgery, this artificial bone will be placed at the missing location of the skull for sealing the skull.

(a) Before the manufacturing, imaging of the skull should be performed to obtain the 3D geometry of the missing part of the bone. Can you suggest a bio-imaging method? Please describe briefly the working principle.
(b) The scaffold was then fabricated by a rapid prototyping machine. The substrate material you could use is the composite of hydroxyapatite and PEG-PLA-PEG. Figure 11.P3 is showing the chemical synthesis of PEG-PLA-PEG. What kind of the polymerization is involved in this process? Please explain.

Table 11.P2 Extension ratios (λ) for different molar masses (MP) of polyisobutylenein cyclohexane

M_P (10^3 g/mol)	97	170	374	643
λ	1.12	1.25	1.46	1.65

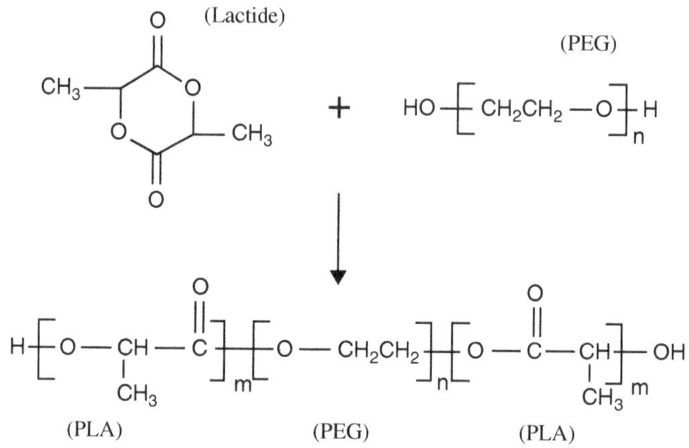

Fig. 11.P3 Polymerization of PEG-PLA-PEG

(c) The composite should include mostly PEG-PLA-PEG, whose viscosity (μ) is ~8.872×10^{-3} kg/m·s. During the rapid prototyping process, the composite should be applied onto the sample for deposition using a syringe. The injection operation was controlled by a computer. The composite solution should flow through a syringe needle with a length of 0.1 mm (L) and an inner radius (R_b) of 0.1 mm. The radius of the plunger (R_p), where the force for fluid motion is obtained through pressure exerted by the thumb, is 1 cm. Assume that the solution flow rate is 5 nL/h. What is the minimum required force applied to support the required rate of the liquid injection?

(d) The missing piece of bone had the shape as a quarter of a circular disc as shown below. Taking into account that a bone cell should have a diameter ~10 μm, the scaffold should have the separating hole/gap widths D_{gap} such that the cells can deposit and migrate inside the scaffold. Please design for scaffold and the corresponding path of the extruder/syringe needle for the fabrication.

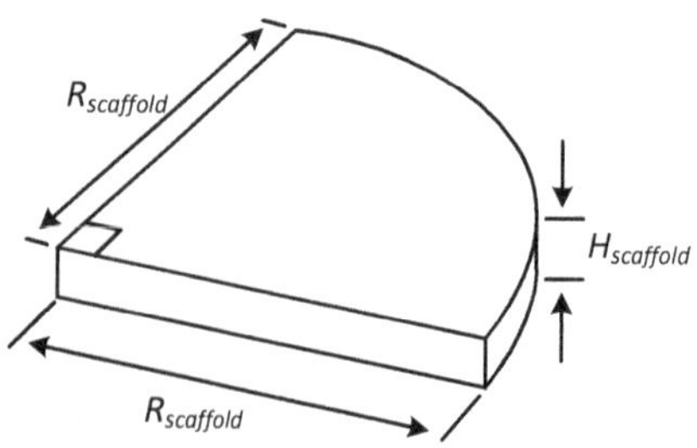

Fig. 11.P4 Geometry of a missing piece of bone

(e) Please express the porosity of the scaffold as a function of D_{gap}.
(f) Considering that the cured scaffold material has the Young's modulus close to cure PLA ($E_{PLA} = 3.5$ GPa), please express the equivalent compressive stiffness of the scaffold as a function of D_{gap}.
(g) Based on the design and fabrication strategy you provided in part (d) and further considering the injection flow rate of the composite (as mentioned in part (c)) is now flexible (no longer fixed as 5 nL/h), denoted as Q, please estimate for the manufacturing time.
(h) To achieve a functional scaffold, the allowable D_{gap} should be ≥ 30 μm such that cells can migrate into the inner scaffold body. The equivalent compressive stiffness $E_{scaffold}$ of the scaffold should be ≥ 0.5 GPa to maintain the structural shape after implantation. The patient's injury site would need the scaffold dimensions of $R_{scaffold} = 2$ cm and $H_{scaffold} = 3$ mm. Furthermore, the flow rate Q can only be set ≤ 10 nL/h by the injection system. Given all these parameter settings/constraints, can you find out a set of feasible operation parameters and manufacturing time for your scaffold design? Please show how you obtain the results.

References and Further Reading

1. Blitterswijk, C.A., Boer, J.: Tissue Engineering. Academic Press, Cambridge, MA, USA (2015)
2. Schwab, M.: Encyclopedia of Cancer. Springer, New York (2008)
3. Brandrup, J., Immergut, E.H., Grulke, E.A., Abe, A., Bloch, D.R.: Polymer Handbook, vol. 7. Wiley, New York (1989)
4. Sperling, L.H.: Introduction to Physical Polymer Science. Wiley, Hoboken (2005)
5. Bosworth, L., Downes, S.: Electrospinning for Tissue Regeneration. Elsevier, Sawston, UK (2011)
6. Burdick, J.A., Mauck, R.L.: Biomaterials for Tissue Engineering Applications: A Review of the Past and Future Trends. Springer, New York, NY, USA (2010)
7. Basu, B., Katti, D.S., Kumar, A.: Advanced Biomaterials: Fundamentals, Processing, and Applications. Wiley, Hoboken (2010)
8. Bártolo, P., Bidanda, B.: Bio-Materials and Prototyping Applications in Medicine. Springer, New York (2008)
9. Subia, B., Kundu, J., Kundu, S.C.: Biomaterial Scaffold Fabrication Techniques for Potential Tissue Engineering Applications. IntechOpen (2010)
10. Minuth, W.W., Strehl, R., Schumacher, K.: Tissue Engineering: Essentials for Daily Laboratory Work. Wiley-VCH (2003)
11. O'Brien, F.: Biomaterials and scaffolds for tissue engineering. Mater. Today. **14**, 88–95 (2011)
12. Hollister, S.J.: Porous Scaffold Design for Tissue Engineering. Nat. Mater. **4**, 518–524 (2005)
13. Woodfield, T.B.F., Blitterswijk, C.V., Wijn, J.D., Sims, T.J., Hollander, A.P., Riesle, J.: Polymer scaffolds fabricated with pore-size gradients as a model for studying the zonal organization within tissue-engineered cartilage constructs. Tissue Eng. **11**, 1297–1311 (2005)
14. Teixeira, S., Ferraz, M.P., Monteiro, F.J.: Biocompatibility of highly macroporous ceramic scaffolds: cell adhesion and morphology studies. J. Mater. Sci. Mater. Med. **19**, 855–859 (2008)

Chapter 12
Process Design Optimization

Abstract In product design and manufacturing process planning, we prefer the "best" product design out of all feasible designs, which is called the optimal design. Optimization techniques have become the major quantitative tools in engineering design and decision-making process. Process optimization is a process that adjusts a current process to optimize some specified sets of parameters without violating constraints. The most common goals are minimizing cost while maximizing profit and efficiency. This chapter introduces the formulation of a design problem as an optimization statement incorporated with the design constraints. Essential computational methods for solving the optimization statements with demonstrated examples are also discussed.

12.1 Introduction

In the previous chapters, we discussed the design, material selection, manufacturing, process parameters, and characteristics of biomedical devices in a range of aspects. Process planning for manufacturing biomedical devices is an important stage in product design and the development cycle. Without systematic determination of optimal designs, engineers in many cases used to conduct process designs by trial and error and/or through case studies. Clearly, this is very time consuming and inefficient. A good process designer should understand the manufacturing process and the effect of each individual parameter of the process plan. A systematic thinking flow will improve a process planner's performance. These procedures define in detail that will transform raw material into the desired product. More precisely, process planning can be defined by a sequence of activities which are mainly comprised of the following:

- Interpretation of the specifications contained in the definition drawing of a part, including dimensions and tolerances, geometrical tolerances, surface roughness, material type, blank size, number of parts in a batch, etc.
- Selection of manufacturing processes and tools which are candidates for processing a part and its features by the corresponding constraints imposed in the definition drawing

R. H. W. Lam, W. Chen, *Biomedical Devices*,
https://doi.org/10.1007/978-3-030-24237-4_12

- Determination of production tolerances and setting dimensions which ensure execution of the design tolerances while choosing production dimensions for reasons of commodity and capability of manufacturing machinery
- Selection of starting surfaces and datum surfaces to ensure precise execution of processing operations simultaneously with a selection of holding fixtures and checking the stability of a part by appropriate clamping
- Sequencing of operations as a function of priorities imposed by accuracy and technological constraints
- Grouping of elementary operations on the same machine so that operation time will be reduced while respecting accuracy requirements
- Selection of machines to execute the technological operations, taking into account the number of workpieces to be produced
- Selection of inspection methods and inspection instruments to guarantee final conformity of the component with functional requirements
- Determination of processing conditions for every elementary operation which enables the computation of working times and costs to carry out an economic evaluation
- Editing of process sheets to be assembled in a comprehensive process planning file which is transferred to the manufacturing department for execution.

A huge amount of preparation work has to be carried out before final decisions about a process plan can be made. The modern approach of computer-aided process planning can off-load some of the manual work of the process planner by using information databanks and computerized algorithms to select proper manufacturing conditions. However, a complete automated process planning system still seems to be ideal, except for special applications with well-defined conditions. Particularly, it is important to emphasize that process planning is required at any product development stage, including the manufacturing plant, regardless of plant size, part complexity, and batch size.

In product design and manufacturing process planning, we prefer the "best" product design out of all feasible designs, which is called the optimal design. Generally, there are multiple solutions to a design problem, and the first selected solution is not necessarily the best. Thus, the need for optimization is inherent in the design process. Optimization techniques have become the major quantitative tools in engineering design and decision-making process. Optimization can be achieved by adjusting the process configuration parameters until desired quantities are maximized and undesired ones are minimized. Process optimization is a process that adjusts a current process to optimize some specified sets of parameters without violating constraints. The most common goals are minimizing cost while maximizing profit and efficiency.

12.2 Optimization Statement Formulation

To formulate a design problem as a precise goal for the optimization process, we need to define key attributes to sufficiently represent the practical problem, including an objective function, a predictive model, and variables. The objective function is a quantitative performance measure that needs to be minimized or maximized, e.g., production cost, expected profit, product durability, etc. The predictive model describes the system under consideration, which includes all the elements and constraints for the specific problem. The problem formulation involves transferring all the elements and constraints into a set of equations and inequalities. The variables that appear in the predictive model must be adjusted to satisfy the constraints. This can usually be accomplished with multiple instances of variable values, leading to a feasible region that is determined by a subspace of these variables. In many engineering problems, this subspace can be characterized by a set of decision variables. The number of these decision variables can be interpreted as the "degrees of freedom" of the optimization problem.

In the typical design optimization situation, we need to establish a general configuration for which the numerical values of the independent variables have not been fixed. We should decide on an appropriate objective function $U(\underline{x}) = U(x_1, x_2, \ldots, x_N)$ that defines the value of the design in terms of the independent variables, where $x_1, x_2, \ldots, x_N$ are N design parameters which can be represent cost, time, profit, energy consumption, etc. Furthermore, constraints aroused from physical laws and limitations, product specifications, user preferences, or from compatibility conditions on the individual variables should be taken into consideration. A constraint is a condition of an optimization problem that the optimal solution and all other feasible solutions must satisfy. There are several types of constraints, primarily equality constraints, inequality constraints, and integer constraints. The set of candidate solutions that satisfy all constraints is called the feasible set. Functional constraints, or equality constraints $H_v(\underline{x}) = H_v(x_1, x_2, \ldots, x_N) = 0$, specify relations that must exist between the variables, where the integer $v = 1, 2, \ldots, N_E$ and N_E is the number of equality constraints. $H_v(x_1, x_2, \ldots, x_N) = 0$ describes the target characteristics of the system (e.g., production rate, material balances). For example, if we optimize the volume of a rectangular storage tank V, where $x_1 = l_1$, $x_2 = l_2$, $x_3 = l_3$, then the functional constraint would be that $H(x_1, x_2, x_3) = V - x_1x_2x_3 = 0$. On the other hand, the inequality constraints $G_w(\underline{x}) = G_w(x_1, x_2, \ldots, x_N) \leq 0$ can define process specifications or constraints for feasible plans and schedules, where the integer $w = 1, 2, \ldots, N_I$ and N_I is the number of inequality constraints.

Most design problems come with a large number of design requirements and limitations. Engineering design problems are rarely unconstrained. The optimization process is a process from mathematical modeling to real-world practical problems, composed of design variables, objective functions, and constraints which can be concluded to the following form:

$$\begin{array}{ll} \text{Minimize} & U(x_1, x_2, \ldots, x_N) \\ \text{Subject to} & G_w(x_1, x_2, \ldots, x_N) \leq 0 \\ & H_v(x_1, x_2, \ldots, x_N) = 0 \end{array} \tag{12.1}$$

Alternatively, we may use a vector $\underline{x} = [x_1, \ x_2, \ \ldots, \ x_N]^T$ to represent design variables. Then, we get the more compact form of the optimization statement as

$$\begin{array}{ll} \text{Minimize} & U(\underline{x}) \\ \text{Subject to} & G_w(\underline{x}) \leq 0 \\ & H_v(\underline{x}) = 0 \end{array} \tag{12.2}$$

Since real numbers are often used in most engineering problems, we may skip defining $\boldsymbol{x} \in R_N$ in the target function. Equations 12.1 and 12.2 are the normal mathematical forms of optimization processes. The normal form should be highly compatible with commercially available software, and therefore, the optimization problems and the corresponding product design process can be solved with computers. When optimization problems have different descriptions, we can transfer the descriptions into this normal form. For example, if the problem requires the maximum value of $U(x, y)$, all we need to do is find the minimum value of $-U(x, y)$. This is because maximizing $U(x, y)$ has the same meaning as minimizing $-U(x, y)$. When the objective function and the constraint function in the mathematical model are all linear functions of the design variables, the problem is called a linear optimization problem or a linear programming problem. When at least one of the objective functions and the constraint functions is a nonlinear function, such a problem is called a nonlinear optimization problem or a nonlinear programming problem. In reality, most product design problems are nonlinear optimization problems.

As a demonstration of the optimization statement formulation, we may consider an example of designing a cylindrical tank (with cover) to store a fixed volume of liquid V, with the minimum material cost. We may assume the tank is made of thin metallic sheets, manufactured by the sheet metal process. The design variables are the tank diameter D and its height h. D and h should be within reasonable ranges, i.e., $D_{\min} \leq D \leq D_{\max}$ and $h_{\min} \leq h \leq h_{\max}$. The tank will be constructed by forming and welding thin steel plates. Then, the cost per unit area of steel plate is C. Therefore, the total cost will depend directly on the area of plate A that is used.

If we formulate the above example as an optimization statement, we may set the material cost of the tank as the objection. Clearly, the surface area (A) of the tank is $A = 2(\pi D^2/4) + \pi Dh$, and the material cost should be $C \times A$. Moreover, we should consider all the relevant functional constraints and regional constraints. A functional constraint exists on the volume requirement: $V = \pi D^2 h/4$. The regional constraints are introduced by the range of acceptable dimensions: $-D_{\min} \leq D \leq D_{\max}$ and $-h_{\min} \leq h \leq h_{\max}$. Therefore, the optimization statement is

$$
\begin{array}{ll}
\text{Minimize} & C\left[\left(\pi D^2/2\right) + \pi D h\right] \\
\text{Subject to} & D - D_{\max} \le 0 \\
& D_{\min} - D \le 0 \\
& h - h_{\max} \le 0 \\
& h_{\min} - h \le 0 \\
& V - \pi D^2 h/4 = 0
\end{array}
$$

Generally, the solution of a constrained optimization problem is the point of multiple variables within the feasible space, inducing the minimum value of the object function. Now, we consider the following nonlinear constrained optimization problem:

$$
\begin{array}{ll}
\text{Minimizing} & U(\underline{x}) = 4{x_1}^2 - x_1 - x_2 - 2.5 \\
\text{Subject to} & G_1(\underline{x}) = -{x_2}^2 + 1.5{x_1}^2 - 2x_1 + 1 \le 0 \\
& G_2(\underline{x}) = {x_2}^2 + 2{x_1}^2 - 2x_1 - 4.25 \le 0
\end{array}
$$

As seen from the corresponding contour plot of $U(\underline{x})$ in Fig. 12.1, the feasible regions are the overlap of different regions of constraints. The optimal solution is the minimum point of $U(\underline{x})$ in the feasible regions. Although there are many numerical solvers available in the market already, we would like to introduce a strategy for solving the optimization problems as discussed later in Sect. 12.3.

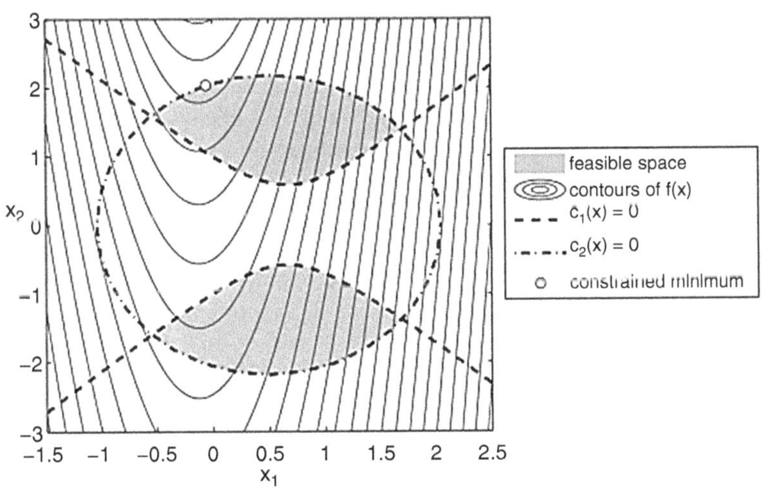

Fig. 12.1 Example contours and feasible regions for a simple constrained optimization problem

12.3 Common Objective Functions

12.3.1 *Machining Process Cost*

The first common objective function mentioned here is about the machining process as we have discussed previously in Chaps. 6 and 10. The total cost of a machined part C_{machine} is the sum of the machining cost C_{mc}, the cost of the tooling C_{t}, and the cost of the material C_{m}:

$$C_{\text{machine}} = C_{\text{mc}} + C_{\text{t}} + C_{\text{m}} \tag{12.3}$$

The machining cost is usually the major component of the unit production cost. It depends on the unit machining time t_{unit} and the total effect of costs of the machine, those being labor and overhead:

$$\begin{aligned} C_{\text{mc}} &= C_{\text{rate}} t_{\text{unit}} \\ C_{\text{rate}} &= \left[M(1 + \text{OH}_{\text{m}}) + W\left(1 + \text{OH}_{\text{op}}\right)\right] \end{aligned} \tag{12.4}$$

where C_{rate} is cost rate, \$/min, M is machine cost rate (\$/min), OH_{m} is machine overhead ratio, W is labor rate for operator (\$/min), and OH_{op} is operator overhead ratio. Typical costs which contribute to the "overhead" are salaries, benefits, maintenance, repair, compliance, etc. The production time t_{unit} is the sum of the machining time t_{m} (i.e., T_{cut} as introduced in Chap. 9) and the nonproduction or idle time t_{i}:

$$t_{\text{unit}} = t_{\text{m}} + t_{\text{i}} \tag{12.5}$$

where the idle time t_{i} includes job setup time, tool change time, hand loading/unloading time, and downtime because of machine/tool failure:

$$t_{\text{i}} = t_{\text{set}} + t_{\text{change}} + t_{\text{hand}} + t_{\text{down}} \tag{12.6}$$

The cost of tooling for one tool (C_{t}) is prorated cost of "the cost of the cutting tools (C_{tool})," dependent on the machine time (t_{m} or T_{cut}) and the tool life (T_{tool}):

$$C_{\text{t}} = C_{\text{tool}} t_{\text{m}} / T_{\text{tool}} \tag{12.7}$$

Thus, the objective function for the cost of a manufactured part can be

$$\begin{aligned} U = {} & \left[M(1 + \text{OH}_{\text{m}}) + W\left(1 + \text{OH}_{\text{op}}\right)\right] \\ & \times \left(t_{\text{m}} + t_{\text{set}} + t_{\text{change}} + t_{\text{hand}} + t_{\text{down}}\right) + \frac{C_{\text{tool}} t_{\text{m}}}{T_{\text{tool}}} + C_{\text{m}} \end{aligned} \tag{12.8}$$

In case the material cost C_m is negligible and the idle time is much shorter than the machining time, $t_{unit} = t_m = T_{cut} = kA_m/(vL_{feed})$, where A_m is the cross-sectional area of cut, v is the cutting speed, and L_{feed} is the feed distance. Then, the cost per part is

$$C_{machine} \approx C_{rate} t_m + \frac{C_{tool} t_m}{T_{tool}} = \frac{kA_m}{vL_{feed}} \left(C_{rate} + \frac{C_{tool}}{T_{tool}} \right) \tag{12.9}$$

The tool life T_{tool} can be formulated by a similar relationship, proposed by F. W. Taylor back in 1905, in which T_{tool} has a power-law relationship as $T_{tool} = \Psi_{tool} v^{-p} L_{feed}{}^{-q}$ and p and q are constants and typically $p > q$. Then, we have

$$C_{machine} \approx \frac{kA_m}{vL_{feed}} \left(C_{rate} + \frac{C_{tool}}{\Psi_{tool}} v^p L_{feed}^q \right). \tag{12.10}$$

12.3.2 Machining Process Time

The machining operation can also be optimized based on minimum time of the operation. There can be still ranges of the cutting speed v and feed distance L_{feed} fulfilling the minimum machining cost; therefore, we can find the machining time within these ranges of parameters. Recall Eq. 12.10. When differentiating with respect to v and L_{feed} for determining the parameters for the minimum machining cost

$$\frac{\partial C_{machine}}{\partial v} = kA_m \left(-\frac{C_{rate}}{v^2 L_{feed}} + \frac{(p-1)C_{tool}}{\Psi_{tool}} v^{p-2} L_{feed}^{q-1} \right) = 0 \text{ and} \tag{12.11}$$

$$\frac{\partial C_{machine}}{\partial L_{feed}} = kA_m \left(-\frac{C_{rate}}{vL_{feed}^2} + \frac{(q-1)C_{tool}}{\Psi_{tool}} v^{p-1} L_{feed}^{q-2} \right) = 0. \tag{12.12}$$

Clearly, $\partial C_{machine}/\partial v = 0$ and $\partial C_{machine}/\partial L_{feed} = 0$ induce the local minimum of $C_{machine}$. Further considering the relationship between the feed distance $L_{feed}{}^*$ and cutting speed v^* for the minimum machining cost from Eq. 12.11:

$$v^* = \left[\frac{\Psi_{tool} C_{rate}}{(p-1) C_{tool} L_{feed}^* q} \right]^{1/p} \tag{12.13}$$

In the practical case, the machine specifications can only offer limited v and L_{feed}. L_{feed} is often the major limiting factor, and $L_{feed}{}^*$ can be simply set as the maximum level that the machine can provide rather than being calculated.

According to the machining process characteristics as discussed in Chap. 10, the machine time for different machining processes share the same form, which is expressed as

$$t_{\mathrm{m}} = \frac{\pi D(L + L_{\mathrm{A}})}{kL_{\mathrm{feed}}^{*} v^{*}} \tag{12.14}$$

Accordingly, the total manufacturing time as the objective function can be set as the production time t_{unit}:

$$t_{\mathrm{unit}} = \frac{\pi D(L + L_{\mathrm{A}})}{kL_{\mathrm{feed}}^{*} v^{*}} + t_{\mathrm{set}} + t_{\mathrm{change}} + t_{\mathrm{hand}} + t_{\mathrm{down}} \tag{12.15}$$

12.3.3 Batch Development Processes

For mass production, process optimization offers a dramatic improvement for the concerned objective. Equally as important, process planning is still suited for small batch sizes. Product development with small batch sizes still requires optimization of the defined objectives. The difficulty here is to find a reasonable compromise between time of preparation (e.g., thinking time) and production time. We may start with a brief definition of the product development cost. Let the total cost to produce a part be the following:

$$C = QC_{\mathrm{d}}T_{\mathrm{d}} + C_{\mathrm{t}}T_{\mathrm{p}} + C_{\mathrm{s}}T_{\mathrm{s}} \tag{12.16}$$

where C is the total cost (\$), Q is the batch quantity, C_{d} is the hourly rate of direct work (\$ per hour), T_{d} is the direct manufacturing time (hours), C_{t} is the hourly rate of indirect labor (\$ per hour), T_{p} is the indirect labor time (hours), C_{s} is the hourly rate for setup time (\$ per hour), and $T_{\mathrm{s}} =$ is the time for setup (hours).

Furthermore, the indirect labor time T_{p} can be divided into two elements: (1) T_{pf} as a fixed time to handle an order and to arrive at an initial process plan and (2) T_{pv} as the time to generate alternative process plans and evaluate them in order to arrive to an optimum process plan. That is, $T_{\mathrm{p}} = T_{\mathrm{pf}} + T_{\mathrm{pv}}$. T_{pv} is a variable representing the thinking time. It seems that the total cost will increase with T_{pv}. However, the longer T_{pv} is, the "better" the process plan can be generated with a shorter direct machining time T_{d}. Importantly, the value of T_{pv} determines the balance between the preparation and direct production of the product. Properly setting T_{pv} can optimize the economics of handling both a small batch size order and a large batch size order. An evaluation of T_{pv} and its effect on T_{d} should be made. If it is assumed that the direct machining time T_{d} depends on the thinking time T_{pv}, T_{d} can be approximated by

$$T_d = K_1 + K_2/T_p = K_1 + K_2/(T_{pf} + T_{pv}), \qquad (12.17)$$

where K_1 and K_2 are constants. K_1 has a value that depends only on the manufacturing process, whereas K_2 depends on the process planner with a smaller value for "better" process planners. Their values can be assigned by two boundary conditions reflecting (1) the minimum direct manufacturing time T_d which can be obtained with a sufficiently long thinking time T_{pv} for the optimal process planning and (2) T_d without any thinking process ($T_{pv} = 0$):

$$\begin{aligned} T_d &= K_1 = T_{d\_min}, \text{ at } T_{pv} = \infty \\ T_d &= K_1 + K_2/T_{pf} = T_{d\_max}, \text{ at } T_{pv} = 0 \end{aligned} \qquad (12.18)$$

where the values of T_{d_max} and T_{d_min} are known and are specific for a particular product development case. These values are related to K_2, and therefore, they depend upon the expertise of the process planners.

Hence, K_2 can be obtained by

$$K_2 = (T_{d\_max} - T_{d\_min})T_{pf} \qquad (12.19)$$

Thus

$$T_d = T_{d\_min} + \left[(T_{d\_max} - T_{d\_min})T_{pf}\right]/T_p \qquad (12.20)$$

Substitute T_d into Eq. 12.16 yields

$$C = QC_d\left[T_{d\_min} + \frac{(T_{d\_max} - T_{d\_min})T_{pf}}{T_{pf} + T_{pv}}\right] + C_t(T_{pf} + T_{pv}) + C_sT_s \qquad (12.21)$$

where the values of C_d, C_t, and C_s are known and are specific for every product development task. Differentiating Eq. 12.21 with respect to the thinking time T_{pv} and setting it equal to zero, we can obtain the minimum total cost of manufacturing a part with the quantity of products as a parameter and the thinking time as the variable:

$$\frac{dC}{dT_{pv}} = -QC_d\left[\frac{(T_{d\_max} - T_{d\_min})T_{pf}}{(T_{pf} + T_{pv})^2}\right] + C_t = 0 \qquad (12.22)$$

Hence, the optimal thinking time T_p^* is

$$T_{pv}^* = \sqrt{QC_dT_{pf}(T_{d\_max} - T_{d\_min})/C_t} - T_{pf} \qquad (12.23)$$

In essence, formulation of this relationship demonstrates that the thinking time T_{pv} is an important parameter in determining a machining process. Without providing an adequate amount of time for process planning, a process planner usually works under heavy pressure and lack of time, so unnecessary loss of production costs may be generated without even being noticed. To shorten T_{pv}, it is extremely important to have an expert as the process planner. Although each process will produce the part as specified, different processes will have different machining times and costs because of the different thinking times required, reflecting the capability of the process planner. On top of this, nowadays there are computer-aided process planning algorithms that further shorten the thinking time required. However, research and development efforts in the field of computer-aided process planning over the past decades have been widely implemented in the industry. Nevertheless, there are still great demands for process optimization capability for new products, especially for novel and custom products such as biomedical devices.

Expectedly, for the same product part, different process planners correspond to different "K_2" values in Eq. 12.18. Consequently, there is always the need to rely on continuous improvements on the process planner's intuition, knowledge, and expertise for developing process configurations which will achieve design/production objectives. Thereby, the "K_2" value of the planner should reduce based on experience. The experienced process planner usually makes better decisions based on what has happened in the past without spending much time to analyzing the problem at hand.

12.4 Solving Optimization Problems

12.4.1 Newton-Raphson Method for Local Optimization

In this section, we introduce a feasible strategy to solve process design-related optimization problems. Yet, this will not be able to offer a comprehensive skill set for general optimization and computation methods. Here, we first consider that (1) the case of problems is without any constraints and (2) there is no local maximum point and only one local minimal point for the objective function of multiple parameters $\underline{x} = [x_1, x_2, \ldots, x_N]^T$, and hence, such local minima must be the global minimum. The object function can be expressed as $U(x_1, x_2, \ldots, x_N)$ or $U(\underline{x})$. At the minimum point of $U(\underline{x})$, the gradient denoted as $\underline{f}(\underline{x})$ should be zero such that

$$\underline{f}(\underline{x}) = \begin{bmatrix} f_1 \\ f_2 \\ \vdots \\ f_N \end{bmatrix} = \begin{bmatrix} \partial U/\partial x_1 \\ \partial U/\partial x_2 \\ \vdots \\ \partial U/\partial x_N \end{bmatrix} = \begin{bmatrix} 0 \\ 0 \\ \vdots \\ 0 \end{bmatrix} \tag{12.24}$$

To solve such multivariate system $\underline{f}$, we may first consider the Taylor series expansion of this multidimensional function about a reference set of parameters $\underline{x}_o = [x_{o\_1},\ x_{o\_2},\ \ldots,\ x_{o\_N}]^T$. For example, when $N = 2$, the solution of the

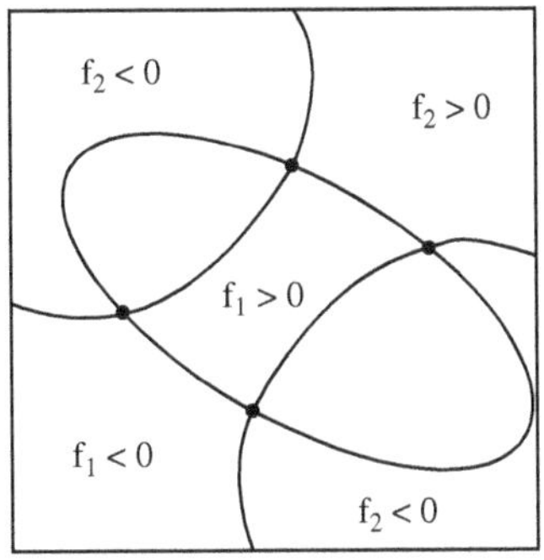

Fig. 12.2 Sample contours of two-dimensional function $f_1(x_1, x_2)$ and $f_1(x_1, x_2)$

equation above can be geometrically explained. The equation $f_i(x_1, x_2) = 0$ represents contour curves in the (x_1, x_2) plane that partition the plane into regions in which the function $f_i(x_1, x_2)$ takes either positive or negative values. The solutions that satisfy both equations are the intersections of the contour curves of both $f_1(x_1, x_2)$ and $f_2(x_1, x_2)$ (Fig. 12.2).

The Newton-Raphson method can also be generalized for solving an N-dimensional system $\underline{f}(\underline{x}) = 0$. We first consider the Taylor expansions of the N functions:

$$
\begin{aligned}
\underline{f}(\underline{x}_o + \delta\underline{x}) &= \begin{bmatrix} f_1(\underline{x}_o + \delta\underline{x})\ 2 \ \vdots\ f_N(\underline{x}_o + \delta\underline{x}) \\ f_2(\underline{x}_o + \delta\underline{x}) \end{bmatrix} \\
&= \begin{bmatrix} f_1(\underline{x}_o) \\ f_2(\underline{x}_o) \\ \vdots \\ f_N(\underline{x}_o) \end{bmatrix} + \begin{bmatrix} \partial f_1/\partial x_1 & \cdots & \partial f_1/\partial x_N \\ \vdots & \ddots & \vdots \\ \partial f_N/\partial x_1 & \cdots & \partial f_N/\partial x_N \end{bmatrix} \begin{bmatrix} \delta x_1 \\ \delta x_2 \\ \vdots \\ \delta x_N \end{bmatrix} \\
&\quad + \underline{O}(\delta\underline{x}^2)
\end{aligned}
\tag{12.25}
$$

where $\delta x = [\delta x_1,\ \delta x_2,\ \ldots,\ \delta x_N]^{\mathrm{T}} = [x_1 - x_{o\_1},\ x_2 - x_{o\_2},\ \ldots,\ x_N - x_{o\_N}]^{\mathrm{T}} = \underline{x} - \underline{x}_o$, $\underline{O}(\delta\underline{x}^2)$ is a vector function as summation of terms in the least order of $\delta\underline{x}^2$, and $\delta\underline{x}^2 = [\delta x_1{}^2,\ \delta x_2{}^2,\ \ldots,\ \delta x_N{}^2]^{\mathrm{T}}$. We further ignore the higher-order terms described by $\underline{O}(\delta\underline{x}^2)$. For small magnitudes of $\delta\underline{x}$, we may approximate $\underline{f}(\underline{x})$ as $\underline{f}^*(\underline{x}_o + \delta\underline{x})$ as N equations. These equations can be written in the linear algebraic form:

$$
\underline{f}^*(\underline{x}_o + \delta\underline{x}) = \underline{f}(\underline{x}_o) + \mathbf{J}(\underline{x}_o)\delta\underline{x} \tag{12.26}
$$

where $\mathbf{J}(\underline{x}_o)$ is an $N \times N$ **Jacobian matrix** defined over the function vector $\underline{f}(\underline{x}_o)$, defined as

$$
\mathbf{J}(\underline{x}_o) = \begin{bmatrix} J_{11} & \cdots & J_{1N} \\ \vdots & \ddots & \vdots \\ J_{N1} & \ldots & J_{NN} \end{bmatrix} = \begin{bmatrix} \partial f_1(\underline{x}_o)/\partial x_1 & \cdots & \partial f_1(\underline{x}_o)/\partial x_N \\ \vdots & \ddots & \vdots \\ \partial f_N(\underline{x}_o)/\partial x_1 & \cdots & \partial f_N(\underline{x}_o)/\partial x_N \end{bmatrix}
$$

with the ijth element being $\partial f_i/\partial x_j$. Note that the ith row vector of $\mathbf{J}(\underline{x}_o)$ is the gradient vector of the ith function $f_i(\underline{x}_o)$.

Now, reconsidering the local unconstrained problem described in Eq. 12.24, we may first propose a guess solution as $\underline{x}^*_0$. This guess has a deviation $\delta\underline{x}^*_0$ from the real solution $\underline{x}_S$, i.e., $\delta\underline{x}^*_0 = \underline{x}_S - \underline{x}^*_0$. Then, $f(\underline{x}^*_0)$ can be viewed as the error of such estimate. If $\underline{f}(\underline{x})$ is a linear function, the error terms $O(\delta\underline{x}^{*2}_0)$ are zero. By assuming $f(\underline{x}^*_0 + \delta\underline{x}^*_0) = f(\underline{x}_S) = 0$, then we can find the roots as $\underline{x}^*_0 + \delta\underline{x}^*_0$, where $\underline{x}_S$ can be calculated by

$$\begin{aligned}\underline{x}_S &= \underline{x}_0^* + \delta\underline{x}_0^* = \underline{x}_0^* + \mathbf{J}\left(\underline{x}_0^*\right)^{-1}\left[\underline{f}\left(\underline{x}_0^* + \delta\underline{x}_0^*\right) - \underline{f}\left(\underline{x}_0^*\right)\right] \\ &= \underline{x}_0^* - \mathbf{J}\left(\underline{x}_0^*\right)^{-1}\underline{f}\left(\underline{x}_0^*\right).\end{aligned} \tag{12.27}$$

If the equations are nonlinear

$$\begin{aligned}\underline{x}_S &= \underline{x}_0^* + \delta\underline{x}_0^* = \underline{x}_0^* + \mathbf{J}\left(\underline{x}_0^*\right)^{-1}\left[\underline{f}^*\left(\underline{x}_0^* + \delta\underline{x}_0^*\right) + O\left(\delta\underline{x}^{*2}_0\right) - \underline{f}\left(\underline{x}_0^*\right)\right] \\ &= \underline{x}_0^* - \mathbf{J}\left(\underline{x}_0^*\right)^{-1}\underline{f}\left(\underline{x}_0^*\right) + \mathbf{J}\left(\underline{x}_0^*\right)^{-1}O\left(\delta\underline{x}_0^*2\right)\end{aligned} \tag{12.28}$$

Hence, if we apply Eq. 12.31 to nonlinear problems, it would be only an approximation of the real root, denoted as $\underline{x}^*_1$, without considering the higher-order terms in $\underline{f}(\underline{x})$. We may consider this approximation as a better estimate than that of the previous $\underline{x}^*_0$. We can then attempt to further improve our approximation by replacing $\underline{x}_{g0}$ with $\underline{x}_{g1}$ in the above equation. This approximation can be applied iteratively:

$$\underline{x}_{n+1}^* = \underline{x}_n^* + \delta\underline{x}_n^* = \underline{x}_n^* - \mathbf{J}\left(\underline{x}_n^*\right)^{-1}\underline{f}\left(\underline{x}_n^*\right), \quad \text{for } n = 0,1,2, \ldots \tag{12.29}$$

where $\underline{x}^*_n = \left[x^*_{n\_1}, x^*_{n\_1}, \ldots, x^*_{n\_N}\right]^{\mathrm{T}}$ is the estimate of $\underline{x}_S$ after n iterations of the approximation. Such iteration searches x_n in the N-dimensional space from an initial position $\underline{x}^*_0$ in such a direction that reduces the error of the new estimate. In the case that the inverse matrix of the Jacobian matrix $\mathbf{J}$ does not exist, the pseudo-inverse $\mathbf{J}^* = (\mathbf{J}^{\mathrm{T}}\mathbf{J})^{-1}\mathbf{J}^{\mathrm{T}}$ can be used in the iteration; this is called the Gauss-Newton method.

The Newton-Raphson method assumes the analytical expressions of all partial derivatives $\partial f_i(x)/\partial x_j$ can be made available based on the functions $f(x)$ so that the Jacobian matrix $\mathbf{J}$ can be computed. However, $\underline{f}(\underline{x})$ is the gradient of the objective function $U(\underline{x})$, and $\mathbf{J}(\underline{x})$ involves further differentiating $\underline{f}(\underline{x})$. Therefore, finding all these analytical expressions can be tedious and challenging. Alternatively, we can estimate the ijth component $J_{ij} = \partial f_i(\underline{x})/\partial x_j$ in $\mathbf{J}$ at $\underline{x}^*$ based on the finite difference approximation:

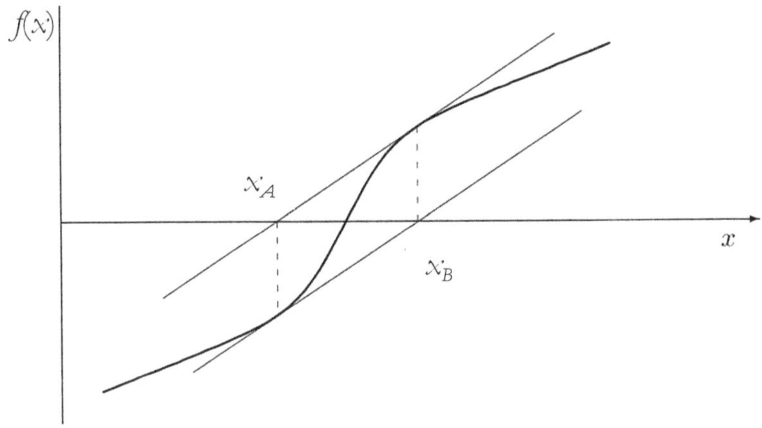

Fig. 12.3 Example of the "cycle" issue during computation using the Newton's method

$$\begin{aligned} J_{ij} &= \frac{\partial f_i(\underline{x}^*)}{\partial x_j} \approx \frac{f_i\left(x_1^*, \ldots, x_j^* + \Delta x_J^*, \ldots, x_N^*\right) - f_i(x_1^*, \ldots, x_N^*)}{\Delta x_J^*}, \quad \text{and} \\ f_i(\underline{x}) &= \frac{\partial U(\underline{x})}{\partial x_i} \approx \frac{U(x_1, \ldots, x_i + \Delta x_f, \ldots, x_N) - U(x_1, \ldots, x_N)}{\Delta x_f} \end{aligned} \tag{12.30}$$

where Δx^*_J and Δx_f should be very small, yet they should also be chosen to be sufficiently large and with signs (positive or negative) such that the calculated J_{ij} would have finite and representative values.

There are special cases that the Newton-Raphson method may not be able to find an optimal solution. Sometimes, the estimates may get stuck in a "cycle." Considering a one-dimensional example as we can see from Fig. 12.3, the function has the same derivative in x_A and x_B. The Newton iterative scheme cycles between these two values x_A and x_B never induce convergence to the solution x_S. One way to handle this "cycle" issue is to alter the recursive scheme. The inverse of Jacobian matrix $\mathbf{J}^{-1}(\underline{x}^*_o)$ in Eq. 12.29 has a magnitude that can directly induce the solution if $\underline{f}(\underline{x})$ is linear. However, this magnitude does not apply for a nonlinear $\underline{f}(\underline{x})$. We may modify Eq. 12.31 as

$$\underline{x}^*_{n+1} = \underline{x}^*_n - K_{\text{damp}}\mathbf{J}\left(\underline{x}^*_n\right)^{-1}\underline{f}\left(\underline{x}^*_n\right), \quad \text{for } n = 0,1,2, \ldots \tag{12.31}$$

with $K_{\text{damp}} \in [0, 1]$. K_{damp} can be set as $1/2^{\xi}$, where ξ is a nonnegative integer at the beginning of computing and updating $\underline{x}^*_{n+1}$. This method is the so-called damped Newton's method. In the estimate of $\underline{x}^*_{n+1}$, ξ can be first set with an initial value of 0. When a calculated $\left|\underline{f}\left(\underline{x}^*_{n+1}\right)\right| \geq \left|\underline{f}\left(\underline{x}^*_n\right)\right|$ in an iteration, we may choose not to update $\underline{x}^*_n$. Instead, we update ξ by $\xi = \xi + 1$ and recalculate $\underline{f}(\underline{x}^*_{n+1})$ until $\left|\underline{f}\left(\underline{x}^*_{n+1}\right)\right| < \left|\underline{f}\left(\underline{x}^*_n\right)\right|$.

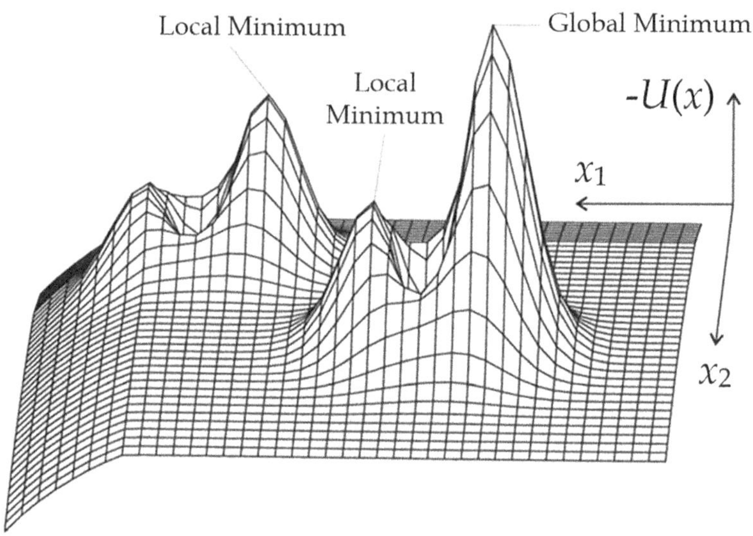

Fig. 12.4 Examples of the global and local minimum points. Note that the vertical axis is $-f(x)$, whose negative sign is added to better show the minimum points

To further introduce the concept of global optimization, we may first consider a function $f(\underline{x})$ defined over a two-dimensional space $\underline{x} = (x_1, x_2)$. As outlined in Fig. 12.4, we distinguish between local and global optima. A global optimum is an optimum of the whole domain of $\underline{x}$, while a local optimum is an optimum of only a subset of $\underline{x}$. The damped Newton's method performs the solution searching based on the gradient of a solution candidate under consideration. This means that when such a candidate locates near a local minimal point, the damped Newton's method will search out the local minimum as the final solution, rather than the global minimum. Obviously, a global minimum can be found when the current solution candidate is close to the global minimum. Whether the damped Newton's can obtain the global minimum would depend on the initial guess of the solution. As a first thought, an intuitive strategy is to compute a global optimal solution a few times from various starting points to avoid local extremes.

12.4.2 Penalty Methods for Constrained Optimization

To handle a constrained optimization problem, we may attempt to approximate it as an unconstrained one and then apply the related search techniques (e.g., Newton-Raphson method) to obtain the solutions. However, the regional constraints are more challenging. The conversion to an unconstrained optimization is accomplished by adding (1) penalty methods with a term to the objective function that prescribes a high cost for violation of the constraints or (2) barrier methods with a term that favors points in the interior of the feasible region over those near the boundary. For a problem with n variables and m constraints, both approaches work directly in the n-dimensional space of the variables. The discussion that follows emphasizes penalty methods recognizing that barrier methods embody these shared principles.

Here, we consider a simple example. Suppose there is a machining operation requiring a total feed length $L = 100$ (mm) operated by a feed rate f. The existing machine supports a maximum f of 5 (mm/s) (Without specifically mentioned, we know $f \geq 0$.). Our object is to minimize the machine time, so the corresponding optimization statement is

$$\begin{aligned} \text{Minimize} \quad & U(x) = 100/x \\ \text{Subject to} \quad & x - 5 \leq 0 \\ & -x \leq 0 \end{aligned} \tag{12.32}$$

Our approach is to embed the constraints in the objective function. The next step is to start by establishing a penalty function for every constraint. For the inequality constraint "$x - 5 \leq 0$," we can define a penalty function $P(x)$ such that $P(x) = 0$ when $x - 5 \leq 0$ or $P(x) = x - 5$ for $x - 5$ is >0. We may further define a more compact function as the quadratic loss $P(x) = [\max(0, x - 5)]^2$, where max( ) returns the maximum of the parameters, i.e., either 0 or whenever $(x - 5)$ is larger. Notably, the penalty function will be integrated with the objective function, which would then be solved by the Newton-Raphson method for continuous functions; therefore, we should configure the penalty function as continuous over x. Likewise, the penalty function for $-x \leq 0$ is $P_2(x) = [\max(0, -x)]^2$. Considering we are always trying to minimize $U(x)$, we add this the quadratic loss function to update the objective function as $F(x) = 100/x + \Phi_I[\max(0, x - 5)]^2 + \Phi_{I,2}[\max(0, -x)]^2$, where Φ_I and $\Phi_{I,2}$ are the correction factors for the inequality constraint. Generally, we would set the correction factors to be a very large value such that the constraints will never be violated during the computation search for the optimal solution. Its higher value increases accuracy, and the solution will less likely violate the constraints. The steep gradient around the penalty function is less of a problem as long as we adopt the damped Newton's method. In practice, the resultant equivalent unconstrained optimization problem is solved repeatedly with the penalty functions until the solutions converge.

The functional constraints are sometimes easy to handle by directly replacing the variables with the constrained values. Alternatively, it is still rather easy to convert these constraints into quadratic loss functions because we do not need to worry about the operation (max, $g(x)$) as the regional constraints. We can convert each of the functional constraints $H(x) = 0$ to a penalty function $P(x)$, which is a quadratic loss function to be added to the objective function with a very large correction factor:

$$P(x) = \Phi_E H(x)^2 \tag{12.33}$$

where Φ_E is a correction factor of the penalty effect. Normally we would set Φ_E as a large positive number. The lowest value of $P(x) = 0$ will occur when $H(x) = 0$, which is exactly what we need. After eliminating all the constraints by modifying the objective function properly, we can obtain an equivalent optimization statement without any constraints. In principle, we can solve the new converted problem using the modified Newton-Raphson method as discussed in the previous section.

Nevertheless, for the cases with a severe nonlinearity in the objective function integrated with boundary conditions, there might be a tendency to obtain a solution right over the boundaries. This is because the penalty functions do not set any strict constraints to the variables. Not matter how large the correction factors are set, violations of the conditions would only be reflected as a higher value of the objective function, rather than any strict indications. For regional boundary conditions, because exterior penalty functions start outside the feasible region and approach it from the outside, they only find extremes that occur on the boundaries of the feasible region, meaning that the computed solution could be located outside the boundaries. Anyways, this sub-optimal "infeasible" solution should be very close to an optimal solution. It is better to use methods which further refine the modified objective function in order to fulfill the problem specifications. Back to our example, the integrated objective function would become

$$\text{Minimize } F(x) = 100/x + \Phi_I(\max(0, \ x - 5))^2 + \Phi_{I,2}[\max(0, \ -x)]^2. \quad (12.34)$$

In order to fulfill the requirements of the constraints, the modified objective functions are often further combined with interior penalty functions. This type of penalty function begins with an initial point inside the feasible region, which is why these procedures are called interior penalty functions or "barrier methods." Before beginning, all constraints must be converted into (expression) ≤ 0 form. This method will not work with equality constraints. The simplest interior penalty function, called an "inverse barrier function" $B(x)$, involves writing a barrier function for each constraint $G_w(x) \leq 0$ as follows:

$$B(x) = -\Phi_{B,w}/\min(0, G_w(x)) \quad (12.35)$$

where $\Phi_{B,j}$ is a correction factor. It has two effects: (1) The negative sign changes the constraint from a negative number (≤ 0) to a positive and (2) the fraction ensures that as $G_w(x)$ approaches the constraint boundary at 0, the barrier function gets infinitely large. Therefore, this barrier creates a positive number that approaches infinity as x nears the boundary. As with exterior penalty functions, the inverse barrier method causes some graphical problems. Because it involves an inverse function $1/G_w(x)$, it creates an asymptotic graph. As long as we stay on the left side of the barrier, the method will work. Each of the exterior penalty functions includes a multiplier $\Phi_{B,j}$. We will begin with $\Phi_{B,j} = 1$ and gradually reduce $\Phi_{B,j}$ by, say, factors of 2. This has the effect of successively reducing the penalty. In the beginning the penalty is large, preventing the minimum point from crossing the barrier of the constraint. This barrier is so effective that the minimum point will not be very accurate. As the penalty gets smaller, the minimum point will approach closer to the boundary. Recalling our previous example described in Eq. 12.32, we will first calculate for the sub-optimal solution x_E^* by integrating the objective function with the barrier functions. We convert the constraint $x - 5 \leq 0$ as the barrier function as $B(x) = -\Phi_B/\min(x - 5, 0)$ and the same for the constraint $-x \leq 0$. The modified objective function becomes

$$\begin{aligned}\text{Minimize } F(x) = 100/x + \Phi_I\left(\max(0,\ x-5)^2 + \Phi_{I,2}[\max(0,\ -x)]^2\right.\\ -\Phi_B/\min(0,\ x-5) - \Phi_{B,2}/\min(0,\ -x).\end{aligned} \tag{12.36}$$

With this expression, we can first set $\Phi_I = 0$, $\Phi_{I,1} = 0$, $\Phi_B = 1$, and $\Phi_{B,1} = 1$ using the damped Newton's method. Importantly, in every computation, we should check whether the component in the Jacobian matrix (Eq. 12.33) has a non-zero value because we need to ensure x will change in the following calculations. If not, we should adjust the corresponding Δx^*_j in Eq. 12.37 until a non-zero Jacobian component is obtained. If the answer falls in the feasible region, we update x_{n+1}^* in Eq. 12.29 and repeat the computation with half of the value of Φ_B and $\Phi_{B,2}$, and use x_{n+1}^* as the next starting point until we first obtain an infeasible solution falling outside the constraint boundaries. Afterward, we increase/refine Φ_I and $\Phi_{I,\,2}$ and repeat the computation until a feasible solution within the error tolerance is obtained.

Furthermore, we need also to find a way to match the solution variables with the functional constraints. A simple way is to add another correction factor Φ_F to the penalty function as described in Eq. 12.30 such that

$$P(x) = \Phi_E(H(x) + \Phi_F)^2. \tag{12.37}$$

Let's consider that there is an optimization problem consisting of only one functional constraint. After finding a "virtual" optimal solution x^* for an optimization problem using the strategy as described in Eq. 12.33, we may update the required correction factor Φ_F by setting $\Phi_F = \Phi'_F + H(x^*)$, where Φ'_F is the value Φ_F in the previous round of computation. Then, we replace the penalty function using Eq. 12.37 and compute again using the damped Newton's method. We repeat the refinement of Φ_F and recalculation of x^* until x^* becomes feasible and falls within the solution tolerance.

Together, we may convert the optimization statement (Eq. 12.2) to the unconstrained one with the form:

$$\begin{aligned}\text{Minimize } U(x) + \sum_v \Phi_{E,v}\left[H_v(\underline{x}) + \Phi_{F,v}\right]^2\\ + \sum_w \Phi_{I,w}\left(\max\left(0, G_w(\underline{x})\right)\right) - \frac{\Phi_{B,w}}{\min\left(0, G_w(\underline{x})\right)}\end{aligned} \tag{12.38}$$

There are many variables involved in these penalty methods, including (1) the initial value of correction factors, (2) how fast the correction factors grow, (3) the initial value of $\underline{x}$, and (4) the step sizes in the damped Newton's methods and Jacobian matrices. In an actual problem, these variables are tried in different combinations. Therefore, it should be very clear that process design requires extensive technical knowledge and experience from the process designer. A competent designer can significantly reduce the amount of thinking time for an optimal process configuration as discussed previously in Sect. 12.3.3.

12.5 Demonstrated Examples

12.5.1 Milling over a Surface of an Artificial Knee

Milling over surfaces of implanted biomedical devices can help define both shape and surface roughness as we previously discussed in Chap. 10. In this section, let's consider a scenario in which a patient has been arranged for an implantation surgery of an artificial knee joint after a severe car accident. The expected outcome after the surgery is described as Fig. 12.5a. The lower tibial stainless steel plate is confirmed to be covered by a polyethylene sheet, whereas the material for the upper femoral replacement component is only stainless steel.

To facilitate the geometric modeling for the following discussion, we may consider the tibial plate as a simplified shape shown in Fig. 12.5b. The side length of the tibial stainless steel plate (L_{plate}) would be a key geometric parameter to define for the shape and area to be milled. The gray region indicates the surface needed to be milled in order to generate an appropriate surface roughness for better bone cell attachment (as elaborated in Sect. 10.2.3). The milling path includes two stages of slab milling as shown in Fig. 12.6. During milling, we consider that the cutter is always set with the same moving velocity throughout both the milling stages. The direction of milling paths on the material surface in either stage is along the same

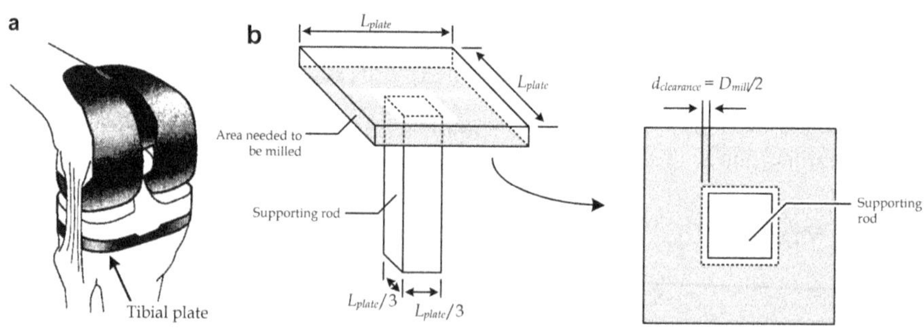

Fig. 12.5 (**a**) Basic structure of an artificial knee, which includes a tibial plate as indicated by arrow. (**b**) Simplified geometry of the tibial plate, with the expected rough surface highlighted in gray

direction. After the first milling stage, a series of grooves are expected to be generated over the surface as shown in the *left* inset of Fig. 12.6. Between the two stages, the milling head would move from the end position of stage 1 to the start position of stage 2 which will also change its orientation. The second-stage milling is similar to the first stage, except that the direction of "grooves" generated is perpendicular to those from the first stage. The resultant surface profile is illustrated in the *right* inset of Fig. 12.6.

As a more practical consideration, we would also take other factors such as tooling cost and labor wages (mentioned in Sect. 12.3.1) into account. The processing time for setting up/finalization of the machine process (T_{set}) is assumed to be 2 min, and the time for reorientation of the milling head between the two milling stages (T_{change}) is set as 10 s. For the other costs, the machine cost (M) is \$0.5/min with an overhead ratio (OH_m) of 0.1; the labor rate (W) is \$3/min with an overhead ratio (OH_{op}) of 0.25.

In terms of the tooling and machine specifications, we assume each tool (milling cutter) costs $C_{tool} = \$100$, and the tool life ($T_{tool}$) is ~3 h of cutting time. We further assume that the tool is not broken during the process. The milling cutter has a diameter D_{mill} (= 5 mm) and a height H_{mill} (= 3 mm) with a number of teeth on the cutter n (= 2). The maximum moving speed/feed rate of the milling head (F_{max}) is 50 mm/s. The maximum rotational speed of the milling cutter is 6000 rpm, yet the suggested rotational speed for this milling operation (N) is only 300 rpm. We would like to ignore the downtime of the machine.

Let's consider that the final tibial plate has a width of 60 mm and the required surface roughness for cell attachment has an upper bound (R_U) of 2 μm and a lower bound (R_L) of 3 μm. We should be able to formulate an optimization statement based on all the above information. First, we may estimate (1) the required distance of moving path for either stage 1 or stage 2 of the milling process and (2) the moving distance between the two stages. For the moving distance of either stage, we could first consider the total area of the process area, which is roughly equal to the plate area subtracted by the rod area (= $(8/9)L_{plate}^2$). The total machined "length"

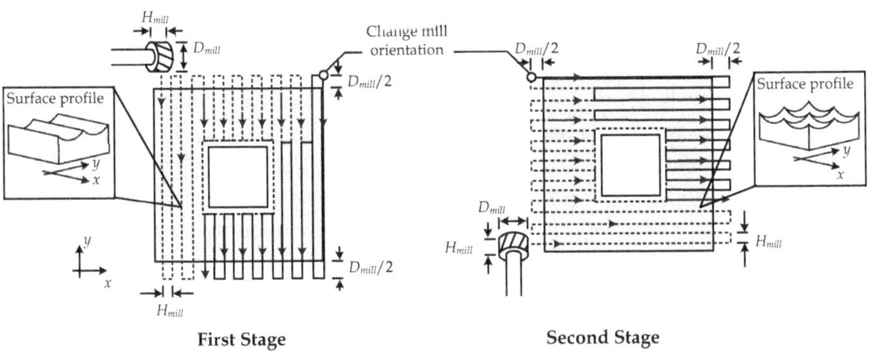

Fig. 12.6 The milling path and expected surface profiles during the two stages of the surface machining process for defining a surface roughness level

($L_{machine}$) should be around such machined area divided by the mill cutting height, i.e., $L_{machine} \approx (8/9)L_{plate}^2/H_{mill}$. The area is machined by multiple parallel lines of milling, and the number of lines is roughly about L_{plate}/H_{mill}. As previously discussed in Sect. 9.2.2, we need to add allowance distances on both sides of each milling line. Yet, for the area above and below the middle rod as shown in Fig. 12.6 (*left*), the middle is blocking the milling paths, and the blocking distance on either side of the rod is $D_{mill}/2$, compensating the allowance distance of those milling paths. Therefore, the additional moving distance caused by the allowance would be (2/3) D_{mill}. The total distance of moving path L_{cut} during stage 1 is

$$L_{cut} \approx \frac{8L_{plate}^2}{9H_{mill}} + \frac{2D_{mill}L_{plate}}{3H_{mill}} = \frac{2L_{plate}}{9H_{mill}}\left(4L_{plate} + 3D_{mill}\right) \tag{12.39}$$

Likewise, there should be a similar total distance for a moving path during stage 2. Besides, the moving distance between the two stages (L_{travel}) is rather straightforward, equal to $L_{plate}/3 + H_{mill}$. Between the two stages, other than the cutting movement, the cutting would also need to alter its orientation. The total machining time would be $(2L_{cut} + L_{travel})/F_{max}$. Adopting the relation given in Eq. 12.8, we may express the objective function U. Recalling Eq. 10.11, we can further obtain the regional constraints. Hence, the optimization statement can be written as

$$\begin{aligned}
&\text{Minimize}\\
&U = \left[M(1+\mathrm{OH_m}) + W\left(1+\mathrm{OH_{op}}\right)\right] \times \left(t_{set} + t_{change}\right)\\
&\quad + \left[M(1+\mathrm{OH_m}) + W\left(1+\mathrm{OH_{op}}\right) + \frac{C_{tool}}{T_{tool}}\right] \times \frac{L_{plate}}{F_{max}}\left[\frac{16L_{plate} + 12D_{mill}}{9H_{mill}} + \frac{1}{3} + \frac{H_{mill}}{L_{plate}}\right]\\
&\text{Subject to } R_L - \frac{\left(F_{max}/nN\right)^2}{8\left[D_{mill}/2 + F_{max}/(N\pi)\right]} \le 0\\
&\qquad \frac{\left(F_{max}/nN\right)^2}{8\left[D_{mill}/2 + F_{max}/(N\pi)\right]} - R_U \le 0
\end{aligned} \tag{12.40}$$

The preset values can be either set as the functional constraints or they can directly replace the corresponding variables in the above optimization statement. For this simple example, we may not need to solve for the optimal cost using the penalty method. In fact, the minimum cost should imply the process with the shortest cutting time and the largest roughness ($R_U = 3$ μm) given that the required constraints are maintained. After plugging in the defined values and calculating, we obtain a minimum cost to be ~\$80. Comparing against the finest roughness (2 μm) with the manufacturing cost of ~\$109, these optimization procedures may help to reduce >20% of the production cost.

12.5.2 *Optimizing Speeds and Feeds for Machining*

In some cases of machining processes, many tools may be engaged simultaneously on one or several identical workpieces. Each of the individual machining processes can be fed along a production line, such that the production time per single machine part can be shortened. As an example for the demonstration purpose, let us consider a product part being fabricated with two spindles; and two tools are involved in each of the spindle, as shown in Fig. 12.7. On the first spindle, two tools (tool "1" and tool "2") are involved in the turning operations on two cylindrical sections of the workpiece with lengths L_1 and L_2 and diameters D_1 and D_2. On the second spindle, tool "3" is applied for the turning operation of the workpiece section with a diameter of D_3 and a length of L_3. Then, tool "4," which is a drill cutter with a diameter of D_4, is applied to drill on the *right* surface of the workpiece with a depth of L_4. Furthermore, as these four machining processes with different tools are applied to different workpieces simultaneously, we may minimize the production time by configuring that the machine time for all the tools $t_{m,i}$ ($i = 1, 2, 3, 4$) is the same.

Now, let's consider that $D_1 = 112$ mm, $D_2 = 60$ mm, $D_3 = 86$ mm, $D_4 = 28$ mm, $L_1 = 40$ mm, $L_2 = 80$ mm, $L_3 = 50$ mm, and $L_4 = 30$ mm. We further assume that the time for changing tools ($t_{change,1}$, $t_{change,2}$, and $t_{change,3}$) is the same for tools "1," "2," and "3" with 5 min and the tool cost ($C_{tool,1}$, $C_{tool,2}$, and $C_{tool,3}$) is also with the same price of \$2 per tool, as they are all the turning cutter. For the drill cutter (tool "4"), we consider the time of tool changing $t_{change,4}$ is 3 min and the cost ($C_{tool,4}$) is \$5 per tool. The tool life equation for tools "1," "2," and "3" is $v_i^3 f_{r,i}^2 T_{tool,i} = 3.2 \times 10^6$, where v_i is the cutting speed (m/min), $f_{r,i}$ is the feed per workpiece rotation (mm), $T_{tool,i}$ is the tool life (min), and i is the tool index of the turning cutters (1–3, and). The upper limit for such feed $(f_{r\_max})_{1,2,3}$ is 0.4 mm. The tool life equation for tool "4" is $v_4^5 f_{r,4}^2 T_{tool,4} = 2 \times 10^7$, and the upper limit for the feed $(f_{r\_max})_4$ is 0.3 mm. As tools "1" and "2" are working with the same spindle, their machine speeds (rpm), denoted as N_1 and N_2, are the same ($N_1 = N_2$). Besides, the machine speeds for tool "3" (N_3) and tool "4" (N_4) are the same ($N_3 = N_4$). Considering the relation $L_1/f_{r,1} = L_2/f_{r,2}$ and the upper limits of the feeds, we obtain $f_{r,1} = 0.2$ mm and $f_{r,2} = 0.4$ mm. Likewise, we also obtain $f_{r,3} = 0.4$ mm and

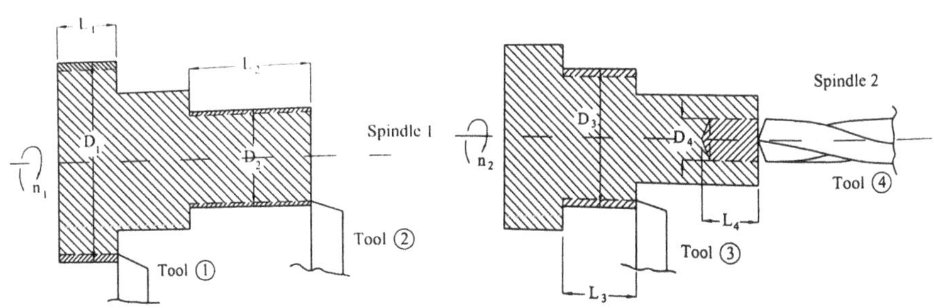

Fig. 12.7 Part of a multi-spindle turning operation. Two spindles are shown with two tools machining in each of the two stations. Speeds and feeds for all tools can be optimized using the constraint dictating that all tools must work for full cycle time

$f_{r,4} = 0.24$ mm. Then, according to Eq. 10.1, the cutting speeds are $v_i = \pi D_i N_i$, where i = 1, 2, 3, and 4. In this particular example, since we wish to configure the operations of tool "1" and tool "3" to have the same machine time (T_{cut} as described in Eq. 10.2), we can obtain the relation of $N_3 = 0.625N_1$ or $N_4 = 0.625N_2$. Hence, the cuttings speeds and the machine speeds would have the relations $v_1 = 0.3519N_1$, $v_2 = 0.1885N_2$, $v_3 = 0.1689N_3$, and $v_4 = 0.0550N_4$.

We may simplify Eq. 12.8 by combining the labor rate and machine cost rate as just the cost rate C_{rate}, which is assumed as $1 per minute. Then, we have the total machine cost per part U_{part}:

$$U_{part} = C_{rate}T_{cut} + \sum_{i=1}^{4}\left(C_{rate}t_{change,i} + C_{tool,i}\frac{T_{cut}}{T_{tool,i}}\right) + C_{fix} \tag{12.41}$$

where C_{fix} is the fixed manufacturing cost including the material cost, which is a constant in this example. We may calculate the machine time T_{cut} by $T_{cut} = t_{m,1} = L_1/(f_{r,1}N_1) = 200/N_1$.

It should be noted that the cost for tool changing is fixed in this manufacturing operation. From the given tool life equations, we can obtain

$$U = C_{fix} + \frac{200}{N_1} + 1.57 \times 10^{-6}{N_1}^2 + 2.32 \times 10^{12}{N_1}^4 \tag{12.42}$$

To find the optimum N_1, the first derivative of U_{part} with respect to N_1 should be zero:

$$\frac{dU_{part}}{dN_1} = -\frac{200}{{N_1}^2} + 3.14 \times 10^{-6}N_1 + 9.28 \times 10^{12}{N_1}^3 = 0; \tag{12.43}$$

and therefore $N_1 = 359$ rpm. To double check on whether this point is a minimum of U_{part}, the second derivative of U_{part} should be positive; and clearly

$$\left.\frac{d^2U_{part}}{d{N_1}^2}\right|_{N_1=359} = \frac{400}{{N_1}^3} + 3.14 \times 10^{-6} + 27.84 \times 10^{12}{N_1}^2 > 0. \tag{12.44}$$

This optimum N_1 implies the cutting speeds (v_1 = 126 m/min, v_2 = 68 m/min, v_3 = 61 m/min, and v_4 = 20 m/min) and tool lives ($T_{tool,1}$ = 40 min, $T_{tool,2}$ = 65 min, $T_{tool,3}$ = 90 min, and $T_{tool,4}$ = 116 min). The cost per part $U_{part} = C_{fix}$ + $0.796.

12.5.3 Face Milling over a Surface of an Artificial Knee

Let's consider another example. Imagine that there is an urgent need for the manufacturing of an artificial bone for the replacement of a bone fragment missing in the skull of a patient who had a serious injury in a car accident. X-ray

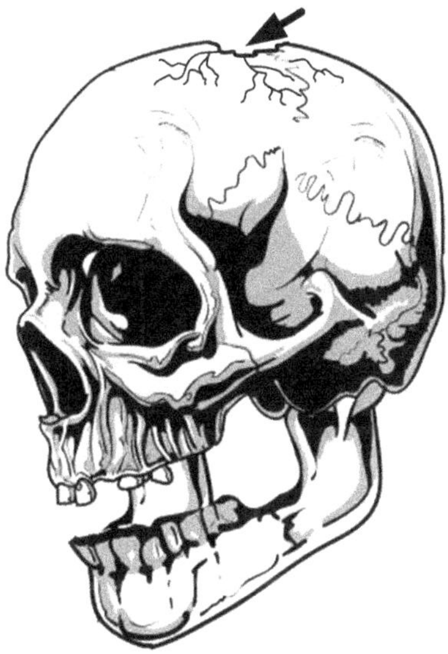

Fig. 12.8 Example of a skull fracture

computerized tomography was applied to visualize the fracture sites shown in the Fig. 12.8. Moreover, a surgery has been previously performed to remove a broken piece of bone. After the primary healing of the patient, the patient has already overcome the critical moment. Now, the next step is to manufacture a scaffold for the subsequent bone cell seeding and culture in order to develop the in vitro artificial bone. In the coming surgery, it is expected that this artificial bone will be placed at the missing location of the skull to seal the skull.

The scaffold has been planned to be fabricated by a rapid prototyping technique called fused deposition modeling (Sect. 7.5.4). The substrate material we could use is a composite of hydroxyapatite and PLA-PEG-PLA, which is assembled from poly-lactic acid (PLA) and poly-(ethylene glycol) (PEG). The composite should include mostly PLA-PEG-PLA liquid, whose viscosity (μ) is ~8.872×10^{-3} kg/m·s. A feasible chemical synthesis of PLA-PEG-PLA as shown in Fig. 12.9 is to first dissolve lactide as lactic acid and generate PLA via a condensation polymerization scheme. Then, by adding PEG into the solution, another condensation polymerization occurs and generates PLA-PEG-PLA.

During the rapid prototyping process, the composite liquid should be applied onto the sample for deposition using a syringe. The injection operation would be controlled by a computer. The composite solution should flow through a syringe needle with a length of 10 mm (L_n) and an inner radius (R_n) of 0.1 mm. The radius of the plunger (R_p) where pressure is created by the user's thumb to create the force for fluid motion is 1 cm. Assume that the process requires a solution flow rate denoted as Q. Recalling Eq. 2.26, the required force F for supporting the liquid flow rate can be expressed as the required fluidic pressure P_p multiplied by the needle cross-sectional area:

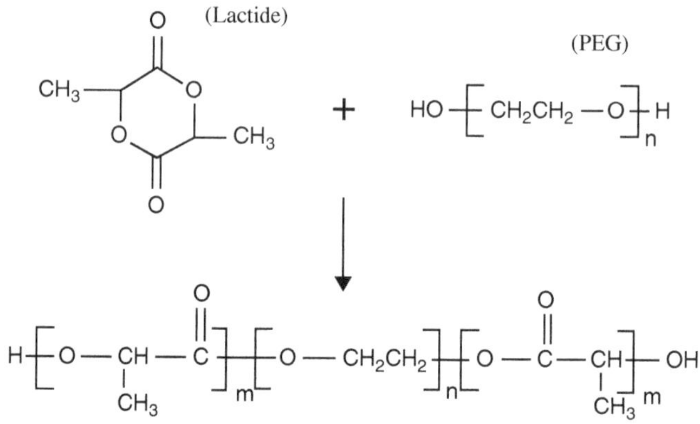

Fig. 12.9 Polymerization of PLA-PEG-PLA

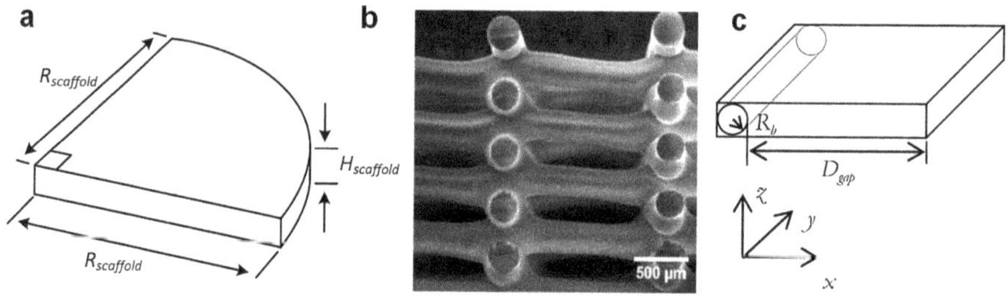

Fig. 12.10 (**a**) Shape of the expected scaffold. (**b**) Design architecture of the scaffold. (Olubamiji [7], Biofabrication. Reproduced with permission from IOP Science.) (**c**) Shape of a repeating unit

$$F = \pi R_{\mathrm{p}}^2 P_{\mathrm{p}} = \frac{8\mu R_{\mathrm{p}}^2 L_{\mathrm{n}} Q}{R_{\mathrm{n}}^4} \tag{12.45}$$

In addition, we assume that the missing piece of bone had the shape of a quarter of a circular disc with a radius R_{scaffold} (=2 cm) and thickness H_{scaffold} (= 3 mm) as shown in Fig. 12.10a. Taking into account that a bone cell should have a diameter ~10 μm, the scaffold should have separating hole/gap widths D_{gap} such that the cells can deposit and migrate inside the scaffold. The scaffold can be fabricated with a designed architecture as a multilayer mesh as shown in Fig. 12.10b. The structural material of the scaffold is PLA-PEG-PLA, whose modulus of elasticity (E_{fiber}) is ~3.5 GPa. A key design parameter of the scaffold is its porosity P_{scaffold}. We may approximate P_{scaffold} by examining a repeating unit of the designed architecture as shown in Fig. 12.10c, which consists only of a straight fiber in an empty volume. We assume that the syringe movement is $Q/(\pi R_{\mathrm{b}}^2)$ such that the generated fiber would have a radius of R_{b}, matching with the inner radius of the needle. Because the scaffold has the hole width of D_{gap}, the repeating unit would have a width of

D_{gap} + $2R_b$ and a height of $2R_b$. The ratio between the material volume and the total volume of the repeating unit should then be $\pi R_b^2/[2R_b(D_{gap} + 2R_b)]$. Hence, the porosity P can be approximated as

$$P = 1 - \frac{\pi R_b}{\left(4R_b + 2D_{gap}\right)}. \tag{12.46}$$

Recalling Eq. 12.1, the effective elastic modulus of the scaffold for designed architectures is

$$E_{scaffold} = E_{fiber}\left[\frac{\pi R_b}{\left(4R_b + 2D_{gap}\right)}\right]^2. \tag{12.47}$$

The majority of the time during scaffold fabrication T_{fab} would be for the polymer deposition. T_{fab} can be estimated by the total fiber length (L_{fab}) of the fiber mesh divided by the average velocity of the polymer outflow, i.e., $T_{fab} = L_{fab}/[Q/(\pi R_b^2)]$. To estimate T_{fab}, we may first consider the number of layers in the scaffold, which is about $H_{scaffold}/(2R_b)$, as well as the fiber length in each layer. In each mesh layer, as we know that for each distance of $2R_b + D_{gap}$ there would be one neighboring fiber, then we know that the fiber length would be around $(\pi R_{scaffold}^2/4)/[2R_b + D_{gap}]$. Hence, the total fiber length in the scaffold is

$$L_f \approx \frac{H_{scaffold}}{2R_b} \times \frac{\pi R_{scaffold}^2}{4\left[2R_b + D_{gap}\right]}. \tag{12.48}$$

Further considering that the bone cells (osteoblasts) have an average diameter of 10–20 μm, we may have D_{gap} which are larger than an acceptable length D_{min} (= 40 μm) such that the cells can migrate into the inner scaffold body. The scaffold should offer an adequate level of effective elastic modulus larger than a minimal acceptable level E_{min} (= 0.5 GPa) to maintain the subsequent bone regeneration and calcification. In the case where we have the extruder set with a maximum injection force (F_{max}) of 20 N on the FDM machine, the optimization statement may be written as:

$$\begin{aligned}
&\text{Minimize} && \frac{\pi^2 H_{scaffold} R_{scaffold}^2 R_b}{8\left[2R_b + D_{gap}\right]Q} \\
&\text{Subject to} && E_{min} - E_{fiber}\left[\frac{\pi R_b}{\left(4R_b + 2D_{gap}\right)}\right]^2 \le 0 \\
& && D_{min} - D_{gap} \le 0 \\
& && -F \le 0 \\
& && F - F_{max} \le 0
\end{aligned} \tag{12.49}$$

Clearly, F should be set as the maximum value for the largest flow rate and fastest needle movement. Then, we should compute D_{gap} for the minimal allowable effective elastic modulus and check if it is larger than D_{min}. Otherwise, we should set $D_{gap} = D_{min}$. With these considerations, the minimal fabrication time and the corresponding manufacturing parameters can be calculated accordingly.

12.5.4 Laser Sintering of a Tibial Plate

In this section, we attempt to formulate an optimization statement for the minimal fabrication time of the tibial plate using metal laser sintering. Sometimes, a custom biomedical implant is required to be fabricated within a short time frame because of the severe immediacy of the surgery. Here, we chose the tibial plate in this discussion because it is a more familiar component as we have initiated some discussions on it in the previous Sect. 12.5.1. We also consider that the tibial plate has the bulk shape as the previous example (a rectangular block) as described in Sect. 12.5.1, with the plate width $W_{plate} \approx 50$ mm and plate thickness $H_{plate} \approx 5$ mm. Furthermore, this tibial plate has a designed architecture as a metallic "scaffold," which features 100% interconnectivity. Taking into account that a bone cell should have a diameter ~10 μm, the scaffold should have separating hole/gap widths D_{gap} between 60 μm and 300 μm such that the cells can deposit and migrate inside the scaffold. The requirements of D_{gap} in horizontal directions (x and y) of the scaffold can be achieved by modifying the path of laser movement during the laser sintering process, whereas that in the vertical direction (z) can be defined by the number of layers repeatedly printed with the same pattern, as illustrated in Fig. 12.11. The metal/alloy wires should have both its width (W_{wire}) and thickness (B_{wire}) ≥ 30 μm for the sufficient adhesion between consecutive welded spots of the stainless steel wire. On the other hand, the effective compressive stiffness and yield strength of the scaffold should be at least higher than those of the cortical bone (E_{bone} and S_{bone}).

Imagine that your supervisor asked you to choose stainless steel (mostly iron, 17% chromium, and 10% nickel) as the scaffold material, which has a melting point

Fig. 12.11 Example of a scaffold with designed architecture, multiple layers of the same path are printed/sintered in order to thicken each layer pattern and to define the pore heights. (Pawar [8], Science Advances)

(T_{melt}) of 1800 K, a boiling point (T_{boil}) of 2800 K, thermal diffusivity $\alpha = 4.4$ mm^2/s, and the product of density and specific heat capacity (ρC_p) of 3.6 N/(mm^{2o}C). The density of stainless steel is denoted as ρ; and the laser absorption of stainless steel is denoted as A. In the sintering process, stainless steel microbeads with the same diameter in the range of ~10 μm are used. As a brief approximation, the microbeads are assumed to be fully packed such that the material volume has a ratio of χ_V (~80%) relative to the bulk volume and the effective thermal diffusivity is about χ_α (~93%) of the intrinsic thermal diffusivity α. Furthermore, you are advised to consider the biocompatibility of the fabricated scaffold from the ion release point of view. The ion release rates of iron (Q_{Fe} = 8.4 μg/cm^2/week), chromium (Q_{Cr} = 0.01 μg/cm^2/week), and nickel (Q_{Ni} = 1.15 μg/cm^2/week) should also be considered together with the wear effect in order to ensure all the ion levels are within the ratio $\Omega_{threshold}$ (= 6% or 0.06) of the maximum allowable intake levels for iron (I_{Fe} = 200 μg/l/day), chromium (I_{Cr} = 35 μg/l/day), and nickel (I_{Ni} = 400 μg/l/day). Basically, you need to ensure the ion release from the stainless steel implant will not cause health problems to the patient.

In addition, the metal laser sintering machine available in the factory can support a scanning speed of the laser beam (V_{laser}) below 10 mm s^{-1}, and each laser-sintered layer should have a thickness of about 30 μm. You may choose either CO_2 laser or ND-YAG laser for the process. The laser can be applied with the pulse-width modulation scheme for repeated pulses with an active laser power P_{peak} (= 10 kW) and a pulse time length (t_p, in the range of 1 ps–0.1 ns) for every repetition duration t_{period} = 1 ns.

To calculate for the minimum manufacturing time, we should first choose the laser type with a higher absorption A for stainless steel. This means we should choose ND-YAG laser for the process such that $A \approx 0.35$. Then, we consider the temperature profile of a moving laser beam $T_m(x, r)$ at the exposure site by recalling Eqs. 8.1 and 8.19:

$$T_m(x, r) = \frac{P_{peak} t_p}{t_{period}} \times \frac{A}{2\pi \chi_V \rho C_p \alpha \chi_\alpha r} \exp\left(-\frac{V_{laser}(r + x)}{2\alpha \chi_\alpha}\right) \tag{12.50}$$

where r is the radial distance from the exposure spot and x is the rear position along the laser sintering path. $T_m(x, r)$ should be above the melting point of steel within the region of the sintering steel wire with both the width and thickness >30 μm. At $x = 0$, $T_m(0, r)$ is reduced as r, implying that the wire width is roughly double the wire thickness; and we should just need to consider the wire thickness B_{wire} or the radius of a welded spot R_{wire} for this constraint. We further approximate the term exp(−£) as 1 − £, then

$$T_{melt} \leq \frac{A P_{peak}}{2\pi \chi_V \rho C_p \alpha \chi_\alpha t_{period}} \times t_p \left(\frac{1}{R_{wire}} - \frac{V_{laser}}{2\alpha \chi_\alpha}\right), \tag{12.51}$$

and the further approximation would be

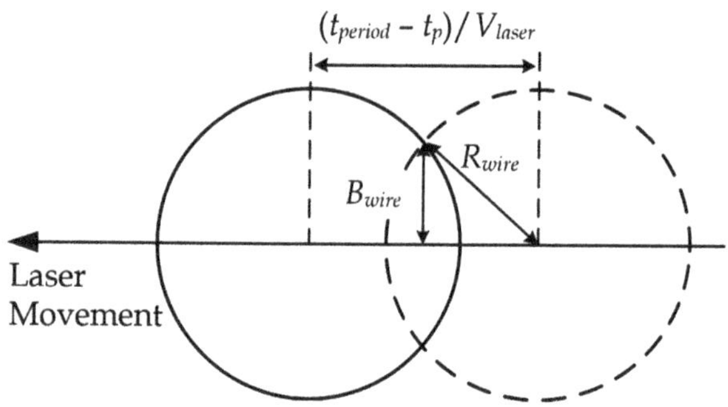

Fig. 12.12 Relationship between laser moving speed (V_{laser}) and minimal laser-sintered wire thickness (B_{wire}), which equals to half of the minimal wire width (i.e. $B_{\text{wire}} = W_{\text{wire}}/2$)

$$\frac{t_{\text{p}}}{R_{\text{wire}}} = \frac{2\pi\chi_{\text{V}}\rho C_{\text{p}}\alpha\chi_{\alpha}t_{\text{period}}T_{\text{melt}}}{AP_{\text{peak}}}. \tag{12.52}$$

The formed wire can be considered as multiple welded hemispherical spots linking up together along the laser path, as shown in Fig. 12.12. The center-center distance between two hemispherical spots is roughly equal to $V_{\text{laser}}(t_{\text{period}} - t_{\text{p}})$, meaning that the minimum wire thickness B_{wire} (= 30 μm) would be at the middle position between any two consecutive spot centers, and B_{wire} can be expressed as

$$B_{\text{wire}} = \sqrt{4R_{\text{wire}}^2 - V_{\text{laser}}^2\left(t_{\text{period}} - t_{\text{p}}\right)^2} \tag{12.53}$$

Hence, Eq. 12.52 can be expressed as

$$\frac{t_{\text{p}}}{\sqrt{B_{\text{wire}}^2 + V_{\text{laser}}^2\left(t_{\text{period}} - t_{\text{p}}\right)^2}} = \frac{\pi\chi_{\text{V}}\rho C_{\text{p}}\alpha\chi_{\alpha}t_{\text{period}}T_{\text{melt}}}{AP_{\text{peak}}} \tag{12.54}$$

$$V_{\text{laser}} = \frac{1}{\left(t_{\text{period}} - t_{\text{p}}\right)}\sqrt{\left(\frac{AP_{\text{peak}}t_{\text{p}}}{\pi\chi_{\text{V}}\rho C_{\text{p}}\alpha\chi_{\alpha}t_{\text{period}}T_{\text{melt}}}\right)^2 - B_{\text{wire}}^2}. \tag{12.55}$$

Apparently, Eq. 12.55 is a function of V_{laser} and t_{p}. We need to find the optimal value of t_{p} such that V_{laser} is maximum, implying that $dV_{\text{laser}}/dt_{\text{p}} = 0$. Differentiating Eq. 12.55, we can find such optimal t_{p}.

$$\frac{dV_{\text{laser}}}{dt_{\text{p}}} = 0 = \frac{1}{\left(t_{\text{period}} - t_{\text{p}}\right)^2}\sqrt{\left(\frac{AP_{\text{peak}}t_{\text{p}}}{\pi\chi_{\text{V}}\rho C_{\text{p}}\alpha\chi_{\alpha}t_{\text{period}}T_{\text{melt}}}\right)^2 - B_{\text{wire}}^2} + \frac{1}{\left(t_{\text{period}} - t_{\text{p}}\right)}\frac{t_{\text{p}}}{\sqrt{t_{\text{p}}^2 - \left(\frac{\pi\chi_{\text{V}}\rho C_{\text{p}}\alpha\chi_{\alpha}t_{\text{period}}T_{\text{melt}}}{AP_{\text{peak}}}\right)^2 B_{\text{wire}}^2}}, \text{ and hence} \tag{12.56}$$

$$t_{\mathrm{p}} = \frac{\pi \chi_{\mathrm{V}} \rho C_{\mathrm{p}} \alpha \chi_{\alpha} T_{\mathrm{melt}} B_{\mathrm{wire}}^{2}}{A P_{\mathrm{peak}}}. \tag{12.57}$$

Substituting Eq. 12.57 into Eq. 12.58, we have

$$V_{\mathrm{laser}} = \frac{A P_{\mathrm{peak}}}{\left(A P_{\mathrm{peak}} t_{\mathrm{period}} - \pi \chi_{\mathrm{V}} \rho C_{\mathrm{p}} \alpha \chi_{\alpha} T_{\mathrm{melt}} B_{\mathrm{wire}}^{2}\right)} \sqrt{\frac{B_{\mathrm{wire}}^{4}}{t_{\mathrm{period}}^{2}} - B_{\mathrm{wire}}^{2}}. \tag{12.58}$$

Subsequently, V_{laser} can be calculated by Eq. 12.55 and then R_{wire} can be calculated by Eq. 12.52. It should be mentioned that R_{wire} should be slightly larger than B_{wire} and R_{wire} defines the thickness of each laser-sintered layer. Requirements of the gap distance ($D_{\mathrm{gap}} > 60$ μm) can be easily achieved by laser sintering two layers with the same laser path and one wire pattern (as previously described in Fig. 12.11).

The manufacturing time (T_{fab}) can be computed by $T_{\mathrm{fab}} = L_{\mathrm{wire}}/V_{\mathrm{laser}}$, where Lwire is the total laser-sintered wire length. Similar to Eq. 12.48, L_{wire} is approximately expressed as

$$L_{\mathrm{wire}} = \frac{W_{\mathrm{plate}}^{2} H_{\mathrm{plate}}}{\left(2R_{\mathrm{wire}} + D_{\mathrm{gap}}\right) R_{\mathrm{wire}}} \tag{12.59}$$

Therefore, the optimization statement, with some of the key constraints, may be written as

$$\begin{aligned} &\text{Minimize} \quad \frac{W_{\mathrm{plate}}^{2} H_{\mathrm{plate}}}{\left(2R_{\mathrm{wire}} + D_{\mathrm{gap}}\right) R_{\mathrm{wire}} V_{\mathrm{laser}}} \\ &\text{Subject to} \quad V_{\mathrm{laser}} = \frac{A P_{\mathrm{peak}}}{\left(A P_{\mathrm{peak}} t_{\mathrm{period}} - \pi \chi_{\mathrm{V}} \rho C_{\mathrm{p}} \alpha \chi_{\alpha} T_{\mathrm{melt}} B_{\mathrm{wire}}^{2}\right)} \sqrt{\frac{B_{\mathrm{wire}}^{4}}{t_{\mathrm{period}}^{2}} - B_{\mathrm{wire}}^{2}} \\ &\qquad\qquad E_{\mathrm{bone}} - E_{\mathrm{steel}} \left[\frac{\pi R_{\mathrm{wire}}}{\left(2R_{\mathrm{b}} + D_{\mathrm{gap}}\right)}\right]^{2} \leq 0 \end{aligned} \tag{12.60}$$

The remaining constraints in Eq. 12.60 would include the biocompatibility requirements based on the ion release and the given upper and lower limits of the design parameters. For values of the minimal manufacturing time and other optimal parameters, interested readers can calculate themselves.

Problems

Problem 12.1

A simple, single-pass, cylindrical turning operation is considered as shown in Fig. 12.P1. Workpiece material is stainless steel, strength = 2000 N/mm^2, length $L = 350$ mm, and mean diameter $d_m = 92$ mm. For the turning operation, cutting speed $v = 120$ m/min, feed per revolution = 0.25 mm, and depth of cut $a = 5$ mm. Determine tangential cutting force (N), cutting power (W), metal removal rate (cm^3/mm), and time of machining (min).

Problem 12.2

A simple, single-pass, cylindrical turning operation for a product part is considered as shown in Fig. 12.P1. Workpiece material is stainless steel, strength = 2000 N/mm^2, length $L = 200$ mm, and mean diameter $d_m = 75$ mm. For the turning operation, cutting speed $v = 150$ m/min, feed per revolution $f_r = 0.25$ mm, depth of cut $a = 5$ mm, machine cost rate is \$0.4 per mm of feed distance, and time for changing tool $t_{change} = 8$ mm. Also given that the cost per cutting tool $C_t = \$4$ and the tool life equation is $v^3 f_r T_{tool} = 8 \times 10^6$ where T_{tool} is the tool life (min), determine the power consumed (kW), machining time, and cost per part.

Problem 12.3

A single-tool, single-pass, cylindrical turning operation for a product part is considered as shown in Fig. 12.P1. Workpiece material has a dimension of length $L = 400$ mm and mean diameter $d_m = 80$ mm. It is given that the tool life equation

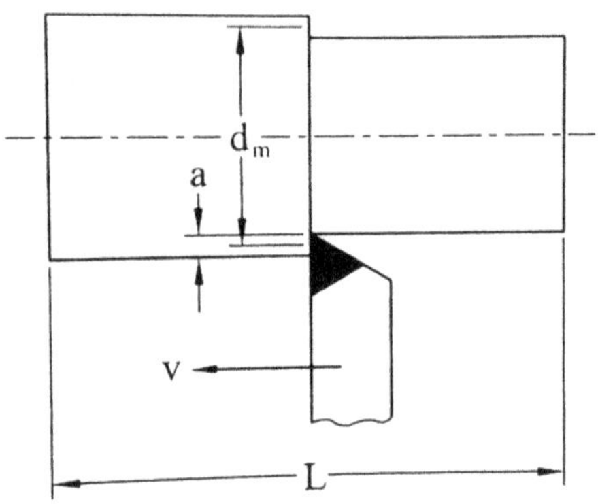

Fig. 12.P1 A single-tool, single-pass, cylindrical turning operation

$v^{3.5}f_r^{2.5}T_{tool} = 15.24 \times 10^6$ where T_{tool} is the tool life (min), v is cutting speed (m/min), and f_r is feed per revolution (mm). The machine cost rate is \$0.5 per minute of machining time, the time for changing tool $t_{change} = 5$ mm, and the cost per cutting tool $C_t = \$2.5$.

(a) The feed f_r is limited by the maximum permissible cutting force of 2516 N. If the cutting force F_C is determined by $F_{cut} = 1400bf_r$ and $b = 5$ mm, what is the maximum feed per revolution?
(b) Express the machining time T_{cut} as a function of v, and determine the optimum cutting speed (m/min) for the shortest machining time.
(c) What is the corresponding machining time T_{cut} and the minimum cost per part?

Problem 12.4

A single-tool, single-pass, cylindrical turning operation for a product part is considered as shown in Fig. 12.P1. Workpiece material is stainless steel, strength = 2000 N/mm^2, density × specific heat capacity = 3.7 N/(mm$^{2\circ}$C), length $L = 250$ mm, and mean diameter $d_m = 80$ mm. For the turning operation, feed per revolution $f_r = 0.25$ mm, depth of cut $a = 5$ mm, time for changing tool $t_{change} = 7$ mm, and the cost rate of machining is \$0.65 per machining time (min). It is given that the tool life equation is $v^3 f_r^2 T_{tool} = 1 \times 10^7$ (where T_{tool} is the tool life (min) and v is cutting speed (m/min)), and the cost per cutting tool C_t is \$3. Determine the optimal cost per part and the corresponding (optimal) machine speed (rpm). Also, determine the cutting force (N), metal removal rate (cm^3/min), and power consumed (kW). Further assuming that all heat is carried off by the chipped material, determine the final average temperature of the chipped material (°C) above the environmental temperature.

Problem 12.5

A two-tool cylindrical turning operation on one spindle for a product part is considered as shown in Fig. 12.P2. Workpiece material has strength = 1800 N/mm^2 and the dimensional parameters $d_1 = 50$ mm, $d_2 = 75$ mm, $l_1 = 100$ mm, $l_2 = 50$ mm. It is given that the tool life equation is $v^3 f_r^2 T_{tool} = 7 \times 10^7$, where T_{tool} is the tool life (min), f_r is feed per revolution (mm), and v is cutting speed (m/min). For the turning operation, depth of cut for region “1” $a_1 = 5$ mm, depth of cut for region “2” $a_2 = 4$ mm, time for changing tool $t_{change} = 6$ mm, cost rate of machining is \$0.8 per machining time (min), and cost per cutting tool $C_t = \$9$.

(a) Assuming machine speed $N = 750$ rpm, calculate the cutting speed, feed per revolution, tool life, and cutting force for region “1” and region “2,” separately. Further, calculate the total machine time (min) and power consumed (kW).

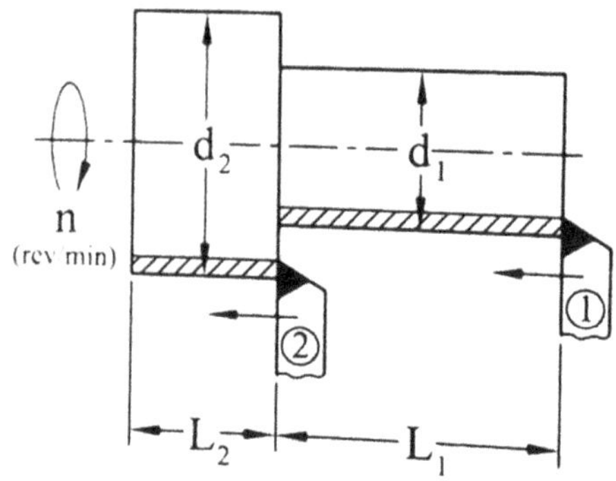

Fig. 12.P2 A two-tool cylindrical turning operation on one spindle

(b) Assuming machine speed $N = 500$ rpm, repeat part (a).
(c) Find the optimum spindle (machine) speed N for the minimum machine cost.

Problem 12.6

Please reconsider the example discussed in Sect. 12.5.2, except that we assume the production of just one product part. This means the machine time for the four different tools can be different, and the total cutting time T_{cut} is the sum of the machine time for different tools ($t_{m,i}$, $i = 1, 2, 3, 4$), i.e., $T_{cut} = t_{m,1} + t_{m,2} + t_{m,3} + t_{m,4}$. Calculate the minimum manufacturing cost of the part, the corresponding optimum cutting speeds (v_1, v_2, v_3, v_4), and tool lives ($T_{tool,1}$, $T_{tool,2}$, $T_{tool,3}$, $T_{tool,4}$).

Problem 12.7

The cervical disc is indicated for implantation in neck joints of skeletally mature patients for reconstruction of a degenerated cervical disc following arm pain and/or a neurological deficit due to a single-level abnormality localized to the disc space. Considering that a commercially available cervical disc is composed of cobalt-chromium (mainly Co, 30% Cr, and 0.8% Fe) endplates and a central polyethylene core, the broad radius of the polyethylene core allows for the required joint motion (as shown in Fig.12.P3).

(a) Please propose a manufacturing process flow for producing the cervical disc, excluding surface finishing steps. You may include drawings to describe the shape changes of materials throughout the process.
(b) Here, for a simple analysis, we may consider the cervical disc as three blocks of materials as shown in Fig. 12.P4. The upper and lower Co-Cr blocks have the same side lengths $L_{Co\text{-}Cr}$ (= 20 mm) and thickness $H_{Co\text{-}Cr}$ (= 2 mm).

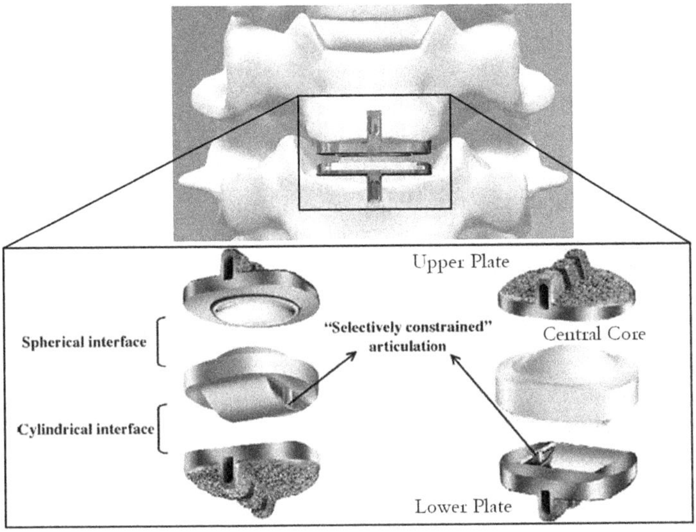

Fig. 12.P3 A cervical disc for implantation in neck joints

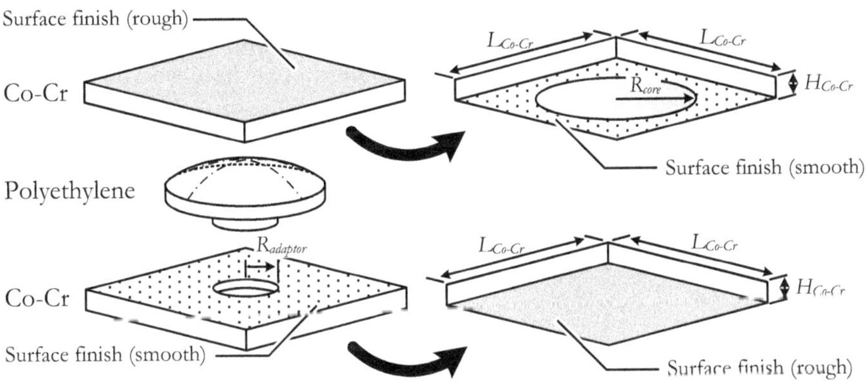

Fig. 12.P4 Simplified geometry of a cervical disc

Assuming the weight of head for a 70-kg man (blood volume ~5 L) is ~5 kg and the neck joint has n_{neck} (≤ 100) times per day of rotations with θ degree ($< 90^\circ$ or $\pi/2$ radian) every day, we may consider the wear only between the upper Co-Cr plate and the polyethylene core with a radius of R_{core} with an equivalent sliding distance of $2\theta R_{\text{core}}/3$ for each the neck rotation. The allowable R_{core} is between 5 mm and 8 mm.

The polyethylene core contains a rod structure with a radius of R_{adaptor} (= 5 mm) for adhesion to the lower Co-Cr plate.

Further, the ion release rates of iron ($Q_{\text{Fe}} = 0.6$ μg/cm^2/week), chromium ($Q_{\text{Cr}} = 0.03$ μg/cm^2/week), and cobalt ($Q_{\text{Co}} = 0.1$ μg/cm^2/week) should also be considered together with the wear effect in order to ensure all the ion levels are within the ratio $\Omega_{threshold}$ (= 6% or 0.06) of the maximum allowable intake levels for iron (I_{Fe} = 200 μg/L/day), chromium (I_{Cr} = 35 μg/L/day), and cobalt (I_{Co} = 28 μg/L/day).

To enable the cell attachment on the uppermost and lowest surfaces (gray surfaces shown in the above figure) of the cervical disc, surface finishing should be performed to obtain a surface roughness σ_{rough} between 2 μm and 3 μm. On the other hand, the inner surfaces (highlighted with dotted shading in the above figure) should be smooth enough ($\sigma_{smooth} \leq 200$ nm) to minimize cell attachment. To achieve these required surface roughnesses, we can implement slab milling on each of the four surfaces with the path, including two stages of slab milling as described in the Fig. 12.P5 (Here, we have only shown the "smooth surfaces" as an example, with an allowance about D_{mill} around the core perimeter or the rod. The "rough surfaces" do not have the circular vacancies.)

The chosen milling cutter has a diameter D_{mill} (= 3 mm) and a height H_{mill} (= 2 mm), with the number of teeth on the cutter n (= 2). The moving velocity has to be set the same throughout the milling stages. Between the two stages, the milling head will change in orientation and move from the end position of stage 1 to start position of stage 2.

The maximum moving speed/feed (F_{move}) rate of the milling head is 50 mm/s. The maximum rotational speed of the milling cutter is 6000 rpm, yet the suggested rotational speed for this milling operation (N) is only 300 rpm. In addition, the milling machine needs other processing time for setting up/finalizing the machine process (T_{set} = 2 min) and reorientation of the milling head/working material between the milling stages (T_{change} = 10 s).

For the costs, the machine cost (M) is \$0.5/min with an overhead ratio (OH_m) of 0.1, and the labor rate (W) is \$3/min with an overhead ratio (OH_{op}) of 0.25.

For each use, the tool (milling cutter) costs C_{tool} = \$120, and the tool life (T_{tool}) is ~3 h cutting time. We assume that the tool is not broken during the process.

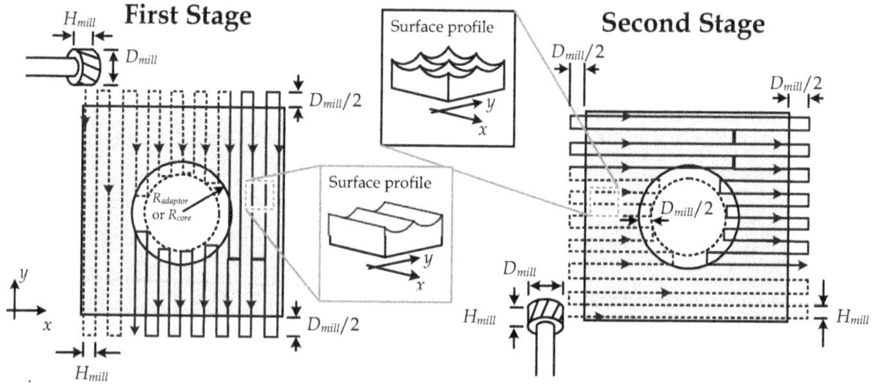

Fig. 12.P5 Two stages of slab milling for defining surface roughness

Now, please propose an optimization statement (including both the objective function and the constraints) for the minimum cost for the surface finish process. (Note: Use symbols only, but not the given values, for this part.)

(c) Based on your answer in part (b), please find the minimized manufacturing cost by considering the above given values provided, with explanations. Furthermore, discuss how much production cost can be potentially saved by comparing this value with the maximum possible cost of a feasible manufacturing process.

Problem 12.8

Imagine that you are responsible for manufacturing an artificial bone for the replacement of a bone fragment missing in the skull of a patient who had a serious injury in a car accident. A surgery was performed previously to remove a broken piece of bone. During the primary bone healing of the patient, your role is to manufacture a scaffold for the subsequent bone cell seeding and culture in order to develop the in vitro artificial bone. The scaffold fabrication is based on the 3D printing of a biomaterial with the "designed" architecture. In the next surgery, this artificial bone will be placed at the missing location of the skull for sealing the skull.

(a) Before the manufacturing, imaging of the skull should be performed to obtain the 3D geometry of the missing part of the bone. Can you suggest a bio-imaging method? Please describe briefly its working principle.
(b) The scaffold was then fabricated by a rapid prototyping machine. The substrate material you could use is a composite of hydroxylapatite and PEG-PLA-PEG. Figure 12.P6 shows the chemical synthesis of PEG-PLA-PEG. What kind of polymerization is involved in this process? Please explain.
(c) The composite should include mostly PEG-PLA-PEG, whose density is 1.14 g/cm^3 and viscosity (μ) is ~8.872 $\times$ 10^{-3} kg/m·s. During the rapid prototyping process, the composite should be applied onto the sample for deposition using a syringe. The injection operation was driven by a computer-controlled pressure supply. The composite solution should flow through a syringe needle with a length of 0.1 mm (L_{needle}) and an inner radius (R_{needle}) of 50 μm. The inner radius of the syringe body (R_{syringe}) is 10 mm, and the length of the syringe body (L_{syringe}) is 10 cm. Assume that the solution flow rate is 5 μL/hr. Should the flow be laminar or turbulent, and why? What is the minimum required pressure applied to support the required rate of the liquid injection?
(d) The missing piece of bone has the shape of a circular disc with radius R_{scaffold} and thickness H_{scaffold}. Taking into account that a bone cell in suspension should have a diameter ~10 μm, the scaffold should have a large enough separating hole/gap widths D_{gap} such that the cells can deposit and migrate inside the scaffold. Considering that the cured scaffold material has a Young's modulus close to cured PLA ($E_{\text{PLA}} = 3.5$ GPa), please express the equivalent compressive stiffness of the scaffold as a function of D_{gap}. (Note: Use symbols only, but not the given values, for this part.)

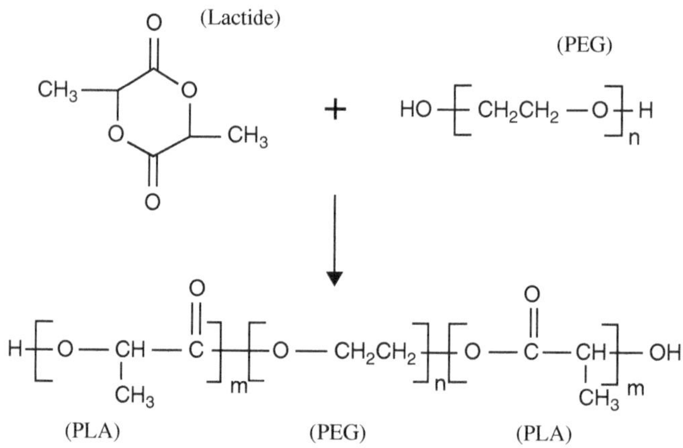

Fig. 12.P6 Polymerization of PEG-PLA-PEG

(e) Based on the design and fabrication strategy you provided in part (d) and further considering the injection pressure P of the composite (in other words, the flow rate is no longer fixed as 5 μL/h), please estimate the manufacturing time. (Note: Use symbols only, but not the given values, for this part.)

(f) Please propose an optimization statement (including both the objective function and the constraints) for the shortest manufacturing time by considering the following additional information.

(g) The disposition is achieved by injection of the PEG-PLA-PEG dissolved in a solvent (tetrahydrofuran) with a volumetric ratio $\Omega = 0.5$ (50%). Driving pressure of the solution can only be set as ≤ 10 Pa by the injection system. The allowable separating distance of the designed architecture D_{gap} should be ≥ 50 μm such that cells can migrate into the inner scaffold body. The equivalent compressive stiffness $E_{scaffold}$ of the "designed" scaffold should be ≥ 0.5 GPa to maintain the structural shape after implantation. (Note: Use symbols only, but not the given values, for this part.)

(h) Given $R_{scaffold} = 2$ cm and $H_{scaffold} = 4$ mm, can you find out the minimum manufacturing time based on your optimization statement above? Should the result correspond to the maximum injection flow rate, the minimum equivalent stiffness, and/or the minimum D_{gap}?

Problem 12.9

Imagine that you are responsible for manufacturing an artificial ear for replacement of an ear cartilage of a patient who has a chronic inflammatory disease (*left* photograph of Fig. 12.P7). Generally, cartilage is composed of specialized cells called chondrocytes that produce a large amount of collagenous extracellular matrix, an abundant ground substance that is rich in proteoglycan and elastin fibers.

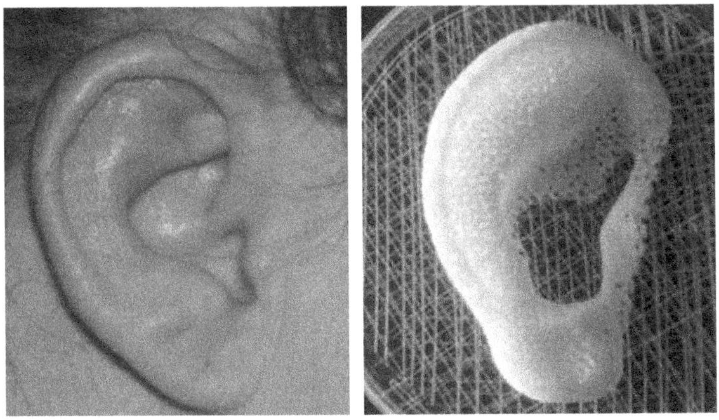

Fig. 12.P7 Chronic inflammatory ear (*left*) and the 3D printed scaffold (*right*). (Lekpa, F.K., Chevalier, X.: Refractory relapsing polychondritis: challenges and solutions. Open Acc. Rheumatol. Res. Rev. 10, 1 (2018))

Throughout the process, your role as a biomedical engineer in the hospital is to design the rapid prototyping manufacturing process for the scaffold and facilitate the subsequent chondrocyte seeding and culture in order to develop the in vitro artificial ear. The scaffold fabrication is based on the 3D printing similar to the fused deposition modeling (FDM) of a biomaterial with the **"designed" architecture** (*right* photograph of Fig. 12.P7).

(a) Before the manufacturing, imaging of the ear cartilage should be performed to obtain the 3D geometry of the missing part of the bone. Can you suggest a bio-imaging method? Please describe briefly the working principle with 30 words or less.

(b) The scaffold was then fabricated by a rapid prototyping machine. The substrate material you could use is the composite of biodegradable polycaprolactone (PCL) and lignin. The chemical synthesis of the poly-PCL-lignin is shown in Fig. 12.P8. What kind of polymerization is involved in this process? Please explain.

(c) The composite should include mostly volumetric ratios of 80% PCL, 2% lignin, and 18% acetone, whose viscosity (μ) is ~180×10^{-3} kg/m·s and density (ρ) is 1.145 g/cm^3. During the rapid prototyping process, the composite should be applied onto the sample for deposition using a syringe. The injection operation was driven by a computer-controlled pressure supply. The composite solution should flow through a syringe needle with a length of 0.1 mm (L_{needle}) and an inner radius (R_{needle}) of 50 μm. The inner radius of the syringe body (R_{syringe}) is 10 mm, and the length of the syringe body (L_{syringe}) is 10 cm. Assume that the solution flow rate is 0.5 nL/h. Should the flow be laminar or turbulent? Please explain. What is the minimum required applied pressure to support the required rate of the liquid injection?

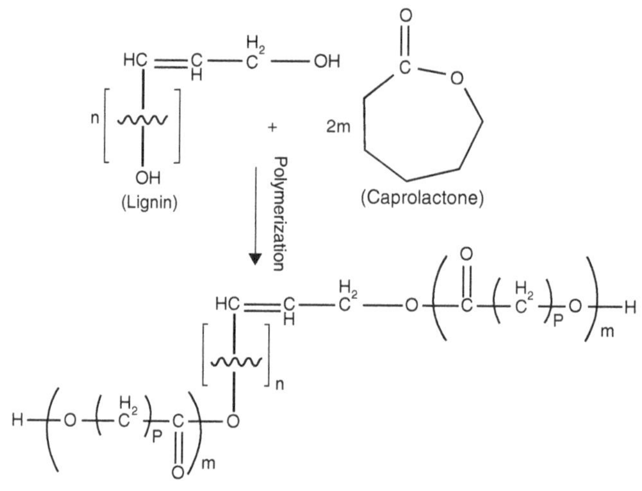

Fig. 12.P8 Chemical synthesis of the poly-PCL-lignin

(d) Let's assume that the soft bone of the ear has the shape as a half circular disc (i.e., a half circle with a certain thickness) with radius R_{scaffold} and thickness H_{scaffold}. Taking into account that a bone cell in suspension should have a diameter ~10 μm, the scaffold should have a large enough separating hole/gap widths D_{gap} such that the cells can deposit and migrate inside the scaffold. We set $D_{\mathrm{gap}} \geq 30$ μm. Considering that the cured scaffold material has a Young's modulus close to cured PCL ($E_{\mathrm{PCL}} = 1$ GPa), please express the equivalent compressive stiffness of the scaffold E_{scaffold} as a function of the printed fiber diameter D_{fiber} and D_{gap}. Please assume that D_{fiber} contains mainly the poly-PCL-lignin without acetone, and therefore, the cross-sectional area of the fiber is ~82% of the inner cross-sectional area of the syringe needle, given that the moving speed of the syringe is fast enough. (**Note:** Use symbols only, but not the given values, for this part.)

(e) The scaffold with a "designed" architecture may be printed as a multilayer mesh. However, the printed poly-PCL-lignin fibers have a diameter smaller than the cell, the gap with one layer of the fibers, i.e., D_{fiber}. In order to fulfill the required D_{gap} as mentioned in (d), please design an IMPROVED printing strategy, and explain it using drawings and descriptions. Please consider that the proposed strategy can still adopt the estimation of E_{scaffold} as described in (d). Please also specify the values of (1) D_{fiber}, (2) D_{gap}, and (3) number of each key feature layers for your design.

(f) Based on the design and fabrication strategy you provided in part (e) and further considering the injection pressure P of the composite (in other words, the flow rate is no longer fixed as 0.5 nL/h), please estimate the manufacturing time. (Note: Use symbols only, but not the given values, for this part.)

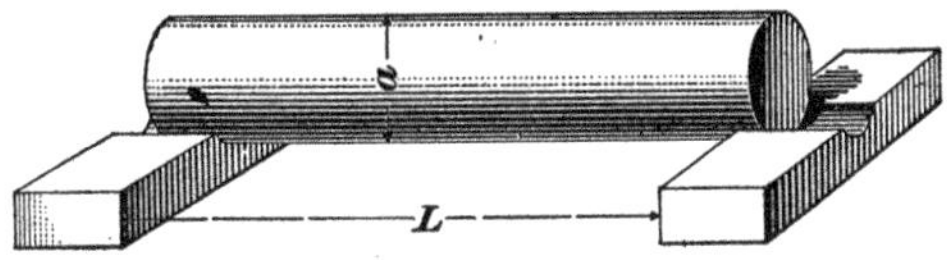

Fig. 12.P9 A simply supported beam with a circular cross-section

(g) We wish to select the optimum material for a sufficiently light and stiff fiber/beam supported by the two previously printed fibers as shown in Fig. 12.P9. It is required that the beam has a fixed length of L_{fiber} where its fiber diameter on the side D_{fiber} is adjustable to fit the design requirement. The mass of the beam is $M = \pi D_{\text{fiber}}{}^2 L_{\text{fiber}}\, \rho_{\text{PCL}}/4$ where ρ_{PCL} is the density of the polymerized PCL material (1.145 g/cm^3). For a simply supported beam with a pressure per length P_{fiber} along the beam length caused by the fiber weight $P_{\text{fiber}} = \pi D_{\text{fiber}}{}^2 \rho_{\text{PCL}} g/4$ where g is the gravitational acceleration, the central deflection δ is

$$\delta = \frac{5P_{\text{fiber}}L_{\text{fiber}}^4}{384E_{\text{PCL}}I_{\text{fiber}}}$$

where the moment of inertia $I_{\text{fiber}} = MD_{\text{fiber}}{}^2/16$.

Determine the appropriate material performance index.

For your final design configuration (after you finish part (i) later), please find the central deflection δ, and check if it is within 0.1 μm.

(h) Please propose an optimization statement (including both the objective function and the constraints) for the shortest manufacturing time by considering the following additional information:

The disposition is achieved by injection of the PCL-lignin dissolved in a solvent (acetone) with a volumetric ratio Ω, which can be set between 0.6 (60%) and 0.8 (80%) for different fiber diameters D_{fiber}, under a fast enough syringe movement as mentioned previously in part (d). A maximum flow rate of 3 nL/h and a maximum pressure of 100 kPa can be offered by the injection system. The allowable separating distance of the designed architecture D_{gap} should be $\geq$30 μm in all x, y, and z directions such that cells can migrate freely into the inner scaffold body. The equivalent compressive stiffness E_{scaffold} of the "designed" scaffold should be larger than the level of the cartilage ($\geq$2 MPa) to maintain the structural shape after implantation. (Note: Use symbols only, but not the given values, for this part.)

(i) Given D_{scaffold} should be ~2 cm and H_{scaffold} should be between 2.95 mm and 3.05 mm, can you find out the minimum manufacturing time based on your optimization statement above? Should the result correspond to the maximum injection flow rate, the minimum equivalent stiffness, and/or the maximum porosity?

Problem 12.10

Please reconsider the example discussed in Sect. 12.5.4, except that now titanium is considered instead of stainless steel. Please solve for values of the optimal manufacturing time and other related parameters.

References and Further Reading

1. Pierre, D.A.: Optimization Theory with Applications. Courier Corporation, New York (1986)
2. Sobieszczanski-Sobieski, J., Morris, A., Van Tooren, M.: Multidisciplinary Design Optimization Supported by Knowledge Based Engineering. Wiley, Chichester (2015)
3. Fries, R.C.: Reliable Design of Medical Devices. CRC Press, Boca Raton, FL, USA (2016)
4. Rao, R.V.: Advanced Modeling and Optimization of Manufacturing Processes: International Research and Development. Springer, London (2010)
5. Pardalos, P.M., Rosen, J.B.: Constrained Global Optimization: Algorithms and Applications. Springer, Berlin/New York (1987)
6. Liberti, L., Maculan, N.: Global Optimization: From Theory to Implementation. Springer, New York (2006)
7. Olubamiji, A.D., Izadifar, Z., Si, J.L., Cooper, D.M., Eames, B.F., Chen, D.X.: Modulating mechanical behaviour of 3D-printed cartilage-mimetic PCL scaffolds: influence of molecular weight and pore geometry. Biofabrication. **8**, 025020 (2016)
8. Pawar, A.A., Saada, G., Cooperstein, I., Larush, L., Jackman, J.A., Tabaei, S.R., Cho, N.-J., Magdassi, S.: High-performance 3D printing of hydrogels by water-dispersible photoinitiator nanoparticles. Sci. Adv. **2**, e1501381 (2016)

Index

R. H. W. Lam, W. Chen, *Biomedical Devices*,
https://doi.org/10.1007/978-3-030-24237-4

T

U

V

W

X

Y

Z